Prevention of Accidents and Unwanted Occurrences

Theory, Methods, and Tools in Safety Management

W0234990

Prevention of Accidents and Unwanted Occurrences

Theory, Methods, and Tools in Safety Management

Second Edition

Urban Kjellén
Eirik Albrechtsen

CRC Press
Taylor & Francis Group
Boca Raton London New York

CRC Press is an imprint of the
Taylor & Francis Group, an **informa** business

First published in paperback 2024

First published 2017 by CRC Press
2385 NW Executive Center Drive, Suite 320, Boca Raton FL 33431

and by CRC Press
4 Park Square, Milton Park, Abingdon, Oxon, OX14 4RN

CRC Press is an imprint of Taylor & Francis Group, LLC

© 2017, 2024 Taylor & Francis Group, LLC

Publisher's Note
The publisher has gone to great lengths to ensure the quality of this reprint but points out that some imperfections in the original copies may be apparent.

ISBN: 978-1-4987-3659-6 (hbk)
ISBN: 978-1-03-291858-7 (pbk)
ISBN: 978-1-315-12097-3 (ebk)

DOI: 10.1201/9781315120973

**Visit the Taylor & Francis Web site
at http://www.taylorandfrancis.com**

**and the CRC Press Web site at
http://www.crcpress.com**

Contents

Preface

Accidents are unwanted events that cause losses to individuals, companies, and society as a whole. They are symptoms of underlying weaknesses within the company. When accidents occur, they represent an invitation to improve by learning from these weaknesses. Near-accidents and unsafe conditions and acts are other examples of such symptoms offering the potential to learn and improve.

There is no simple answer to the question of how to use the experiences from unwanted occurrences in order to prevent accidents. Accident risks cannot usually be engineered away. Neither is it sufficient to train and motivate people to avoid the risks or to establish and enforce strict work procedures. Only a combination of well-balanced measures will be effective to prevent incidents. These measures need to be adapted to changing conditions inside and outside a company.

This book emphasises the central role of experience feedback in any management system for the prevention of accidents. We define experience feedback as the process by which information on the results of an activity is fed back to decision-makers as new input to modify and improve subsequent activities (Melnick and Everitt 2008). There are many channels for experience feedback – such as reporting of unwanted occurrences, workplace inspections, safety audits, and accident investigations. These methods will provide experience of past and present safety performance that is used to support decisions in controlling hazards and preventing loss. The associated mechanism, feedforward, is applied for the same purpose and is treated in this book in the same manner as feedback. Anticipation of future conditions in risk assessments represents a typical example of such a mechanism. In addition to the formal channels of experience feedback, the tacit knowledge of the members of an organisation is an important source of experience. Participation of personnel is thus a key part in the prevention of accidents through experience feedback.

This book addresses the question of how to improve experience feedback inside companies in order to prevent unwanted occurrences. It presents principles and methods to accomplish effective experience feedback and decision support to manage the risk of accidents. It is neither a prescriptive nor a normative book but intends to help the reader in analysing systems for experience feedback from different scientific perspectives to demonstrate how these perspectives contribute to an overall understanding. This book underlines the need for interdisciplinary approaches to safety research and practice. It also gives a comprehensive overview of different methods and tools for use in safety practice, and the scope, merits, and shortcomings of these methods and tools are discussed. This book thus helps the reader to understand the strengths, weaknesses, and potential pitfalls of these different methods and tools and to decide on which to use.

This book combines theory and principles from safety science with one of its author's professional experience from 30 years of safety management practice in different industries. This professional experience has been useful in providing examples and in demonstrating the applicability of different methods and tools.

This book is not targeted at one domain. The principles and methods presented here are applicable to all organisations that want to manage accident risk. In the literature, accidents are subdivided into different categories such as occupational, major, and traffic accidents. Among major accidents, the so-called process accidents are an important subset involving loss of containment of hazardous materials and chemicals. This book is relevant for all these types of accidents. The target group of this book is thus cross-sectorial and also cross-disciplinary. It has been written with a mixed readership in mind, including safety professionals within industry, insurance companies, authorities, safety researchers, and students in the area of health, safety, and environment (HSE) management. The goals are to provide the reader with:

- Knowledge of principles of accident control through experience feedback
- Knowledge of various methods and tools for use in safety practice and an ability to develop skills to put them into use as well as an ability to reflect about their merits and weaknesses
- Knowledge of principles and approaches for developing measures to control hazards and prevent loss
- Proficiency in assessing potentials and limitations of existing approaches and methods as well as new trends in HSE management

In approaching these goals, the authors have maintained a careful balance between the need to comprehensively list alternative approaches presented in the safety research literature and the need to give precise recommendations. Whereas the experienced professional may benefit most from the former, others with less experience may in a short-time perspective be best served by a handbook approach. We have tried to solve this dilemma by presenting alternative solutions while at the same time giving recommendations regarding our preferred alternatives.

The term HSE management is used for the formal and informal processes of steering an organisation to prevent injury or illness to people, material damage, or environmental losses. Similar terms that do not include environmental issues are occupational health and safety (OHS) management and safety management. In this book, we emphasise the *safety* part of HSE. We also use the associated term 'management of safety', which is a wider concept referring to an organisation's total efforts to control hazards. It includes, but is not limited to, safety management systems.

This is the second edition of *Prevention of Accidents through Experience Feedback*, which was originally published in 2000. One of the drivers behind the second edition has been updating the contents with new knowledge, experience, legislation, standards, and so on that have evolved during the 16 years since the first edition was published. Some parts of this book have undergone major changes based on research directions as well as on our practical experiences. Important changes in the content from the first edition have been noted in the description of the book contents that follows.

Contents of the book

As indicated previously, this book is a combination of a handbook (an overview of principles, methods, and tools in the management of safety) and a scientific book about the foundations, merits, and shortcomings of methods, tools, and principles. Therefore, here we provide a thorough overview of the main contents of this book to help the reader find whatever he or she is looking for.

Section I introduces the reader to the concepts of experience feedback and safety information systems. Chapter 1 introduces some basic concepts and reviews the historical development in the area of accident control through experience feedback. The focus has shifted from safety information systems to experience feedback in general, because the former concept does not grasp the variety of the safety management principles described in this book. Chapter 2 gives an overview of different framework conditions for a company's feedback systems for the management of safety. These are internal and external conditions that influence the systematic management of safety within a company. New in the second edition are updates of regulatory requirements and relevant standards of the International Organization for Standardization (ISO). Chapter 3 gives an example of an application of experience feedback and how it can contribute successfully to the improvement of a company's performance by nearly eliminating accidental emissions to the environment.

The theoretical foundations for the various methods and tools presented in subsequent sections of the book are presented in Section II.

Chapter 4 covers a range of different accident models and accident theories. Some new models have been included in the second edition (e.g. the Swiss cheese model, AcciMap, and the so-called systemic accident models such as Systems Theoretic Accident Model and Processes [STAMP]). At the end of the chapter, some key aspects of safety culture are included, which is an important complementary dimension to the structural safety management approaches that include feedback control systems.

Chapter 5 provides a framework for accident analysis, which is a combination of elements from the accident models presented in Chapter 4. This framework is based on a simple input–process–output model, where the output is loss. The process is the incident sequence created by deviations and barrier failures that lead to loss of control of energies in the system. Loss occurs when a victim is exposed to the uncontrolled energy flow. This accident analysis framework is the key foundation for later chapters on methods for collection of accident data. Relevant and updated classification systems for information about accident risks are presented (e.g. of deviations, consequences, and root causes).

Chapter 6 summarises the basic statistical theory of Poisson processes. These statistics are applied in Section IV as the theoretical foundation for monitoring HSE performance.

Chapter 7 reviews basic principles for experience feedback. Two central principles applicable to experience feedback are presented: diagnosis and feedback control. The chapter reviews some concepts that are important foundations of the systematic management of safety: Deming's cycle, Juran's feedback cycle for the control of anything, Ashby's law of requisite variety, and Van Court Hare's hierarchy of order of feedback. Organisational learning is another important part of this chapter, focusing on how learning based on experience is facilitated and also on potential obstacles to learning.

Chapter 8 summarises issues dealt with in Chapters 4 through 7 by addressing the question 'How can we distinguish adequate feedback control systems from those that are less than adequate?' A set of requirements for feedback control systems is presented related to data collection, distribution, and presentation of information, and the system as a whole. Additionally, requirements for safety performance indicators are presented. In the second edition, requirements for incident investigation and risk assessment have been added.

Chapter 9 presents accident countermeasures based on barrier theory, which has been updated based on new scientific and practical contributions. Barrier theory is a central part of this book and an important foundation for later chapters. We define a barrier as 'a set of system elements (human, technical, organisational) that as a whole provide a barrier function with the ability to intervene into the energy flow to change the intensity or direction of it'. We thus interpret a barrier as a function to prevent or stop an incident sequence. This function is realised by a barrier system consisting of human, technological, and/or organisational elements. Designing for safety of machinery and the permit-to-work system are other important parts of the chapter. Emergency response planning has been added in the second edition.

Chapter 10 addresses the role of humans in both accident control and as a source of deviations that can lead to accidents. The chapter is a major revision from the first edition. It demonstrates how individuals are a part of an active barrier system that ensures the execution of the barrier function. Theories on macro-cognition are used to explain the role of humans in accident prevention, which also demonstrates how the organisational context influences human actions. The chapter also addresses different approaches to improve the quality of human performance in the execution of hazardous work.

Section III presents methods and tools for the collection of data on accident risks. Chapter 11 has an overview of the data-collection methods presented in this part of the book: workplace inspections, safety audits, accident and near-accident reporting and investigation, and reporting of unwanted occurrences. This chapter shows how data collection methods are linked to the accident analysis framework discussed in Chapter 5 and to the safety information system outlined in Chapter 7.

Chapter 12 presents methods and principles for hazard identification, workplace inspections, and safety audits. The second edition of this chapter has been updated by including a new topic: hazard identification. A checklist of hazards is given, which is helpful for other parts of feedback control as well. The chapter section about audits has been updated from the first edition based on ISO 19011.

Chapter 13 is a comprehensive discussion about accident and near-accident reporting and investigation. The chapter gives details on accident investigation methods and reporting systems. It distinguishes among three levels of accident investigation based on severity and frequency. At Level 1, all accidents and near-accidents are reported and analysed by the immediate line management. Level 2 involves investigations by an in-house team and through the use of man, technology, and organisation (MTO) analysis, which may include a description of the incident sequence, barrier analysis, and deviation analysis. In exceptional cases (Level 3), an independent team investigates the accident or near-accidents in depth by using an audit technique supported by checklists such as the safety management and organisation review technique (SMORT). This chapter has undergone major changes from the first edition based on research and practical experiences.

Chapter 14 reviews basic principles for the establishment and application of a database on accident risks. The chapter also demonstrates different methods for statistical analysis of incident data. The content has changed from the first edition due to the developments in information technology.

In Section IV, we present safety performance measures or indicators for use in the types of HSE management activities that involve goal setting and feedback control. Here, we define a safety performance indicator as a metric used to measure an organisation's safety performance in terms of its ability to control the risk of accidents in its activities.

Chapter 15 gives an overview of safety performance indicators based on the accident analysis framework in Chapter 5. This overview shows performance indicators from all three parts of the framework–loss-based indicators, process-based indicators, and causal factor–based indicators. Chapter 15 also presents a review of different interpretations of the concepts of lagging and leading indicators, which is a new addition in the second edition.

Chapter 16 discusses loss-based indicators. These are indicators using data about accidents and losses as the main input. Accident statistics will fluctuate from period to period due to pure chance, and control charts are introduced to evaluate whether there are significant changes in safety performance. Process-based safety performance indicators are presented in Chapter 17. This chapter presents various indicators based on data about incidents and deviations. Chapter 18 looks into safety performance indicators based on data about the underlying contributing factors and basic causes. This chapter includes an overview of rating systems for the performance of general and HSE management. The section on measuring safety climate has been updated from the first edition. Section IV ends with Chapter 19, which gives recommendations on the selection of safety performance indicators, including examples of the various choices.

Section V presents different methods for assessment of occupational accident risk. Risk assessment differs from other methods and tools presented in this book because they are proactive (i.e. they support prevention of accidents at the planning stage). Chapter 20 outlines the basic steps in a risk assessment process, including hazard identification, analysis of risk (i.e. frequency of occurrence and consequences), and evaluation of acceptable risk. An overview of different types of methods for risk assessment is presented. This chapter has been updated from the first edition by including requirements in ISO 31000, *Risk Management*.

A particular type of risk assessment – the coarse analysis or preliminary hazard analysis – is presented in Chapter 21. The coarse analysis is a simple method that generates a crude risk picture. It can be used for most situations and most systems. Chapter 22 demonstrates the steps of job safety analysis (JSA), which is an important tool for operative risk assessment. Merits and limitations of JSA are discussed. Chapter 23 is dedicated to the risk assessment of machinery and is updated from the first edition based on ISO 12100 (machinery safety). It shows steps and checklists for the assessment of the use and risks of machinery and links these to previous sections in the book about machinery safety (Chapter 9). Chapter 24 is dedicated to comparison risk assessment. Steps of comparison risk assessment are presented, including schemes and examples. Comparison risk assessment is used during the design of a new plant to predict the occupational accident frequency rate after start-up (i.e. during operation and maintenance). This is also applied in Chapter 27 on road transportation risk assessments.

Section VI illustrates applications of the different methods and tools presented in Sections III to V on the management of safety in three different industries. Chapter 25 is built around one particular case study, the Ymer offshore oil and gas platform. The chapter describes how experience feedback is accomplished in different phases of a platform's life cycle (i.e. design, construction, and operation). It illustrates both the prevention of occupational accidents and the prevention of major accidents. Chapter 26 is new to the second edition and is organised in a similar way to Chapter 25. It uses a large infrastructure project in the hydropower industry as a case study that is brought through the different project phases into operation. Applications of safety management principles and practices in each phase are demonstrated. Chapter 27 brings us to the area of road transportation safety. Compared to the first edition, the perspective is shifted from the trucking industry to corporate safety management of road transportation. We will illustrate the application of different principles and methods of experience feedback in this new setting.

Acknowledgements

Many people have supported us in the development of this book. We especially want to mention and thank the following people:

- Professor Emeritus Jan Hovden at the Norwegian University of Science and Technology (NTNU) reviewed several of the chapters in this book. He also provided important support to the first edition.
- Professor Trond Kongsvik at NTNU reviewed the chapters about safety performance indicators and principles for feedback control.
- Professor Stein Haugen at NTNU reviewed the section on risk assessment.
- Associate Professor Karin Laumann at NTNU provided valuable input to Chapter 10 on the human element in accident control.
- Senior Research Scientist Knut Øien at SINTEF did a thorough job of reviewing and suggesting updates to Chapter 25 (the case study of the oil and gas industry).
- Dr. Kjellén's colleagues at Statkraft, Jan Arild Berget and Kemal Cihan Şimşek, reviewed Chapter 14 (incident database), Chapter 26 (hydropower industry), and Chapter 27 (work-related road transportation).

We also want to thank Stian Antonsen (SINTEF), Kjetil Vindal Forslund (Statoil), Stian Lydersen (NTNU), and Stig Winge (NTNU and the Norwegian Labour Inspectorate) for their comments on certain parts of this book. In some parts of this book, we have used examples from the master's theses of some of our students at NTNU. Here, we especially want to mention Gro Blindheim, Karsten Boe, Ola Sand Koren, Jørn Lindtvedt, Ingvild Solberg, Eva Svensli, and Elisabeth Wendel.

Kjetil Strand has produced several of the illustrations in this book. We are thankful for his professionalism and his ability to transform our complex ideas into simple illustrations.

We also thank the editorial and the production staff at CRC Press/ Taylor & Francis Group for their professional work.

Urban Kjellén
Oslo

Eirik Albrechtsen
Trondheim

Authors

Dr. Urban Kjellén is a principal advisor at Statkraft, Oslo, and an Adjunct Professor of safety management at the Norwegian University of Science and Technology, Trondheim, Norway. He has about 30 years of industrial experience primarily in various health, safety, and environment (HSE) management positions in investment projects and at the corporate level in the oil and gas, light metal, and hydropower industries. He has published 35 papers in refereed international journals and books on risk analysis and HSE management of design and construction. His books include *Prevention of Accidents through Experience Feedback* (Taylor & Francis, 2000) and *Occupational Accident Research* (Elsevier, 1984). He has been a member of the editorial board of Safety Science and of various standardisation committees.

Dr. Eirik Albrechtsen is an Associate Professor in safety management at the Department of Industrial Economics and Technology Management, the Norwegian University of Science and Technology (NTNU), Trondheim, Norway. He has several years of experience in safety research with NTNU and SINTEF Safety Research within the petroleum and construction industries as well as in research on information security management and societal safety. He has published several papers in refereed international journals within these topics.

section one

Introduction

In Chapter 1, we introduce some basic concepts and review the historical development in the area of accident control through experience feedback. Chapter 2 gives an overview of different framework conditions for the systematic management of safety, both inside and outside a company. In Chapter 3, we will look into a case on the prevention of an accident in the environmental field. It represents a successful application of some basic principles of experience feedback in the reduction of emissions from a fertiliser plant. We use this example to demonstrate the application of some of the issues detailed in later parts of this book and show how they form a coherent whole.

chapter one

Introducing some basic concepts

In pre-industrial society, human muscles accounted for much of the energy used in the production of goods and services. The individual and the work group had hands-on control of the hazards associated with this energy. At the same time, humans were at the mercy of nature and its hazards, as represented by weather extremes, landslides, flooding, wild animals, and so on. Accidents were viewed as phenomena beyond human control.

Since the mid-nineteenth century, industrialisation has involved the harnessing of various energy sources such as coal, hydrocarbons, and hydropower for production purposes. Mechanisation and automation, together with new principles for management and organisation, contributed to the separation of individuals from the direct control of the energies used in production. At the same time, the energy flow increased – as did the potential for accidents with severe consequences. Safety was identified as a specific activity area needing dedicated management attention. Accident prevention was accomplished through the development and enforcement of safety regulations, standards, and rules (such as those put in place in the UK since the first half of the nineteenth century). These were based on accumulated experiences from investigations into accident cases. This 'trial-and-error' approach is still valid today but has become insufficient to master the rapidly changing technology and means of organisation of modern industrial society. Today, safety management is accomplished through systematic feedback and assessments of accident risks at different levels of an organisation. Compared to the early trial-and-error approaches, today's feedback loops are much more diverse.

This book will illustrate the application of methods and tools that are used primarily within – but that are not limited to – the field of occupational accident prevention. Occupational accidents still represent a significant public health problem. It has been estimated that, in 2010, about 350,000 people died in occupational accidents worldwide (Nenonen et al. 2014). The likelihood of experiencing a fatal accident at work is very unevenly distributed by region and by type of industry (Pearson 2009). High accident fatality rates occur in low- and middle-income countries in parts of Asia, Oceania, and sub-Saharan Africa. Those industries with high fatal accident rates are transportation, construction, mining, and basic production industries such as agriculture, forestry, and fishing.

The situation is far more benign in high-income countries in northern Europe and North America. However, even in a country like Norway, about 40 people per year die in occupational accidents, resulting in a fatal accident rate of about 1.7 per 100,000 employees in 2014 (Arbeidstilsynet 2015). This figure has remained somewhat stable during the last decade after a reduction of about 50% between 1970 and 2000. The risk of fatal occupational accidents is unevenly distributed. Those employed in the agriculture, forestry, and fishing industries run a ten times greater risk of dying in an occupational accident than does the average Norwegian employee.

In the UK, the fatal accident rate dropped by 60% in the 20-year period from 1995 to 2015 (HSE 2015). The UK rate of 0.67 fatalities per 100,000 employees for the period of 2009–2011 was about 40% of the European Union's average rate of 1.55. In the United States, the rate of fatal work injuries in 2014 was 3.4 per 100,000 employees (U.S. Bureau of Labor Statistics 2016).

In this book, we will analyse accidents as a sequence of events evolving from lack of control to loss of control, primarily in terms of energy in an industrial system, and finally to where loss occurs through personal injury or damage to the environment or to material assets. Although the consequences are different, the processes that lead to such losses may be analysed in similar ways for occupational accidents, major accidents, and accidents resulting in environmental or material damage. We also will touch on the prevention of major accidents – especially in the oil and gas industries, where major accidents due to fires and explosions also represent a significant occupational safety concern. We will also present examples related to occupational health and environmental concerns, where the consequences of accumulated emissions due to acute events may be severe.

In this volume, health, safety, and environment (HSE) are used in relation to management, which goes beyond the prevention of unwanted occurrences due to loss of control. HSE management includes the prevention of negative effects on people's health or on the environment from planned emissions and discharges. Although there are often separate laws and regulations relating to occupational health and safety (OHS), environmental care, and major accident prevention, different branches of industry, such as those of oil and gas, prefer to develop integrated management systems covering all three areas. We will use HSE in the context of HSE management systems and the term 'safety' in connection with methods, tools, and principles for the prevention of accidents.

This book rests on the basic assumption that accidents at work, and severe accidents in particular, are preventable. The scope for improvements is largest in areas where the risk of accidents is greatest. The last part of this book is dedicated to case studies dealing with the application of various methods and tools in the prevention of accidents in three industries – oil and gas, hydropower, and transportation. Some of these cases build on experiences from large investment projects in developing countries, where the validity of this basic assumption is demonstrated in practice.

This book is based on an empirical tradition in which experience feedback from accidents and incidents plays a significant role. These unplanned and unwanted events are all relatively common in industry. Production stoppages, substandard product quality, and material damages are more frequent consequences of these events than, for example, injury to personnel. Management and workers are used to handling unplanned events on a daily basis and to taking actions to prevent recurrence. With improvements in the planning and control of operations and maintenance, a reduced risk of accidents will follow – although this is not the primary aim. In most industries, the management of safety is dependent on coordinated efforts involving proper planning and control of the operation and good HSE management. These synergetic efforts will result in positive impacts on areas such as cost, progress, safety, and quality (Grimaldi 1970; Kjellén et al. 1997).

The history behind the development of safety management applied in this book goes back almost a century, when H.W. Heinrich in 1931 first published his book on industrial accident prevention, in which he transfered some of the principles of scientific management to the field of accident prevention (Heinrich 1959). Scientific management involves a rational and systematic approach to the analysis and improvement of productivity in manufacturing (Taylor 1911). Heinrich pointed out the importance of basing accident prevention on experience – not only from accidents but also from unsafe acts and conditions. His five steps for the prevention of accidents, which involve data collection, analysis of remedies, remedy selection, implementation, and evaluation – are equally valid today. Many of Heinrich's other propositions have been re-examined in light of new insights, but he nonetheless established an important standard for future developments in safety management based on empirical data and a systematic approach.

The significant impact of scientific management on early developments in safety management was later overtaken by that of quality management. Mass production could not be accomplished without quality control. To speed up production without compromising quality or safety, the U.S. military introduced statistical quality control techniques during World War II. What began as a focus on quality management of product inspection and testing was widened after the war by Joseph Juran and Edwards Deming. They proposed methods for supplying continuous feedback on product quality to those in the higher echelons in order to improve the organisation and management of production. Deming's wheel, or the plan–do–check–act (PDCA) cycle for continuous improvements, which is so central to the fields of safety and quality management today, has its origin in this period (Deming 1993).

Rockwell (1959) pioneered the implementation of statistical quality control techniques in the field of safety management. At the time, the so-called 'problem of safety performance measurement' received increasing attention. Because accidents are infrequent events, the field of safety management was, and still is, hampered by a lack of good measures of safety performance. Rockwell applied statistical sampling techniques in the development of safety performance measures where the statistical basis was determined not by random events such as accidents but by decided sampling frequency of behaviour observations. Today, we recognise this initiative in the so-called behavioural-based safety (BBS) programs designed to reinforce safe behaviour (Komaki et al. 1978).

Contributions from the field of epidemiology represent another important milestone that is central to this book. Gibson (1961) introduced the concept of unwanted or uncontrolled energy transfer as an injury agent in most accidents. The full impact of this perspective was only achieved when Haddon (1980) defined 10 generic principles for the prevention of harm (injury) from transfer of energy. Here, the contributions of Gibson and Haddon are integrated in the so-called energy model. By combining the principles of the energy model with those of quality management, researchers at the U.S. Atomic Energy Commission (AEC) developed the Management Oversight and Risk Tree (MORT) safety management concept (Johnson 1980).

Haddon (1968) also introduced a phase model of accidents that was further elaborated by the Swedish Occupational Accident Research Unit in the beginning of the 1980s, when they introduced the so-called initial phase of an accident, which is characterised by lack of control (Kjellén and Larsson 1981). The concept of deviation analysis was also introduced as a method for identifying and assessing departures from the planned and faultless production process, thereby indicating an increased risk of accidents. This approach became increasingly accepted especially through standardisation work in quality management from the 1980s onwards.

After World War II, sociotechnical studies emerged in response to the developments within scientific management and bureaucratic organisations and the underlying human viewpoint (e.g. Trist 1981). These studies have contributed to our understanding of how human and organisational factors affect the ways in which work is done and technical systems are used. This perspective has been applied in the analysis of accident causes as an interplay among human, technical, and organisational aspects. Another important contribution from sociotechnical studies is the emphasis on employee participation in the development of work systems. These sociotechnical studies have had a major influence on the democratic traditions evident in Scandinavian working life, including participative approaches to OHS management. Worker participation in different arenas resulting in experience exchange is a central concept in this book.

In the late 1970s, the Norwegian Petroleum Directorate (NPD) made two strategic decisions regarding their regulatory regime in the area of HSE: (1) going forward, regulations should primarily be risk-based and of a goal-setting type as opposed to prescriptive, and (2) oil companies were required to develop their own management systems to ensure compliance with regulations in accordance with the principles of internal control. The NPD worked closely with consultants affiliated with the MORT concept in developing these regulations. UK authorities followed a similar approach in the Health and Safety at Work Act of 1974, implying that society should set overall safety goals and companies must decide on how to meet these. Lord Cullen's inquiry into the 1988 Piper Alpha oil platform disaster resulted in a further boost to interest in this risk-based and goal-setting regime, which has further been adapted by the European Union in their directives relating to HSE. The so-called Seveso Directive is one such example (European Commission 2012).

These converging trends in regulatory requirements for HSE management run parallel to developments in management system standards on quality and HSE. The International Organization for Standardization's ISO 9000 on quality management system standards (ISO 2015a), ISO 14001 on environment management systems (ISO 2015b), ISO 31000 on risk management (ISO 2009), and the draft ISO 45001 on OHS management systems (ISO 2016), all apply a management system model integrating Deming's PDCA cycle.

The structural approach to the prevention of accidents through Heinrich's 'five steps in accident prevention' and further development was balanced by safety culture initiatives, especially after the Chernobyl nuclear accident in 1986. We now realise that safety performance cannot be managed adequately without considering the shared values within an organisation that define the safety attitudes and behaviour of its members (Cooper 2000). Although safety culture will not be a central theme, we will touch upon it – especially in the case studies presented in the last section of the book.

The aforementioned developments and disciplines represent the basis for the central theme of this book, which is the prevention of accidents through experience feedback. Experience feedback is defined here as the process by which information on the results of an activity is fed back to decision-makers as new input that is used to modify and improve subsequent activities (Melnick and Everitt 2008). This book presents many channels for accident prevention through experience feedback, such as reporting unwanted occurrences, workplace inspections, safety audits, and accident investigations. Risk assessments use an associated mechanism (i.e. feedforward). And worker participation is an important means of capturing experience and ensuring involvement in the decision-making process for most of these channels.

chapter two

Framework conditions

The design, organisation and operation of a company's feedback systems for the management of accident risks are determined by a number of framework conditions. These are conditions inside and outside a company over which we do not have immediate control but that influence the systematic management of safety in that company. These framework conditions influence the opportunities that an organisation, group, or individual has to control major accidents and occupational accident risks (Rosness et al. 2012). The importance of considering framework conditions has been demonstrated by several researchers, but the notion and operationalisation of the term vary, representing both opportunities for and threats to our ambitions for accomplishing efficient solutions.

2.1   Conditions inside the company

2.1.1   Size, type of technology, and resources

Let us start by looking at three important structural variables within a company's structure: its size, type of technology, and its resources. These internal framework conditions are visible in practice. Size affects the communication patterns within the company. In a small company, the organisational members know each other, and communication links are often direct, informal, and spontaneous. When the company grows, there is an increased need for communication and reporting through formal channels. These principles apply not only to communication in general but also to communication on safety issues. If, for example, an accident happens in a small company, the manager who divides his/her time between customer contacts and supervision will soon know about it. In a large company, management will be dependent on formal accident-reporting routines to obtain the necessary information.

Typically, we find well-functioning formal health, safety, and environment (HSE) management systems in large and resourceful companies in high-risk industries (Skaar 1994). Here, a company's HSE policy and goals are often integrated as part of their overall criteria for success. However, large companies (>500) will have challenges in establishing successful informal exchange of information on safety issues. At the other end of the spectrum, small companies often have an informal HSE management system (Arocena and Nunez 2010), or they implement systems that are tailor-made for large companies but that utilise other resources (Champoux and Brun 2003). Furthermore, small-sized companies may find that HSE management is too formal and that the requirements for documentation are too demanding (Bråten et al. 2012).

Technology is another important structural variable. High-risk industries run production processes involving substantial amounts of hazardous energy such as flammable or toxic substances, nuclear materials, and so on. The potential consequences of an accident are catastrophic and may threaten a company's existence. Often, these industries introduce new and complex technologies, which make safety management more complicated and demanding (Perrow 1984). In these cases, it becomes increasingly difficult for management and workers to understand and anticipate the behaviour of the production processes and the related hazards. The company will be dependent on advanced feedback systems to avoid unwanted events. Not only must all sorts of accidents, incidents, and deviations be reported and followed up, but risk-analysis methods have also to be applied to identify and assess potentially catastrophic events before they happen. Companies in branches of industry where the hazards are obvious and the consequences limited, on the other hand, do not deal with these advanced needs. They may rely on simple incident reporting routines that meet the minimum regulatory requirements.

Resources – in particular economical resources – play a decisive role. Resourceful companies will be able to afford solutions that pay off in the long run. They will be able to acquire the necessary equipment, manpower, and expertise for the design and management of efficient HSE systems. Resources are also needed in order to close the loop from the identification and assessment of accident risks to the design and implementation of remedial actions. When adequate resources are marshalled to solve safety problems, employees will experience this as a sign of management's positive attitudes towards safety. They will in turn be more inclined to report accident risks and to participate in problem-solving activities. If the opposite situation is the case, even a well-designed feedback system for accident occurrences may not be allowed to develop into an efficient tool. Empirical research shows that there can be a higher frequency of incidents during economic recessions (e.g. Lander et al. 2016). Or, in periods of economic crisis, cost cuts may affect the resources allocated to safety activities in a way that has detrimental effects on the commitment of management and workers. Not only economic resources affect safety performance. To complete the picture, there are internal resources that need to be in place in order to maintain efficient feedback systems such as a commitment to safety among managers and workers, organisational safety skills and knowledge, sufficient time to perform tasks related to feedback systems, formal and informal systems for communication, and coordination among the different participants (Hale 2003).

2.1.2   The organisational context

HSE management is influenced by the organisational context within which it operates. Here, we will touch upon Bolman and Deal's four organisational perspectives (Bolman and Deal 1984; Hale and Hovden 1998) to show how an organisation frames HSE management. Each perspective represents a way of looking at an organisation and what goes on inside it. A holistic understanding of organisations is achieved by combining the four perspectives. These are then applied to emphasise different aspects of an HSE management system and the preconditions necessary for that system to work effectively.

The *structural perspective* emphasises goal accomplishment. With respect to safety issues, there are company goals (lost-time injury rates, risk acceptance criteria), operational goals for individual jobs (compliance to rules, procedures), or public goals (safety first, vision zero). For each organisation, there exists a structure that suits the goal, the context, and the organisational members. Goals are achieved through safety result feedback conveyed to management for follow-up and through rationality in the decision-making process. The organisational structure, its distribution of responsibility, and its authority are central parts of this perspective. Development of and compliance with formal rules and procedures are other typical aspects associated with the structural perspective. Using this perspective, problems in the organisation are understood as inadequate structures that should be solved by restructuring. Most of the HSE management systems applied in industry today have been designed on the basis of this perspective: monitoring and control, procedures and rules, rational decision models, and structural allocation of responsibility. The classical bureaucratic approach to safety management according to Heinrich (1959) fits this perspective.

The *human resource perspective* considers human needs as its point of departure. Organisations need individuals, and individuals need organisations. This perspective focuses on the interactions among people, empowerment through delegation of authority to the employees, and employee participation in the decision-making process. Development of informal group norms, how much management cares, and various concerns are emphasised. The socio-technical research tradition (Trist 1981) fits well into this perspective. Socio-technical studies have had a major influence on democratic traditions in Scandinavian organisations as well as on participative approaches to organisational development and occupational health and safety management. Occupational health and safety legislation includes requirements for worker involvement; this is also an important aspect of the draft standard, ISO 45001 on occupational health and safety management systems (ISO 2016). Managers who are influenced by this perspective will seek to facilitate the solving of safety problems as close to the source as possible. From this perspective, we look for organisational solutions, where arenas are established for worker participation in decision-making. These arenas are part of the feedback system and will facilitate the informal exchange of information and experiences regarding safety problems.

The *political perspective* emphasises the fact that decisions in an organisation are made in order to allocate scarce resources. Conflicts of interest will arise among different interest groups (management, workers, experts, and so on) as to what constitutes an acceptable solution to a safety problem. Conflicting objectives among safety, production, and efficiency issues may also occur (Rasmussen 1997). These will be settled through bargaining processes with the participation of the different interest groups. An HSE management system has to provide organisational solutions that allow bargaining to take place. The political situation inside a company and the degree of mutual trust between management and labour are important framework conditions. A reasonable degree of trust and good relations are preconditions for HSE management to be effective. A well-functioning HSE management system will also support processes inside a company that improve management–worker relations.

The *symbolic perspective* emphasises the symbolic value of promoting safety for the organisation and its stakeholders. The 'cultural' element in the management of safety and how people in the organisation interpret and attribute a meaning to accidental events are important aspects. The concept of a feedback system mainly belongs to the structural perspective. The symbolic perspective pays attention to the way things are actually being done in organisations, while the structural perspective pays attention to work as planned. We find that most of the systematic HSE management efforts fall under the structural perspective, including feedback systems. The symbolic perspective thus represents an important complementary understanding of how HSE management operates. (The symbolic perspective is further elaborated in Section 4.8 on safety culture.)

Commitment to safety among workers (and, in particular, managers) is part of the symbolic perspective. Management recognises the value of displaying the company's social responsibility by developing and promoting HSE policies and objectives. To maintain credibility inside and outside the company, top management's commitment has to permeate all levels of the organisation. HSE management will serve as an important tool in this respect. It will have an adequate impact only when it supports openness and debates, where different individual experiences and interpretations of why accidents occur and how to prevent them are shared. Feedback systems on safety must not be tools for transferring secret and/or distorted data or for unilateral top management control.

2.2   Conditions outside the organisation

2.2.1   Regulatory requirements for the management of safety

The employer is the primary body responsible for accident prevention at the workplace. Although safety legislation varies from country to country, and from industry to industry, typical duties of the employer are to ensure that:

- Design and organisation of the workplace meets the requirements defined in laws and regulations.
- Hazards are identified and evaluated and that the necessary remedial actions are implemented.
- Safety performance is monitored regularly.
- Safety work is carried out in co-operation with employees.
- Employees are informed about workplace hazards and about safe work practices and are given necessary safety training.

These different requirements have clear implications for the design of companies' feedback systems on accident risks. Employers have to establish and maintain administrative systems for the collection and use of data relating to workplace hazards. Authorities also bring the employer or the responsible manager to court in cases where violations of regulatory requirements have caused severe accidents with the intention that such actions have deterrent effects and counteract complacency. However, this may, on the contrary, result in self-protective actions – by which the employer hides sensitive information on accident causes.

Regulations play an important role in society's efforts to control hazards and thus prevent injuries to citizens. Regulating mechanisms follow a hierarchical structure: laws and regulations are transferred to and implemented in companies by employers through internal rules and procedures that frame employees' behaviour when they are close to a hazard (Rasmussen 1997). Regulators and authorities need knowledge about safety performance as a part of their regulatory tasks (regulations, audits, and supervision). Thus, experience feedback at a macro level is an important part of risk management in society. Next, we look into some of the most important regulations for safety information systems.

Following a number of major accidents involving dangerous substances in the 1970s and 1980s, authorities in Europe and the United States enacted legislation to prevent such occurrences in the future. This legislation extends further than the general working environment and environmental care legislation and makes explicit what employers have to do to manage risks adequately. The so-called Seveso directives in Europe and the Clean Air Act Amendments in the United States specify HSE management system elements that highly hazardous industries must implement (EPA 1990; European Commission 2012).

Central to these regulations are requirements for performing formal risk assessments. Employers must also carry out investigations of incidents and accidents to determine causes, and they must implement remedial actions to prevent recurrences.

In some countries, enacting regulations that specify HSE management system requirements has been extended to apply to HSE systems in general, including ordinary occupational accidents. The so-called internal control regulations in Norway and Sweden require companies to develop and implement adequate management systems to ensure that their activities are planned, organised, and maintained in accordance with HSE legislation (Arbeidstilsynet 1996; Arbetarskyddsstyrelsen 1996).

In the United Kingdom, the Management of Health and Safety at Work Regulations 1999 have a similar objective. They require employers to carry out risk assessments, to implement measures to remedy the risks, to appoint qualified personnel, and to provide for adequate information and the training of employees.

In the United States, safety and health program management guidelines require employers to set up programs to manage workplace safety and health (OSHA 1989). The aim is to reduce injuries, illnesses, and fatalities by systematically achieving compliance with occupational safety and health regulations.

Let us look closer at the Norwegian internal control regulations (Arbeidstilsynet 1996), which are a typical example of the types of requirements that employers have to meet. According to this legislation, there must be routines inside the company for:

- Maintaining an overview of the HSE regulations
- Ensuring that employees have adequate safety knowledge and skills
- Ensuring that employees follow safety procedures in order to utilise their collective experiences
- Establishing safety goals
- Maintaining an overview of the company's safety organisation and distribution of responsibilities
- Systematically identifying and assessing hazards and implementing action plans to control them
- Identifying nonconformities in relation to regulatory requirements
- Performing audits and reviews to monitor the effectiveness of the company's HSE management system

The requirements in internal control regulations demonstrate that feedback systems need to be in place to ensure that these requirements are fulfilled.

2.2.2 Regulations on record keeping and on reporting injuries and incidents to the authorities

Legislation usually requires employers to investigate incidents and accidents in order to plan preventive measures and to report to the authorities. In the United States, employers must keep records of work-related fatalities, illnesses, and injuries (OSHA 1971/2015). According to the minimum requirements within the European Union, employers must keep a list of occupational accidents resulting in employees being unfit for work for more than three working days (European Council 1989). Similar requirements exist in other countries. The aim is to ensure that employers keep an overview of accidents that have occurred – for preventive purposes and to ensure that the information is available for possible later compensation claims.

The requirements for reporting accidents to the authorities vary among countries. The differences mainly concern the classes of employees to be included and the severity thresholds (e.g. number of days of absence). Usually, there are two levels of reporting to the authorities – immediate reporting of severe accidents and later reporting of less-severe work-related accidents.

Usually, severe accidents involving fatalities, serious injury, fires and explosions, and so on must be reported to the enforcing authorities immediately. In the United States, accidents resulting in fatalities must be reported to the Occupational Safety and Health Administration (OSHA) within eight hours. Any inpatient hospitalisation, amputation, or eye loss must be reported by employers within 24 hours of their learning about it (OSHA 2015). In Great Britain, the enforcing authority must be notified immediately about accidents resulting in death or major injury and about dangerous occurrences (HSE 2013). This notification is then followed by a report within 10 days. Similar requirements exist in the Scandinavian countries.

National laws vary concerning the requirements for reporting less-severe accidents to the responsible authorities. In the United States, the so-called OSHA-recordable accidents involving absence from work, medical treatment, loss of consciousness, restrictions on work or motion, or transfer to another job are reported to the authorities on a sampling basis.

In Great Britain, the Reporting of Injuries, Diseases, and Dangerous Occurrences Regulations (RIDDOR) stipulate that accidents resulting in more than three days' absence must also be reported within 10 days to the enforcing authorities (HSE 2013). In Sweden and Norway, the reporting threshold is absence from work beyond the day of injury.

Reporting accidents to the authorities fulfils different aims. One is to supply the authorities responsible for the development of safety regulations with the necessary knowledge for prevention. A second aim is to provide input to the programs for supervision of compliance with workplace regulations.

2.2.3 Workers' compensation systems

Most countries have compensation systems for work-related disabilities (injury or illness) or premature death (Stellman 1998). Usually, these cover loss of income, medical payments, and rehabilitation expenses. There are different means of organising the workers' compensation system:

- The employer has to provide insurance coverage through an insurance company. Such companies are subject to regulations and supervision by a governmental agency.
- Workmen's compensation is part of the social insurance provided by the government.
- The system is operated by a special government agency – a so-called workers' compensation board.

Usually, it is the duty of the employer to report a disability or fatality to the insurer. When possible, the worker is also required to notify the employer about an injury or illness. Compensation is paid for disabilities and fatalities that result from events and circumstances of employment. Usually, it is irrelevant whether there is any fault on the part of the victim, the employer, or any other party.

The worker or the employer has to provide a factual report about the circumstances related to the disability or fatality for compensation to be paid. In the case of accidents, the question of whether the event resulted from employment usually is not problematic.

Usually, the employer pays the costs of workers' compensation as a percentage of the payroll. It is common practice for premiums to be adjusted so that expenditures and revenues are equalised over time, but a partial subsidy by the government also may occur. Usually, the premium is established for each class of industrial activity based on historical data related to expenditures for compensation. Experience rating is also practised, meaning that the premium to be paid by an employer is adjusted based on previous experience with that employer. Some insurance companies also give premium incentives for improvements in HSE standards and HSE management systems that are made by employers. To summarise, reporting accidents to the insurers provides a basis for compensation to the victim and for the determination of insurance rates.

2.2.4 International standards and guidelines

Managers need to demonstrate their commitment to safety, health, and environmental issues to interested parties such as stakeholders, customers, the community at large, voluntary organisations, and the authorities. One way of doing so is to implement voluntary standards for HSE management and to demonstrate proof of compliance from an independent certifying body. In order to receive such a certificate, the company's HSE management system has to be audited by the independent body to demonstrate that it is operating effectively and that it is in conformance with the standard.

The most important example is the draft ISO 45001 on occupational health and safety management systems, which is expected to be published as an ISO standard in 2017. Another example is Health and Safety Guidance (HSG) 65 on health and safety management, which is published by the UK Health and Safety Executive (HSE 2013). ISO 45001 will replace the British Standard Occupational Health and Safety Assessment Series (BS OHSAS) 18001:2007, *Occupational Health and Safety Management Systems,* which is an internationally applied standard (OHSAS Project Group 2007). The draft version of ISO 45001 (ISO 2016) provides requirements for safety management systems to enable an organisation to improve its safety and health performance by:

- Developing and implementing policies and objectives for its systematic safety management
- Ensuring top management's commitment to the safety management system
- Establishing systematic safety processes that consider the context of the organisation
- Determining hazards and risks associated with organisational activities
- Establishing operational controls to eliminate or minimise risks
- Increasing awareness of hazards and risks and associated operational controls through information, communication, and training
- Evaluating safety performance and seeking to improve it
- Establishing and developing the necessary competencies
- Developing and supporting an OSH culture in the organisation
- Ensuring that workers and their representativeness are informed, consulted and that they participate in safety management systems

The plan–do–check–act (PDCA) circle, which is discussed in detail in Chapter 7, is applied as the basic principle of occupational health and safety management through experience feedback of the draft ISO 45001 standard as well as the HSG65 guidelines. This implies that diagnosis and continuous improvement are key principles in systematic safety management according to the standard. We find the same foundation in BS OHSAS 18001:2007 and in the ISO 9000 standard family on quality management.

There are several other ISO standards relevant to systematic HSE management. The development of the draft ISO 45001 is based on certain elements, including the PDCA circle, that are common to ISO management system standards such as ISO 9001 (quality), ISO 14001 (environment), and ISO 31000 (risk management). These are all important standards for HSE management. ISO 19011 gives guidance on audits that are especially relevant to safety (see Chapter 12). Another important standard for the management of safety is ISO 12100, *Safety of Machinery* (see Chapter 23). There are also sector-specific ISO standards – for example, ISO 39001 on road traffic safety management systems.

The ISO 9000 standard family addresses different features of quality management. There are many similarities among the principles applied in quality management and in occupational health and safety and environmental management, and many companies have developed integrated HSE and quality management systems.

2.2.5   Other non-governmental organisations

There are numerous organisations at the national and international level that are active in the area of safety, such as:

- Associations of safety professionals
- Employers' and workers' organisations
- Joint labour–management councils
- Associations of producers, manufacturers, and operators
- Voluntary organisations

These organisations influence corporate safety practices in different ways. Some provide professional support and services by arranging conference and training programs for safety professionals and by publishing educational materials, professional journals, and textbooks. Other associations represent arenas for experience exchange and for the development of common policies and practices. They represent, for example, a particular branch of industry such as chemical manufacturers, oil companies, or primary aluminium producers. There are also interest groups taking care of members' interests in the area of safety, such as various worker and management organisations.

chapter three

Case study

3.1   Reducing emissions into the air from a fertiliser plant

In this chapter, we will present a case about a fertiliser plant. This case involves the successful introduction of a new system for feedback control of accidental emissions from the plant to the environment. This system illustrates some basic qualities of accident prevention resulting from experience feedback. We have selected a case involving environmental emissions even though the main focus of this book is on occupational safety. However, this case illustrates many principles of accident prevention through feedback control that are the same for all areas of health, safety, and environment (HSE).

An integrated process control system was introduced at this plant. The goal was to reduce emissions of nitrogen compounds into the air by controlling the process through measurement of disturbances (deviations) rather than through measurement of emissions (output). The system provided the control room operators with timely information on process deviations that had the potential to cause environmental pollution (see Figure 3.1). Furthermore, the timely information resulted in faster interventions by the control room operators and thus improved regulation of the system.

Before the integrated process control system was introduced, emissions were registered continuously by measuring the nitrogen content in the exhaust pipe's air. This information, collected once every 24 hours, was summarised in internal reports and in those reports sent to the authorities. In the case of major accidental emissions, the causes were analysed and remedial actions were taken, usually with a delay of more than 24 hours. There was no closed feedback loop operating on a daily basis. Figure 3.2 shows the output from the plant during the two years prior to the introduction of the new system and 1.5 years after the introduction.

Plant management analysed the causes of typical emissions and used this information as input for the new design. They found that accidental emissions could be traced back to process upsets in the form of deviating values for different parameters such as pressure, temperature, and flow. A new system was designed so that sensors continuously measured these parameters, and the results were fed into a process control system.

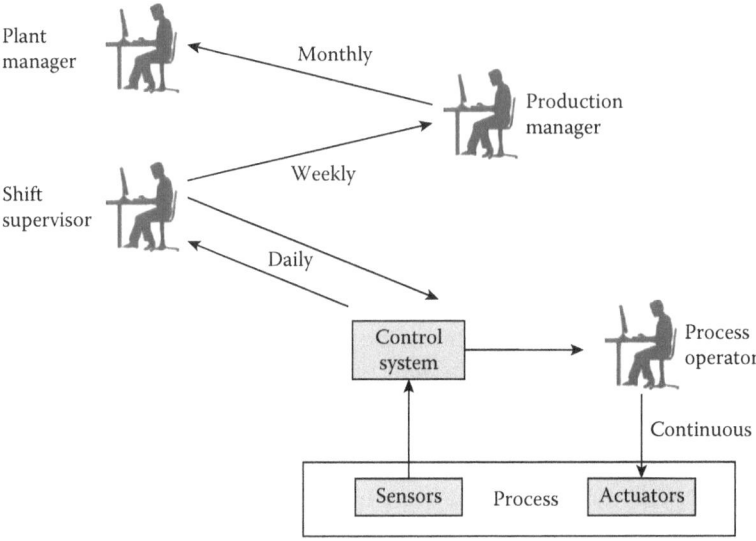

Figure 3.1 Integrated process control system.

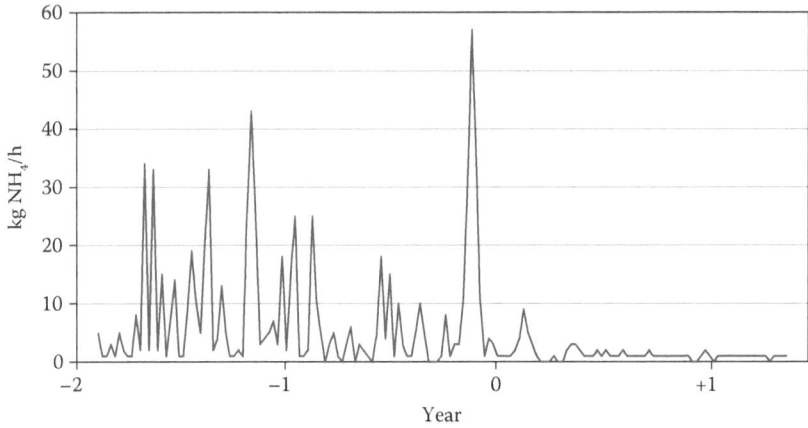

Figure 3.2 Nitrogen compound emissions from the fertiliser plant into the air during a period of 3.5 years. The integrated control system was introduced at Time 0.

Operators monitored the system but normally did not intervene. The new system made it possible for operators to adjust the process in a timely manner and thus prevent emissions into the air. In Figure 3.2, we see the effects of the new system as evidenced by the reduced emissions after it was introduced at year 0.

The new process control system also had some other interesting qualities. It provided management at different levels with filtered and summarised information according to their needs (see Figure 3.1). The shift supervisor, for example, received daily reports on causes of emissions and was responsible for measures to improve the control system and the production process in general.

Using this example, we will look into the contents of the remaining parts of this book, which will help us in answering such questions as:

- *How can the causes of accidental events involving the releases of pollution be analysed?* Chapter 4 gives an overview of accident models that have been developed to support analyses of this type. Chapter 5 presents one particular analysis framework that suits this purpose. The roles of the operators at the sharp end, who are responsible for controlling the process, are discussed in Chapter 10. Chapter 13 presents practical methods used in investigations for determining the causes of accidental releases. Accumulated incident data play a key role in the establishment of periodical accident statistics, and incident databases support this purpose. Chapter 14 provides an overview of the design and application of such databases.
- *How can accidental releases be prevented from occurring?* Chapter 9 introduces the concept of barriers against accidents. Although the focus is on barriers against accidents resulting in personal injury, the same principles are applicable to the prevention of environmental pollution through accidental discharge of chemical compounds.
- *What principles should be applied in the distribution of information on accidental releases and underlying factors to the different levels of an organisation for use in decision-making?* Chapter 7 outlines a number of aspects relevant to this issue. Feedback control and diagnosis are two central themes of this chapter. Typically, the process control system is based on feedback control. Diagnosis is a central concept in problem-solving. It is applicable in two respects. One is in the analysis of causes of accidental releases. The second application is illustrated by the earlier example, where diagnosis was a part of the development of the process control system.
- *When to react with respect to increases or decreases in the frequency of accidental releases?* Accidents are random phenomena. Chapter 6 gives a short review of the basic statistical theory necessary for us to be able to analyse the occurrence of accidents in consecutive time periods. We will use formulas based on the Poisson process in the presentation of safety performance indicators in Section IV of this book.
- *What types of sensors should be used in monitoring the process?* This is a specific question on the design of measurement devices for process control.

A general presentation of different methods for the collection of data on accident risks is given in Chapters 11 through 13.

- *How can the occurrence of accidental releases be prevented without first waiting for the accidents to happen?* Section V presents different methods of risk assessment. These are already applicable at the design stage and will help to identify the need for barriers against accidental releases. The use of risk assessments during the operations phase will speed up experience feedback and the learning process with respect to future incidents.
- *How can the process control system be examined in order to determine whether it functions as planned and is effective and suitable?* Chapter 12 presents the basic principles of quality audits. Audits may be used as tools for answering these questions.
- *What can we learn from other applications of feedback control systems?* Section VI presents different cases showing how the design and management of HSE systems based on feedback control are handled in practice.
- *What specifications should we use when designing a feedback control system?* Chapter 8 presents a list of criteria or requirements for feedback control systems. These criteria are intended for use as a support in answering this question.

section two

Theoretical foundation

This part lays the theoretical foundation for the presentation of various safety management methods and tools that will be discussed in Sections III through V. Chapter 4 gives a presentation of different accident theories and models. Initially, we will focus on accident models that are necessary in order to ask the right questions in an accident investigation. At the end of the chapter, we will highlight some important aspects related to safety culture, an important dimension that is complementary to the more structural approach used in health, safety, and environment (HSE) management. Chapter 5 summarises many important aspects of the understandings of accidents presented in Chapter 4 within an accident analysis framework. We use this framework when we analyse the different types of information about accident risks used in the management of these risks. Chapter 6 summarises the basic statistical theory of the Poisson process. This theory is applied in Section IV on the monitoring of safety performance. Chapter 7 reviews some basic principles for experience feedback. We will introduce two important concepts – feedback control and diagnosis. We will also discuss some important obstacles to efficient learning from experience. Chapter 8 presents a set of criteria for the evaluation of methods in the management of safety based on the concepts of feedback control and diagnosis. Chapter 9 introduces the concept of safety barriers. It also presents applications of this philosophy in the prevention of ordinary occupational accidents as well as major accidents. Chapter 10 concludes Section II by introducing us to the important area of the role of the human element in accident prevention and human errors.

chapter four

Accident theory and models

The main purpose of this chapter is to lay the theoretical foundation for the analytic framework presented in Chapter 5. This framework will be used as a main building block in our presentation of safety management methods and tools in subsequent chapters. The main focus will be on accident models (i.e. simplified representations of the processes in the real world that result in accidental loss). Many of these models include portions that are substantially theoretical and reflect prevailing theories of accident causation. For our purposes, accident theory means a supposition or system of general principles and ideas intended to explain why accidents occur (cf. Oxford University Press 2012).

We end this chapter with a short review of safety culture. This review is not consistently theoretical. We make some selections by focussing on models of safety culture that explain an organisation's informal processes and beliefs and how they affect the risk of accidents.

4.1   On the need for accident models

Accident models play a vital role in the design of methods for accident investigation and of feedback systems for safety management in general. Each accident model has its own characteristics as to the types of 'causal factors' that it highlights. In an accident investigation, for example, accident models support the investigators by:

- Creating a mental picture of the accident sequence
- Asking the 'right' questions and defining the types of data to collect
- Establishing stop rules (i.e. rules for when to terminate the search for new causes further away from the accidental event)
- Checking that all relevant data have been collected
- Evaluating, structuring, and summarising the data into meaningful information
- Analysing relations among pieces of information and seeing interrelations
- Identifying and assessing remedial actions
- Communication among people by providing a common frame of reference

The models are also important in risk assessments by supporting in the identification of hazards, deviations and contributing factors that can lead to accidents.

An important aim of introducing an accident model is to establish a shared understanding within the organisation of how and why accidents happen. It is especially important that those parts of the organisation responsible for the collection of information on accident risks and those responsible for using the information in decision-making use a similar frame of reference. Accident models will thus have a direct influence on safety practices, both consciously and unconsciously.

Old accident 'models' or perceptions explained accidents as a result of fate (Hovden and Larsson 1987). Accident research during the early part of last century studied the relation between accidents and personal traits of the victim. 'Accident-proneness' was a commonly accepted theory at that time (McKennan 1983). It stated that, due to their personal traits, certain individuals are more susceptible to accidents than others. It follows from this theory that accident risk may be reduced substantially by removing 'accident-prone' persons from hazardous jobs. Today, 'accident-proneness' is considered responsible for only a number of accident incidents. It follows that, in most cases, a preventive strategy based on this theory will have only minor effects. It is relevant, however, for specific types of jobs (e.g. pilots) that involve high mental and/or physical demands and where an error may have a significant safety impact. We do not find accident models that focus solely on personal factors in use in industry today.

The first model considering accidents and their context in a systematic way was established in the 1930s. The so-called domino theory in particular has significantly contributed to current theories and practices in safety management. There are several subsequent models that focus on the sequence of causes in a way that is similar to the domino theory.

Since the 1960s, the field of quality management has been an important contributor to the development of safety management as it is practised today. This contribution has resulted in accident models that emphasise processes and deviations; these are known as process models.

Energy transfer is another important aspect for understanding accidents. The energy model is based on injury epidemiology. It explains an injury as resulting from an energy transfer to the victim's body in excess of the 'body injury threshold' (Gibson 1961; Haddon 1968).

Over the years, accident models also have incorporated and highlighted factors pertaining to management, organisation, and the individual. This has resulted in system models that highlight different parts of the organisational context of accidents (Khanzode et al. 2012).

In the subsequent sections, we present a selection of accident models from the research literature that are relevant to the design of methods applied in safety management. This presentation will help us understand the framework for accident analysis in Chapter 5 and will form the basis for our presentation of safety management methods and tools in Sections III through V of this volume. The intentions of Chapters 4 and 5 are twofold. One is to help the reader develop an understanding of the nature of accidents. The second is to build a pedagogic structure to organise the distinct characteristics of the different methods and tools. We will also include models of historical significance to help the reader understand developments in the general understanding of accident phenomenon.

There are many different models presented in the research literature, representing the needs of individual researchers to highlight specific aspects. Fortunately, there are converging trends in accident research in that different subsets of models have important aspects in common. We will focus on these common aspects rather than on the particularities of each model. We will also see how different 'schools' contribute to an overall understanding of accident phenomenon. We have categorised six types of accident models, five of which are presented in this chapter:

- *Causal-sequence models* present an accident as a chain of events that finally ends in some kind of loss.
- *Process models* explain an accident as deterioration from a normal state to deviations that lead to loss of control and injuries.
- *Energy models* explain accidents as transfers of energy and show how barriers can prevent or stop the energy flow and thus protect victims.
- *Logic tree models* present causes of accidents in terms of logical relations among events and conditions in the affected system.
- *System models* pay particular attention to human, technical, and organisational factors or to the interplay among them. These include management models and so-called systemic models that consider the dynamics and complexity of accidents.
- *Cognitive models* analyse human errors in terms of failures in cognitive function – how these are caused by contextual variables and how these variables lead to accidents (presented in Chapter 10).

Many accident models are combinations of these six types of accident models. In this chapter, we have categorised the accident models as to the most dominant type.

4.2   Causal-sequence models

An early and historically very important example of a causal-sequence model is the *chain of multiple events* or *domino theory* (Heinrich 1959) as shown in Figure 4.1. In a causal-sequence model, an accident is described as a chain of conditions and events that culminate in an injury. A link in

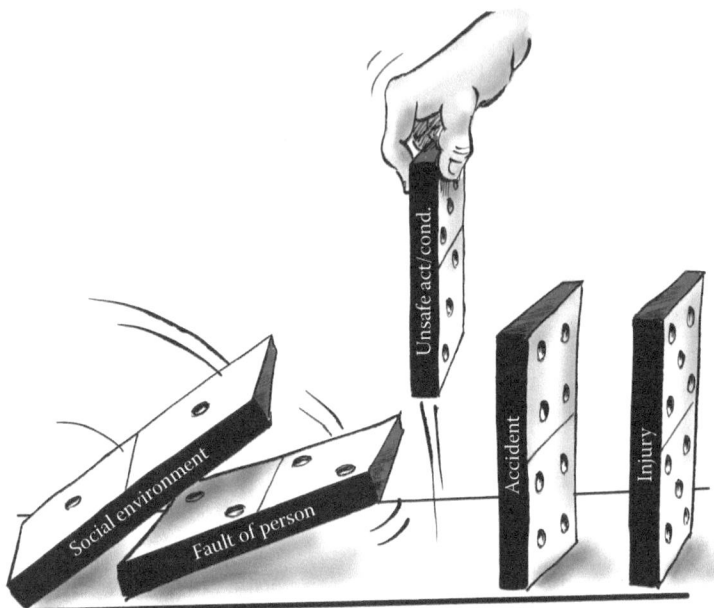

Figure 4.1 The domino theory. (Adapted from Heinrich. 1959. *Industrial accident prevention* [4th edition].)

this chain is an unsafe act or an unsafe condition at the workplace. It is suggested that accidents can be prevented through the reduction of unsafe acts and conditions. By removing the unsafe act/condition, the sequence of events is stopped as is illustrated in Figure 4.1.

In many countries, the domino theory has had a very great influence on the development of procedures for the classification of accidents. Classification is used to standardise the collection of accident data and to reduce the complexity of that data to a manageable level for statistical purposes. An early and important example of a classification system based on the domino theory is the American National Standards Institute's system for the classification of accidents, which dates from 1926. This has been replaced by the record-keeping standard of the Occupational Safety and Health Administration (OSHA 1971/2015). In this system, the following facts are recorded about the accident sequence:

1. Injury
   a. *Nature of injury:* cut, fracture, burns, and so on
   b. *Part of the body affected:* head, neck, upper extremities, and so on
2. Accident
   a. *Accident type:* strike against something, fall, slip, and so on
   b. *Agency of accident:* machine, vehicle, hand tools, and so on

3. Hazardous condition: unguarded, defective tools, unsafe design, and so on
4. Unsafe act: failure to secure, operating at unsafe speed, and so on

These different factors are also used in many of the standard accident reporting classification systems in Europe. Today, classification of accidents with respect to unsafe acts and hazardous conditions is regarded as obsolete, since it does not take into account the multiple causality of accidents.

The domino theory is the predecessor of many current accident models. In these models, the early part of the causal chain (fault of person and social environment) has been replaced by management factors. The aim is to evaluate how these factors affect the likelihood of unsafe acts and conditions. The International Loss Control Institute (ILCI) has modified the original domino theory in their ILCI model to include these types of factors. The last three 'events' in the sequence are similar to those of the domino theory. The 'injury' event has been replaced by the more general category of 'loss' (i.e. injury or fatality to people, property, or environmental damage) and process disruptions. 'Fault of person' has been replaced by 'personal or job factors', and the so-called root causes (i.e. the 'lack of control') include failures in safety management factors similar to those found in quality assurance programs. An investigation according to the ILCI model is carried out from right to left in a *five-way analysis* to identify root causes.

The ILCI model has had a great influence on safety practices in many countries through the International Safety Rating System (ISRS), a safety management auditing and rating system. It also serves as a basis for the categorisation of information in the accident and near-accident reporting systems that are used by many companies.

One problem with causal-sequence examples such as the domino and ILCI models is that no clear distinction is made between the observable facts about the accident sequence on the one hand and the more uncertain 'causal' relationships at the personal, organisational, and management levels on the other hand. The user is thus led to believe that information on, for example, personal factors such as mental stress have the same objective status and are as unambiguous as information on observable facts about the sequence of events. There is thus a risk of misunderstandings and false interpretations – especially at the higher management levels, where detailed knowledge about the accident occurrences is lacking.

The Tripod model relies on causal sequences that are somewhat similar to the logic principles of the ILCI model; see Figure 4.2 (Reason 1991). It has had a great influence on current thinking because it models how 'erroneous' decisions at different management levels lead up to circumstances that result in an accident. We will apply aspects from the Tripod model when we illustrate how an accident occurrence is affected both by

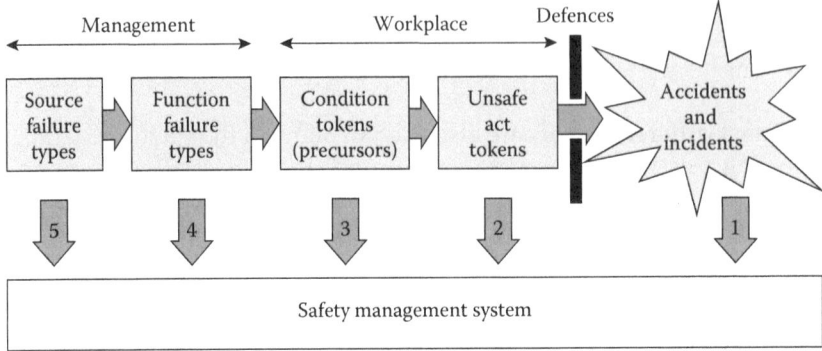

Figure 4.2 The Tripod model of accident causation. (Reprinted from Near-miss reporting as a safety tool, Reason, J., Too little and too late: A commentary on accident and incident reporting systems, 109–120, Copyright 1991, with permission from Elsevier.)

operational decisions at the work system level immediately before the accident and by supervision and higher management decisions.

In the Tripod model, distinctions are made among different classes of failure, mainly coinciding with the levels of organisational hierarchy:

1. *Unsafe act tokens* are specific failures made by individual operators at the sharp end (i.e. the work system level).
2. *Condition tokens* are error-provoking conditions at the local workplace.
3. *Failure types* are general classes of organisational and management failures. There are two failure types:
    a. *Function failures* are latent failures resulting from decisions made by line management, designers, planners, and so on.
    b. *Source failures* are fallible top management decisions at the strategic level.

We recognize a sequence of causes in the model. The 'defences' in the Tripod model shown in Figure 4.2 are similar to the safety barriers outlined by Haddon (1980) in Section 4.4 of this chapter.

The Tripod model also helps us in understanding the relationship among different feedback channels in a safety information system. In Figure 4.2, we identify five different 'channels' for feedback of information on accident risks:

1. Accident and near-miss reporting ideally will cover both the accident sequence and the underlying erroneous decisions.
2. Unsafe-act reporting will cover both the actual acts and their results, such as poor housekeeping.

3. Reporting on unsafe-condition tokens will uncover poor workplace design, high workload, inadequate training, and so on.
4. Reporting on function-failure types at the department level will identify faulty layout, maintenance, procedures, housekeeping, communication, training, and so on.
5. Reporting on source-failure types will uncover substandard strategic decisions due to lack of top-level commitment/competence and to a poor safety culture.

4.3   Process models

Process models help us understand how a production system successively develops from a state of adequate control through increasing lack of control into a state where an accident occurs. Thus, time is a basic factor. In contrast to causal-sequence models, process models make a clear distinction between the accident sequences on the one hand and the underlying causal or contributing factors on the other hand. Figure 4.3 shows the phases of different process models for accidents referred to in this chapter.

One example of a process model is the Occupational Accident Research Unit (OARU) model that distinguishes among different phases of the accident sequence (Kjellén and Larsson 1981). Here, we divide the accident sequence into three phases: the initial phase, the concluding phase, and the injury phase. There are four transitions among the different phases:

1. Transition from normal conditions to a state of lack of control.
2. Transition from lack of control to loss of control.

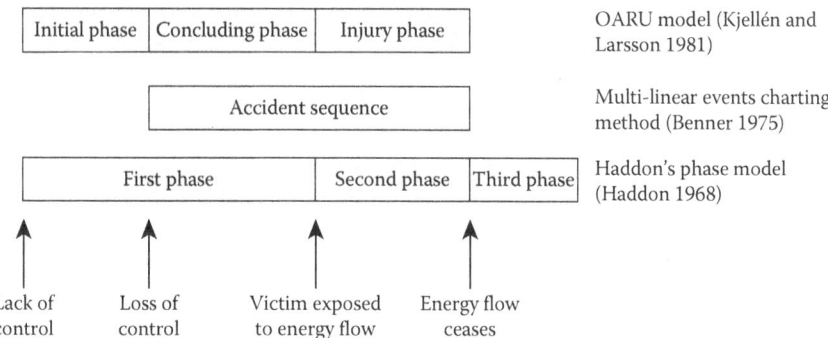

Figure 4.3 The phases of different accident process models.

3. The victim or other target (the body, in the case of personal injury) starts to absorb energy.
4. Energy absorption ceases.

The state of lack of control is characterised by the presence of *deviations* in the system. These are events or conditions that depart from the *norm* for the faultless or planned processes of the system. In the accident research literature, we find different terms used to illustrate the concept of deviation – such as 'critical incident', 'unsafe act', 'unsafe condition', and 'disturbance'. Various types of norms appear in the literature:

* Standard, code, rule, or regulation
* Adequate or acceptable
* Normal or usual
* Expected, planned or intended

A checklist of deviations has been developed to support accident investigations. Deviations are divided into three main categories: work system, environment, and safety systems (see Table 5.10 in Section 5.5). The checklist primarily is based on an industrial engineering systems view and underlines the relationship between accident control and production control.

The OARU model uses the term *determining* or *contributing factors* rather than causal factors. These are human/social, technical, and organisational properties of the work system and department that affect the accident sequence but that only change slowly in comparison with it. Figure 4.4 shows an example of an accident analysed by means of the OARU model. Here, the distinction between the accident sequence and the underlying, determining factors is made clear. Figure 4.4 is based on the following description of an accident.

Example: Ola Andersen worked on a building site installing concrete slabs in a gallery. The team consisted of Ola, a trainee, and a crane driver. Ola's ordinary workmate was sick. Another work team was ready to begin to pour concrete and called for the crane. The installation work had to be finished soon, and a slab was not put in place correctly. Ola moved out on a beam about 3 m above the ground level to use a crowbar to move the slab. He slipped, lost his balance, and fell 3 m to the ground. Ola suffered a back injury and broke a rib. From the accident report, we read that Anderson lost his balance, slipped, and fell one level down.

The significance of deviations as a risk-increasing factor is supported by empirical evidence. The risk of accidents increases when the production system develops into a state of lack of control, where there are production disturbances, defective equipment, non-ordinary personnel, and so on (Kjellén 1983). These deviations represent symptoms of a bad

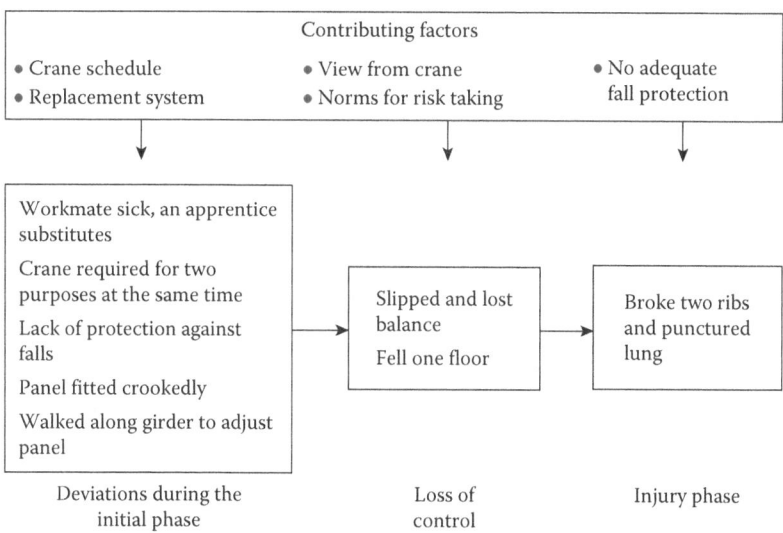

Figure 4.4 Analysis of an accident at a construction site using the OARU model. (Adapted from Kjellén, U. and Larsson, T.J., *J. Occup. Accidents*, 3, 129–140, 1981.)

'control climate' at the plant, that is, deficiencies in the management system for production control (Grimaldi 1970). A bad control climate will not only result in an increased risk of occupational accidents but also in an increased risk of property damages, unscheduled production stops, accidental contamination of the environment, and production outside quality norms.

This does not mean, however, that all types of deviations will increase the risk of accidents. Some cases, for example, deviations from work instructions, represent a means used by personnel at the sharp end to adapt their behaviour to better suit the actual conditions at the workplace. Rather than changing the behaviour, it may be a question of changing the norm. Each type of deviation has to be analysed with respect to its significance to the risk of accidents.

Haddon's phase model was developed for the purpose of studying traffic accidents, and the phases were denoted 'pre-crash', 'crash', and 'post-crash'. The model was later generalised in order to apply to studies of other types of accidents (Haddon 1968). The so-called *Haddon matrix* (Haddon 1980) also includes a second system ergonomics or a human-factor dimension with human, equipment, and environmental aspects. Haddon's matrix has been applied as a useful tool in interventions to prevent accidents in the health sector and in road traffic in particular (Runyan 2015).

Events and causal factors charting (ECFC) models an incident as a sequence of interlinked events (DOE 2000). For each event, causal factors are determined by first considering the direct causes of each single event. In the next step, the root causes of each single event are analysed. The man, technology, and organisation (MTO) investigation method combines the use of checklists similar to those for contributing factors in the OARU model with the ECFC technique (Bento 1999). This method was originally developed for the Swedish nuclear industry but has been applied more widely (e.g. in the oil and gas industry).

The Multi-Linear Events Chartering Method is based on the view that an accident is a process involving interacting events and conditions (Benner 1975). An accident begins when a system is transformed by a disturbance and ends with the last injurious or damaging event. An event is described in terms of an actor (animate or inanimate) and an action. The sequence of actions of each actor and the interrelations among the actions are displayed in a diagram. Also, conditions that influence the different actions are displayed. This model is a central part of the sequentially timed event plotting (STEP) method for accident investigations (Hendrick and Benner 1987).

4.4   The energy model

The *energy model* is rooted in epidemiology. It represents an effort by the medical discipline to systematise the analysis of accident causes in a way similar to that of analysing causes of diseases. Gibson (1961) pioneered this development. He based his model on the fact that a transfer of energy in excess of body injury thresholds causes injury to a person. The injury agent is energy exchange. A central part of the model is the hazard (i.e. an energy source with the potential of creating injury to personnel or damage to the environment or material assets). There are different types of hazards – gravity (potential energy), motion (kinetic energy), mechanical electrical, pressure, temperature, chemical, biological, radiation, and sound. Using this model, Haddon systematised known principles of accident prevention into 10 strategies. These are related to the different points of intervention shown in Figure 4.5 (Haddon 1980).

The energy model and Haddon's strategies have had a significant influence on European legislation and standardisation work such as that related to hazardous chemicals (European Council 1998) and machinery safety (European Council 2006). They are central components in accident investigation such as in the OARU, Management Oversight and Risk Tree (MORT), Safety Management and Organisation Review Technique (SMORT), and Tripod models presented in this chapter.

Table 4.1 shows examples from different areas where safety measures have been classified in relation to Haddon's strategies.

The energy model has three distinct merits. One lies in the support it offers in checking that all possible preventive measures have

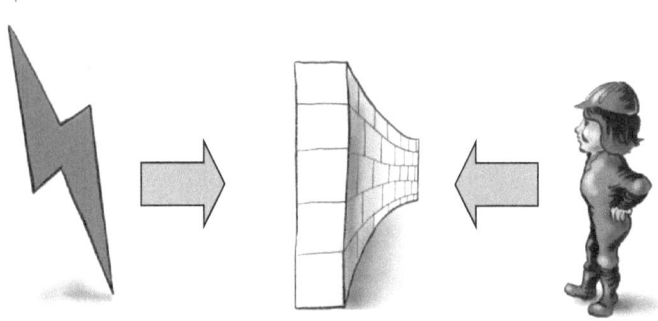

Related to the energy source:	Related to the separation of the energy source from the target:	Related to the vulnerable target:
1. Prevent build up of energy	6. Separate, in time or space, the energy source and the vulnerable target	8. Make the target more resistant to damage from the energy flow
2. Modify the quality of the energy		9. Limit the development of loss (injury or damage)
3. Limit the amount of energy	7. Separate the energy source and the target by physical barriers	
4. Prevent uncontrolled release of energy		10. Stabilise, repair, and rehabilitate the object of damage
5. Modify rate and distribution of the energy		

Figure 4.5 Haddon's 10 accident prevention strategies. (Adapted from Haddon, W., *Hazard Prev.*, 16, 8–12, 1980.)

been identified. In applying the energy model, different priorities have been given to the three main types of strategies. Hence, the primary strategies are those related to the energy source. If it is not possible to eliminate or reduce the hazard to an acceptable level, barriers such as fixed guards are installed. Personal protective equipment is introduced as a last resort. These priorities have been implemented in the European legislation on machinery safety.

It should be remembered, however, that the energy model puts the focus on certain types of measures, whereas others are de-emphasised. Strategy No. 4, prevention of uncontrolled release of energy, is the primary approach in many cases and includes a large array of measures. The energy model is not detailed enough to yield support in identifying and evaluating the specific types of measures that apply in a given situation.

A second merit of the energy model is the support it offers in anticipating the consequences of accidents. After the uncontrolled release of energy, the accident sequence basically follows the laws of physics. The consequences are to a large extent determined by the amount of energy involved. Accident statistics from the Norwegian Directorate of

Table 4.1 Examples of safety measures relating to Haddon's 10 strategies

Type of strategy	Examples of hazards and safety measures		
	Rotating machinery (circular saw)	Toxic chemical (oil vapour and mist from drilling mud in shale shaker)	Motor vehicle (car traffic)
Prevent build-up	Eliminate use of a circular saw by ordering pre-cut pieces of wood	Eliminate oil in mud by using water-based mud	Avoid driving
Modify the qualities	Modified saw blade teeth	Use low-toxicity oil	Softening of hard objects in the cabin
Limit the amount	Limit rotational speed	Smaller evaporation area	Speed limits
Prevent release	Design of start button that prevents accidental start	Not applicable	Sanding and salting of roads
Modify rate and spatial distribution	Emergency stop	Ventilation	Cars with shock-absorbing zones, safety belts
Separate in time or space	Automatic sawing machine	Remote control	Separate lanes for meeting traffic
Separate by barriers	Machine guarding	Air curtain	Cars with safety cage
Make the victim more resistant	Eye protection	Respirators	Helmet
Counter damage	First aid	Not applicable	First aid
Rehabilitation	Depends on type of injury	Depends on type of illness	Depends on type of injury

Labour Inspection show that fatalities are more common among certain types of events with 'high energy content'. Of the 48 work fatalities occurring in 2013, 25% were related to accidents involving motorised vehicles (Arbeidstilsynet 2015). Other dominant types of accidents were 'caught in or between' (23%), falls (13%), and violence (13%). These four categories accounted for three-quarters of all fatalities that year. In Section III, we will use information about the energy content of incidents (i.e. events involving loss of control) to assess their potential for causing loss.

A third important merit is the support the energy model offers in identifying hazards. Each type of industry is associated with certain types of hazards or 'energies' that show up in accident records. This fact is an important input to hazard recognition and to risk assessments.

4.4.1   The Swiss cheese model

Reason (1997) gives an important contribution to our understanding of accidents by analysing how the occurrence of incidents, barrier failures, and human errors are influenced by organisational factors. He explains failures in barriers as active failures and latent conditions (see Figure 4.6). Active failures are errors and violations committed at the sharp end of the system. Latent conditions have existed for a period of time but have gone undetected because the barrier has not been activated. A classic example of the latter is the Concorde flight that crashed in Paris in 2000. Due to the inadequate design of the barrier between the fuel tank and the aircraft tyres, debris from an exploding tyre managed to puncture the tank and cause an explosion. The poor design had existed since the Concorde was put into operation in 1976.

The Swiss cheese model in Figure 4.6 illustrates Reason's theory of organisational accidents in a pedagogical way (Reason 1997; Underwood and Waterson 2014). This model has received significant attention and is

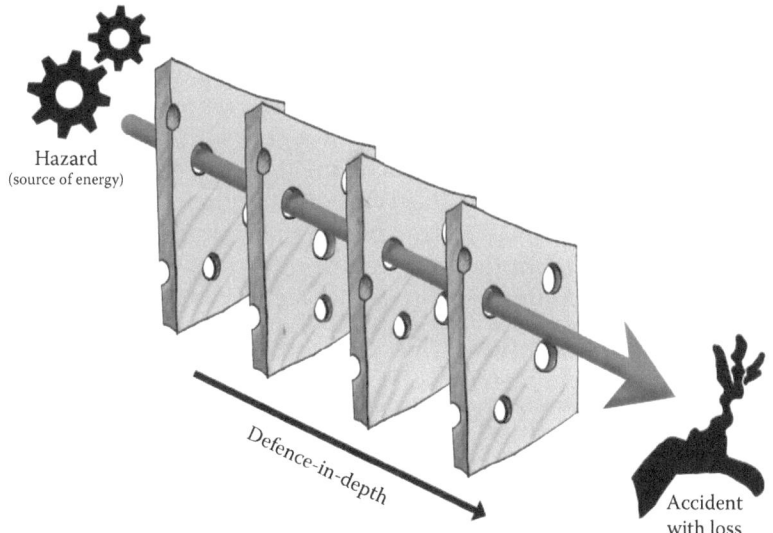

Figure 4.6 Factors leading to accidents. Some of the holes in the barriers are due to active failures, and the others are due to latent conditions. (Adapted from Reason, J., *Managing the Risks of Organizational Accidents*, Ashgate, Aldershot, UK, 1997.)

often referred to in academia and by practitioners (Lundberg et al. 2009). Reason et al. (2006) explain its success as a simple metaphor that encompasses what might be a complex story of accidents caused by failure of several defences generated by organisational factors.

The concept of organisational accidents is based on the notion that accidents have become more challenging to analyse due to the increasingly complex interactions and tighter couplings in modern industrial systems, which we will discuss in more detail in Section 4.6 (Reason 1997; Dien et al., 2004). The origins of accident causes can be traced back to decisions made in many parts of an organisation.

The Swiss cheese model originally illustrated the use of multiple layers of barriers based on the principle of defence in depth to prevent accidents in highly hazardous, complex production systems. The barriers were of the type we know from Haddon's strategies (Section 4.4), which intervene in energy transfer. In later developments, the concept of a barrier has been given a wider interpretation and includes almost any type of measure to prevent accidents. A typical example is the Australian Transportation Safety Bureau (ATSB) model developed for use as a practical tool in investigations (ATSB 2007). Here, the barrier metaphor is used for conditions such as procedures, training, and supervision.

The Tripod beta model is a software tool for accident analysis that is based on organisational accident theory and the Swiss cheese model. It has many similarities with the Tripod model for accident causation described in Section 4.2, but it has failed defence (barrier failure or missing barriers) as its point of departure (Reason 1991). (See also Figure 13.13.) The Tripod beta accident model provides a diagram where all barriers that could have prevented or stopped the incident are identified. Then, each barrier is analysed in light of Reason's theory on organisational accidents by successively identifying causes for the missing or failing barriers in active failures, preconditions for the active failures, and root causes. It thus uses the same principles as the causal-sequence models when explaining each barrier flaw.

4.5   Logical tree models

Logical tree models aim at analysing the causes of accidents in terms of logical relations between events and conditions in the affected system. The models are based on the systems concept, which means that a production system is understood in terms of its components and the relationship among these. Fault tree analysis is an example of a logical tree model applied to assess the risk of accidents. Logical tree

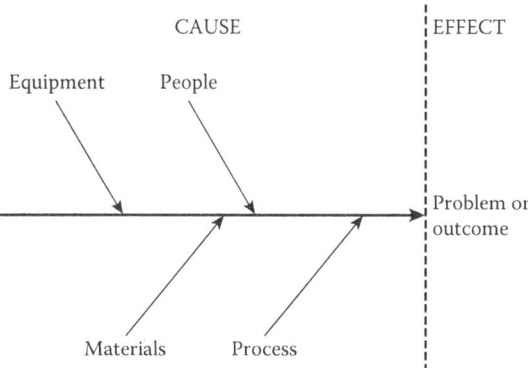

Figure 4.7 Fishbone diagram, modelling causes of a problem or outcome.

models may also be used in diagrams to display the results of an accident investigation.

An example of a logical tree is the *fishbone diagram* (see Figure 4.7), also called the Ishikawa diagram (Ishikawa 1990). This diagram has its roots in quality management and production management but can also be used to analyse causes of deviations and barrier failures in an accident sequence. There are different categories of causes applied – for example, the five Ms used by the Toyota Production System: machine, method, material, manpower, and measurement.

Parts of the Construction Accident Causation (ConAC) model, which was developed for and has been applied in studies of accidents in the construction industry, have similarities to the fishbone model (Haslam et al. 2003, 2005). The origin of the causal sequence is modelled similar to that of the fishbone model, where immediate causes are explained by the following factors: work team, workplace, materials, and equipment. The causal sequence leading to these factors is explained by two levels: shaping factors and origination factors, which are similar to Reason's (1997) active and latent failures as well as to the Tripod beta and the ILCI models.

4.6 System models

System models represent a further development of some of the accident models described so far in this chapter (Khanzode et al. 2012). The accident models previously described also pay attention to factors in management and organisation, even the domino theory from the 1930s has 'social environment' as the first domino chip. System models develop this further by highlighting factors related to management and organisation more systematically and extensively than other accident models. Here, we distinguish among four different types of system models, of which three are further detailed as follows:

- *Hierarchical root-cause analysis models* follow the same principles as the causal-sequence accident models, that is, explaining accidents as results of layers of causes following the organisational hierarchy with root causes in the higher echelons of the management systems and safety management.
- *Human factors models* emphasise the interplay among human, technical, and organisational factors (Chapter 10).
- *Safety management models* emphasise failures in the safety management system, including decision-making.
- *Systemic models* focus on an organisation's ability to be flexible and adaptive to varying conditions. Such models explain accidents in terms of departures from a dynamic equilibrium maintained through safety constraint and feedback control loops.

4.6.1   Hierarchical root-cause analysis models

The Tripod beta and Swiss cheese models discussed in Section 4.4 illustrate how the causes of barrier failures/missing barriers are analysed in layers, following the organisational hierarchy from the work system to the workplace and further to the general organisation and management system. Other models have been developed along the same principles, such as the Human Factors Analysis and Classification System (HFACS). This model does not consider barrier failures but provides a framework for analysing human failures (active failures) by identifying latent conditions (Shappel and Wiegman 2000). The first level of categories in this model is unsafe acts (errors and violations). Underlying causes of the unsafe acts are explored by studying preconditions for unsafe acts, unsafe leadership/supervision, organisational influences, and outside factors. The HFACS offers a simple checklist of factors that can generate the causes, such as resource management – human resources, monetary resources, and equipment/facility resources.

4.6.2   Safety management models

The MORT program was first developed in the 1960s by the U.S. Atomic Energy Commission. In the years since, it has had a great influence on developments in safety management. The thought at that time was to formulate an ideal safety management system from a synthesis of the

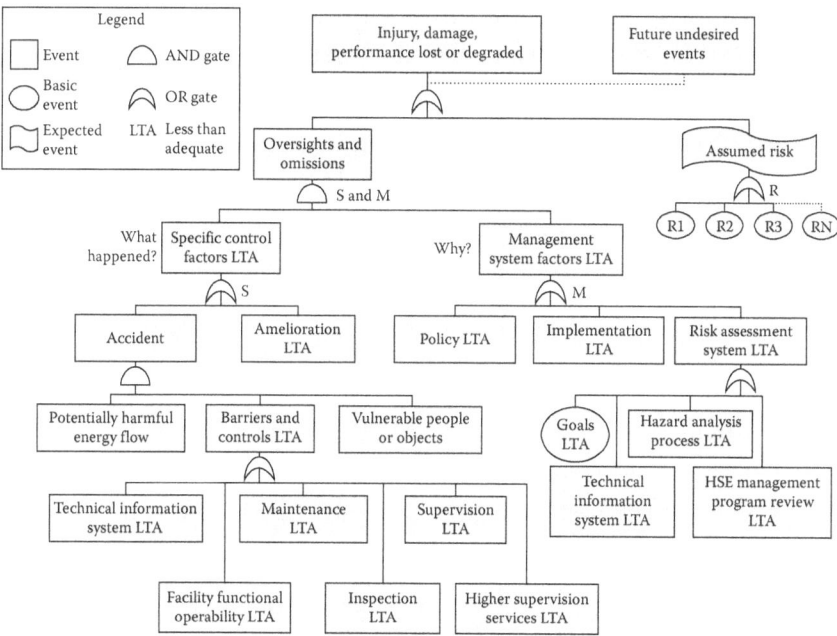

Figure 4.8 The main elements of MORT. (Adapted from Johnson, W.G., *MORT Safety Assurance System*, Marcel Dekker, New York, 1980.)

best accident models and quality assurance techniques then available (Johnson 1980).

Figure 4.8 shows the main elements of MORT (Johnson 1980; Knox and Eicher 1992). They are displayed as events in a logical tree of, in total, about 1500 basic events. Each event corresponds to a question to be addressed in accident investigations and safety audits. The MORT diagram is based on some basic safety management principles from the 1960s and on practical experience from accident investigations.

The top event of the tree is made up of the accidental loss (experienced or potential) to be analysed. The branches below the top event are built up around four principles:

- Risk assessment and assumed risks
- The energy model
- The feedback control cycle according to Juran (1989)
- A system's life cycle

First, the analyst has to address whether the accident represents an assumed risk or whether it is the result of oversights and omissions. If the risk is assumed, the analyst will terminate the investigation. *Assumed risks* are accident risks that have been identified, evaluated, and accepted at a proper management level prior to the MORT analysis of an accident. This corresponds to the R-branch in the MORT diagram.

In all other cases (i.e. when the risk is not assumed according to the afore-mentioned criteria), the analysis will continue along two branches – specific oversights and omissions (S-branch) and management oversights and omissions (M-branch). The *S-branch* of the MORT diagram focuses on the events and conditions of the accident occurrence (actual or potential). Here, time develops from left to right. Causal influences go from bottom to top.

We recognise Haddon's energy model as a key element in the S-branch (Haddon 1980). An event is denoted as an accident when a target (a person or object) is exposed to an uncontrolled transfer of energy and sustains damage. Accidents are prevented through barriers. There are three basic types of barriers: (1) barriers that surround and confine the energy source (hazard), (2) barriers that separate the hazard and the target physically or in time or space, and (3) barriers that protect the target. We find these different types of barriers in the branches below the accidental event. Amelioration relates to the actions taken after the accident to limit the losses (rescue, firefighting, and so on).

At the next level of the S-branch, we recognise factors related to the different phases of the life cycle of an industrial system: the project phase (design and plan), start-up (operational readiness), and operation (super-vision and maintenance). The idea here is to link barrier failures to when they first occurred in this life cycle.

When proceeding along the *M-branch*, the analyst is interested to know why these inadequacies have been allowed to occur. The queries are directed at the main elements of the management system:

- The standard (i.e. policy, goals, requirements, and so on)
- Implementation
- Measurement and follow-up

These are the same basic feedback control elements that we find in the quality assurance principles of, for example, the International Standards Organization (ISO) 9000 series. Influences can be traced back to Juran's work in the 1960s (Johnson 1980).

This part of the focus helps the analyst to generalise the specific find-ings from one specific accident investigation or safety-program evalua-tion. Events and conditions of the S-branch thus often have counterparts in the M-branch. By querying the M-branch, the analyst is able to move his or her focus from the specific circumstances at the accident site to the total management system. The recommendations from this analysis will impact many other accident risks as well.

On the bottom line, MORT has a collection of questions whether specific events and conditions are satisfactory or 'less than adequate'. In the detailed questions, we find elements from such different fields as risk analysis, human-factor analysis, safety information systems, and organisational analysis. The judgements made by the analyst are partly subjective. Training and certifying analysts have proven necessary to ensure an adequate quality.

The literature on evaluations of MORT is sparse. Johnson (1980) reports better coverage of supervisory and management deficiencies in accident investigations after the introduction of MORT. Stephans (2004) claims that the value of MORT is that it is comprehensive, leading accident investigators to ensure that the root causes of all accidents are identified. He also describes complaints from users of MORT that it is time-consuming and too comprehensive for many situations. Experiences have also been gained from evaluations of MORT applications within Finnish industry (Ruuhilehto 1993). These point to MORT's significance in building bridges between safety and production management. The application of MORT also assists in general planning and control and affects the frequency of production disturbances as well. Some limitations also have been identified in the Finnish studies. MORT is not well suited for the identification of immediate risks due to failures and disturbances. Another problem is that there is no scheme for building priorities into the MORT concept.

TapRoot is an accident investigation method that is based on an accident model that has similarities with MORT. The TapRoot accident model is a logical tree that includes branches on human performance, equipment, and management (Paradies and Unger 2002).

SMORT was developed from MORT in order to also make it usable as a tool in the data-collection phase of accident investigation (Kjellén et al. 1987). The basic elements of MORT have been re-arranged in order to provide for a logical sequencing of the accident investigation process in successive tiers (see Section 13.2). SMORT contains a number of checklists for use during the investigations at each tier, see Appendix B.

The *ILCI model* (Section 4.2) identifies three health, safety, and environment (HSE) management elements at the most basic end of the causal hierarchy of the model: inadequate program, inadequate program standard, and inadequate compliance with standard. These elements are built around the three basic elements of administrative systems for feedback control (i.e. goal, implementation, and follow-up). A factor that has contributed to the success of the ILCI model in industrial practice is that it prescribes a standard HSE program with the necessary elements and activities. For each element, there is a set of detailed criteria against which the actual situation must be compared. The ILCI auditing system integrates detailed guidelines, and the value judgements as to what is 'less than adequate' are (to a lesser extent) left to the discretion of the analyst rather than as in MORT and SMORT analyses. The downside is the uncertain validity of the findings for the industrial organisation.

AcciMap is a graphical presentation of decisions and actions made at different levels in a company and in society and how they relate to each other and influence the risk of accidents (Rasmussen 1997; Svedung and Rasmussen 2002). The decisions and actions are displayed as a network of interactions inside and among the six levels shown in Figure 4.9. The merit of AcciMap is that it considers decisions and failures outside the company and includes decisions and actions of regulators and government in the analysis. AcciMap considers the context of safety management at companies. This is relevant in the analysis of major accidents in industries with a strong governmental involvement, such as the nuclear industry, where the scope of an investigation includes the quality of the governance authority as well. The resulting graphical presentation may become very complex.

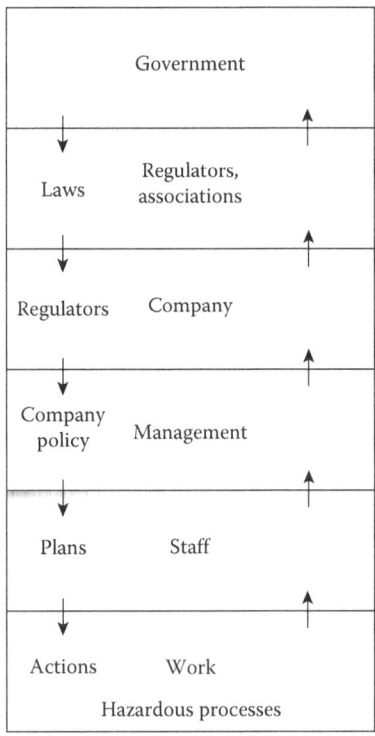

Figure 4.9 Feedback mechanisms in risk management in a dynamic society. (Adapted from Rasmussen, J., *Saf. Sci.*, 27, 183–213, 1997.)

4.6.3   Systemic accident models

In recent years, adaptive safety management approaches such as resilience engineering literature have received considerable attention (Hollnagel et al. 2006). Also, the high reliability organisation (HRO) literature from the 1990s onwards has highlighted the processes necessary in an organisation for it to adapt to changing conditions (e.g. Laporte and Consolini 1991). These safety management approaches emphasise dynamic adaptation and flexibility with regard to both anticipated and unanticipated incidents as a critical means of sustaining normal system functions. In the area of resilience engineering, systemic accident models have been developed. Such models analyse accidents in terms of failures in organisations' ability to be flexible and adaptive to changing conditions. Both adaptive safety management approaches and systemic models have been developed based on arguments that traditional accident models and safety management approaches are mainly suited to analysing simple systems and not to analysing complex systems characterised by the following (Leveson 2004/2011):

- Technology that changes faster than safety engineering techniques
- Technologies that are rapidly introduced into the market, limiting the possibility of testing and assessing them
- New types of hazards (e.g. new artificial chemicals)
- Increased complexity and coupling in systems
- Decreased tolerance in society to accidents with regard to costs and destruction, especially in the case of ripple effects (i.e. situations in which one event causes a series of other events to happen)
- More complex relationships between humans and automation
- Changing regulatory and public views of safety

Systems Theoretic Accident Model and Processes (STAMP) is an example of a systemic accident model that is based on many of the principles on using feedback in decision-making (Leveson 2004/2011). STAMP views safety as a dynamic control issue rather than as a component failure issue. A system is kept in a state of dynamic equilibrium by feedback control loops, and safety is managed by a hierarchical structure of those loops (Leveson 2004). A key part in STAMP is that this control structure enforces constraints on system development and operation with the aim of producing adequate safety performance.

In contrast to other accident models, in STAMP, accidents are not understood as a chain of events but as a result of lack of constraints. Examples of safety constraints are minimum separation in distance between aircrafts while cruising and that aircrafts must maintain a certain over-capacity in lift and power. Some of the constraints have many similarities with what we will describe as barrier functions in Chapter 9.

To enforce the constraints, control structures are required at many levels. These control structures are based on the same principle as Juran's (1989) feedback loop. A human or an automated system controls the physical process by initiating actions based on information feedback about the process. The operating process is controlled by a hierarchy of feedback systems similar to AcciMap (Svedung and Rasmussen 2002). Accidents can thus be explained by information feedback and control actions that are not suited to the state of the operating process.

A merit of this model is that it combines many of the key theories and principles in the management of safety: feedback control including hierarchical layers of control and barriers. The application of STAMP results in a rather complicated model that might be too intricate for applications in the area of occupational accidents. Salmon et al. (2012) has made an evaluation of the model stating that STAMP requires in-depth knowledge about the system being analysed and also requires more data than simpler accident models. Therefore, it is time and resource demanding. Furthermore, it is comprehensive in terms of including the entire socio-technical system of an accident. On the other hand, it results in a deeper understanding of the system. Underwood and Waterson (2014) follow the same line of argument and state that STAMP may be too comprehensive for practitioners and better suited within research.

4.7   Use of accident models in investigations of accidents and near accidents

A primary use of the models presented in this chapter is in methods for accident and near-accident (hereafter called incident) investigations. Accident models help to delimit the analysis, support the search for data, structure the information, develop remedial actions, and provide a framework for presentation of results.

There is extensive literature on investigation methods and associated accident models. Sklet (2004), the Energy Institute (2008), and Hollnagel and Speziali (2008) present overviews of methods and models. Sklet (2004) assessed 14 accident investigation methods based on certain criteria – including directions regarding analyses of the sequence of events and barriers, inclusion of organisational factors in the analysis, analytic approach, and training needs. He concluded that no single method is comprehensive and that a combination of different methods is needed to investigate complex accidents. Dien et al. (2012) give an overview of 'root cause' models and assess their applicability in organisational analysis of events to improve operating feedback. Some of the models are used primarily by scholars, but MORT, MTO, and Tripod have reached industrial application. Okstad et al. (2012) gives an overview of typical investigation methods and associated accident models in use in the Norwegian oil industry, notably MTO, Tripod, TapRoot, and STEP. The authors conclude that the MTO method is favoured by the Norwegian Petroleum Inspectorate, whereas the oil companies' internal investigations are based on common sense rather than on systematic methods based on accident models.

Salmon et al. (2012) compare and assess three methods and associated models developed for use by human-factor experts in accident investigations, AcciMap, HFACS, and STAMP. The authors conclude that the output of the three methods, as applied on the same data from one outdoor accident, varies significantly. This is a good illustration of the well-known phenomenon that Hollnagel (2014) names What-You-Look-For-Is-What-You-Find (WYLFIWYF); in this case, the accident model used in the investigation will direct the inquiry into accident causes and act as a filter against alternative interpretations. Salmon et al. (2012) favour a modified version of AcciMap because of its ability to capture the contributions to the accident from six organisational levels, including governmental decisions on policy, budgets, and decisions at the sharp end, and environmental conditions at the time of the accident. Hollnagel and Speziali (2008) have analysed 21 investigation methods on their applicability in investigating accidents in different types of industrial systems. The authors applied a two-dimensional interaction-coupling chart, according to Perrow (1984), to characterise different types of industrial systems. By plotting the different methods in this chart, the authors concluded that methods based on systemic models such as STAMP are most suitable for the analysis of accidents in industries characterised by complex interaction and tight coupling, such as in the nuclear power industry.

4.8   Safety culture

Safety culture is complementary to the structural approach in HSE management. Whereas the HSE management models are prescriptive and focus on the way things should be done, including the 'cold' aspects of feedback control, safety culture is about the way things really are being done in practice and on 'hot' variables of shared beliefs, attitudes, and norms within the industrial organisation. Safety culture addresses different organisational aspects such as safety-related values, attitudes, perceptions, competencies, and resulting patterns of behaviour at the different organisational levels. We include provisions for workers' involvement and organisational learning in the safety culture perspective. This interpretation of the concept of safety culture has a clear parallel in the human resource and symbolic perspectives of organisations.

In the 1980s, investigations of major accidents, such as the nuclear accident at Chernobyl in 1986, pointed at safety culture as an important contributing factor to accidents. Since then, safety culture has become an important factor in the analysis of accidents and has contributed to accident prevention among practitioners. However, to quote Reason (1997, p. 191): 'Few phrases occur more frequently in discussions about hazardous technologies than safety culture. Few things are so sought after and yet so little understood'.

Based on experiences from the Chernobyl accident, the Advisory Committee on the Safety of Nuclear Installations (ACSNI) has advocated the necessity of including safety culture in the analyses of accidents. The ACSNI proposes a definition of safety culture that has been widely accepted (ACSNI 1993): 'Safety culture is the product of individual and group values, attitudes, perceptions, competencies, and patterns of behaviour that determine the commitment to, and the style and proficiency of, an organisation's health and safety management'. This also is the definition applied by the Health and Safety Executive in the UK and has served as the basis for much research (Antonsen 2009).

There are numerous lists in research literature on the characteristics of a good safety culture; failure to meet these criteria may thus be regarded as accident causes. One example is presented here (cf. Hale 2000):

- The importance given by all employees, but particularly top managers, to safety as an integral part of the business goals and work practices. This means that safety is seen as an inseparable, but explicit, part of the way to do business.
- The engagement and involvement felt by all organisational members in the process of defining, prioritising, and controlling accident risks.
- The creative mistrust that management has in HSE systems, which means that they never show complacency and always are on the alert for new problems.
- The caring trust that all organisational members have in each other. This involves accepting the responsibility to check and to be checked for unsafe practices. All organisational members need a watchful eye and helping hand to cope with the inevitable slips and blunders that can always be made.
- An openness in communication about failures as learning experiences and in imaging and sharing information about new dangers.
- The belief that causes for incidents and opportunities for safety improvements should be sought not just in individual behaviour but in the interaction of many causal factors.

To this list, we would like to add 'compliance culture', the acceptance by the organisation as a whole that regulations from authorities and internal rules and procedures shall be adhered to. If a rule is experienced as unsuitable, it has to be changed not tacitly trespassed. The exception is in

Figure 4.10 HSE culture ladder. (Reprinted from *Safety Sci.*, 45, Hudson, P., Implementing a safety culture in a major multi-national, 697–722, Copyright 2007, with permission from Elsevier.)

the rare emergencies situations where the formally correct procedure is determined to accelerate loss.

The Tripod model includes HSE culture elements (Reason 1991; Hudson 2007). In an accident investigation, the model (see Figure 4.10) is used to classify top management commitment on a scale from 1 (pathological) to 7 (generative-proactive).

There is a relationship between the 'organisational culture' and an organisation's ability to learn from accidents to prevent recurrence. *Lucas' framework of organisational cultures* distinguishes among three different types of organisations in this respect (Van der Schaaf et al. 1991). His framework analyses the types of accident or human-error models that are predominant within an organisation. Such shared models will determine an organisation's ability to learn from experience and to prevent the recurrence of accidents. Lucas distinguishes among three different types of organisational safety cultures and associated human-error models:

1. Traditional 'occupational safety management' culture (i.e. a culture where the causes of errors and accidents are attributed to inattention and carelessness on behalf of the workers). Disciplinary measures will dominate the remedial actions.

2. 'Risk management' culture, where an engineering view of human-error causation is dominant. Errors and accidents are analysed in terms of mismatches between the operators and their environment. Remedial actions typically include design changes and provisions of procedural support.
3. 'Systemic safety management' culture, where the causes of errors are analysed in relation to the total work context. Not only are traditional causes such as poor design and procedures considered but also such aspects as unclear responsibilities, lack of knowledge, and low morale. These in turn are traced back to management liability issues.

It is relevant to apply the approaches by Reason and Lucas to cases where an independent body (e.g. an accident commission) carries out the accident investigation. The commission will ask questions about previous incidents of a similar type and about the organisation's ability to learn from them. The commission may judge the failure to learn from previous experience as a root cause of the incident.

chapter five

Framework for accident analysis

5.1 Characteristics of the accident sequence

This chapter presents a framework for the collection and analysis of data on accident risks as shown in Figure 5.1. In this chapter, we borrow aspects from the different accident models presented in Chapter 4 – especially from the process model, the energy models, and the system models. For pedagogic reasons, we will illustrate the analytic framework mainly through its application in incident investigations. We will also apply this framework in a review of variables and classification systems used in the collection and analysis of data on accident risks. The framework will represent a common basis in our review and analysis of the different safety management methods and tools that will be presented in Sections III to V of this volume.

The framework is suited primarily for applications in industrial and other work systems that Rasmussen (1997) characterises as loosely coupled and as having a relatively high accident frequency but a low magnitude of loss. These types of work systems are suited for an empirical safety management strategy. This framework emphasises the following theoretical concepts (cf. Chapter 4):

- Conceptually, the framework is based on an input–process–output model, which is applied in many areas. Contributing factors and root causes in the human, technical, and organisational systems (input) generate an accident sequence (the process) that produces a loss (outcome). Basically, these are the same human, technical, and organisational systems that, under normal circumstances, produce the wanted output. The input–process–output model has similarities to the causal-sequence (Section 4.2) and logical tree models (Section 4.5).
- The accident sequence is modelled as a process developing through four phases characterised by: (1) lack of control, (2) loss of control, (3) 'target' being exposed to an energy flow, and (4) emergency handling. The last two phases are often overlapping. We recognise this theoretical concept from process models (Section 4.3).
- Causal factors are divided into two classes in the work system and the workplace, deviations and contributing factors. (These are based on the process models presented in Section 4.3.)

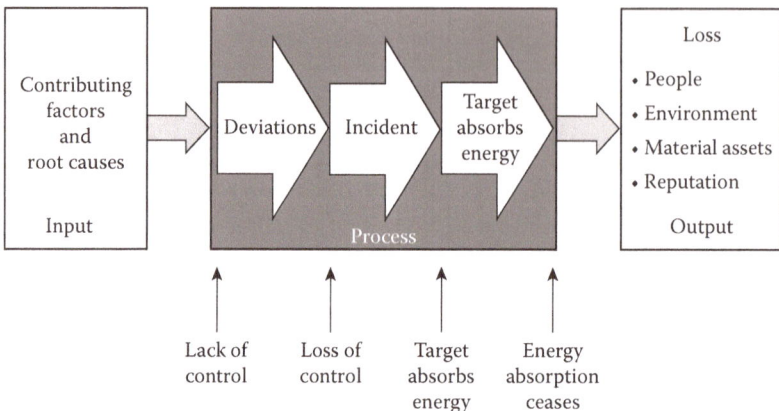

Figure 5.1 Accident analysis framework.

Deviations and contributing factors are further classified based on industrial engineering and human-factor system views (which will be explained in this chapter).

- Barrier functions and their performance are based on the energy model and on Haddon's accident prevention strategies (Section 4.4). Barrier failure as the violation of a norm will also show up as a deviation.
- Higher management causal factors (root causes) are categorised into three classes – project management; health, safety, and environment (HSE) management; and corporate management. A root cause will also constitute nonconformity, if it represents the non-fulfilment of a requirement.

The method includes theoretical concepts similar to the man, technology, and organisation (MTO) and tripod beta analysis methods. The analysis of root causes is based on the Safety Management and Organisation Review Technique (SMORT). It includes a hierarchical analysis similar to the AcciMap approach, although it does not exceed the company level. Sequentially timed event plotting (STEP) is integrated in the method as an alternative to event charting (timeline) for more complex accidents with interacting factors. Barrier analysis differs significantly from that of Tripod beta. It limits the analysis to barriers with functions that intervene in the sequence of events to reduce or eliminate losses in contacts between people and energy flow. Barrier analysis excludes, for example, toolbox talks (informal discussion groups).

The analytic framework is based on accident theory and models presented in Chapter 4. It emphasises that accidents are the result of many of the aspects of the work system and industrial organisation that are designed to produce goods and services. This is why human-factors and industrial engineering system views are introduced. The human-factor (also called systems ergonomics) discipline, according to the International Ergonomics Association (IEA), is concerned with the understanding of interactions among humans and other elements of a system (Wilson 2014). In this book, human factors are about knowledge and methods for analysing and improving the interplay among humans, technology, and organisation (see further details in Chapter 10).

Industrial engineering is a discipline concerned with the optimisation of industrial systems to improve quality and productivity (Salvende 2007). It applies models similar to human factors in analysing work systems.

Let us analyse the framework by starting at the output side. Four different types of losses associated with accidents are included in the framework:

- Losses to people, such as injury or ill health, fatality, and psychological trauma
- Damage to the environment
- Damage to material assets, including loss of production
- Damage to the reputation of the responsible organisation (company), including associated immaterial losses such as a reduced quality of life for people affected by an accident

We recognise the energy model in the parts of the framework starting with the loss of control. Let us focus on injury to people. The energy flow originates from a *hazard* (i.e. a source of possible harm to people, the environment, or to material assets). With very few exceptions (e.g. drowning), the hazard is a source of energy and that will be our focus when we discuss hazards here. Similarly, we define an *incident* as a loss of control of energy in the system or of body movements in relation to an energy source (e.g. rotating machinery). Development of loss occurs when the victim's body comes in contact with the energy flow and the resulting absorption of energy exceeds the body-injury threshold.

There are different types of loss of control:

1. Loss of control resulting from the accidental release of a natural hazard or from a purely technical event without any human intervention at the time of the event.

    Example: The crane load is lost due to a structural failure of the crane boom.

    Example: A robot makes uncontrolled movements due to a programming error.

> *Example:* A large rock or boulder breaks loose during heavy rain and hits a road worker.
>
> 2. A system operator loses control of 'his' system, which is powered by an energy source.
>
>   *Example:* A typical example is driving, where the driver may lose control over the vehicle. The energy in this system is generated by the motor.
>
>   *Example:* A welder loses control of a grinder, which kicks back and cuts his face.
>
> 3. A human operator loses control of a system that is powered by his/ her muscular energy.
>
>   *Example:* A carpenter misses the nail with his hammer and hits his thumb.
>
>   *Example:* A cyclist skids and falls while bicycling.
>
> 4. A person loses control of his/her own body movements.
>
>   *Example:* A person slips and falls on the floor.
>
>   *Example:* A person slips and falls against a rotating saw blade.
>
>   *Example:* A person lifts a heavy box and strains a muscle.

Incidents rarely occur by pure chance. An incident is usually preceded by deviations at the workplace that increase an incident's frequency and/or consequences. Safety measures that eliminate existing deviations (e.g. repair of faulty safety equipment) will have an immediate effect on the risk of accidents. They will, however, not have lasting effects if the deviation may occur again.

The input to the accident process usually consists of the same types of resources at the workplace and in the company that have been established and managed for production purposes. We name these resources contributing factors and root causes when they result in an increased risk of accidents by affecting the sequence of events and extent of loss. By changing such factors, more lasting risk-reducing effects will be achieved.

Contributing factors are human, technical, and organisational conditions in the work system or at the workplace. They differ from deviations in that they generally change only slowly in time and that they represent what is planned or accepted as 'normal'. In practice, it is sometimes difficult to separate contributing factors (input) from deviations in the accident sequence (process). The design and maintenance of machines, for example, will affect the likelihood of technical deviations and incidents. Safety barriers such as guards will, on the other hand, affect the consequences of such incidents.

Root causes are the most basic causes of accidents/incidents (i.e. a lack of adequate management control resulting in deviations and contributing factors). Here, we will limit the analysis to root causes in the company (i.e. causes that are subject to corporate management control). Analytic methods, such as AcciMap, extend the analysis to include the authorities and government as well (Svedung and Rasmussen 2002).

Time is an important variable in the accident process as shown in Figure 5.1, where the accident develops through consecutive phases rather than as a single event or as a chain of causal factors. The accident sequence (i.e. the process of the input–process–output model) consists of transitions through different phases from a lack of control, through loss of control of energies in the system (incident), and further to development of loss (Kjellén and Hovden 1993):

1. The *initial phase* of the accident sequence is described in terms of deviations from the normal and/or faultless production process.
2. The next phase (*incident phase*) starts with a loss of control of energy in the system or of a person's body movements in relation to the danger zone of the energy source.
3. The *injury phase* is characterised by the victim being exposed to the energy flow, resulting in loss. In the case of injury to a person, this occurs when the energy flow exceeds body-injury thresholds.

Figure 5.2 illustrates the phases of the accident process. Our analytic framework helps us in defining the start and the end of the accident sequence. We differentiate between active and latent deviations,

DEVIATIONS	INCIDENT	DEVELOPMENT OF INJURY
A valve lever blocks the gangway Oil spill on the floor Safety gloves not used	The worker slips when he leans forward to pass below the lever He looses his balance and falls	The worker's right hand is squeezed between the falling pipe and the floor

Figure 5.2 Illustration of the accident sequence phases in the analytic framework using the example of a process plant accident.

which is similar to the distinction between active failures and latent conditions according to Reason (1997). The accident sequence starts with an *active deviation* (i.e. an action or event that is logically and chronologically linked to the outcome – injury or damage). It ends when the victim is no longer exposed to the energy flow. *Latent deviations* are failures that have gone undetected for a long time. The focus on deviations also helps in making priorities in the accident investigation.

It is often relevant in investigation practice to map the sequence of events from when the activity started where the accident occurred. An alternative approach is to start with the beginning of a shift. It is sometimes necessary to trace back to earlier shifts to identify when and how the deviations first arose. Stop rules have to be applied in the investigation in order to avoid looking for causes of the accident all the way back in time to the 'garden of Eden' (Rasmussen 1993). This is illustrated by the following case.

Example: The accident occurred in a production cell for car crash-boxes. An operator entered the fenced-in cell through a gate to perform routine correction of a production disturbance. The robot started unexpectedly when the operator corrected the position of the box in the grip of the robot. He was hit by the robot and squeezed between the robot arm and the fence. The immediate cause of the accident was an unexpected start of the robot when the operator corrected the disturbance. This was due to a combination of deviations in the design of the robot's control system and to changes introduced in the production cell one day before the accident. The interlocking guard to the gate of the crash-box cell had been compromised during test production the day before the accident, and the guard had not been restored when normal production was resumed the next day.

The framework also helps us in defining *near accidents*. This is a sequence of events that includes an incident, but where injury or loss was avoided due to pure chance.

In an accident investigation, the first step is to map all relevant facts. This step of the investigation will focus on the losses and the accident sequence (i.e. the extent of injury and damage, the incident, and preceding deviations). In the next step, the investigator interprets the evidence and examines human, technical, and organisational circumstances around the accident in the work system and at the department in order to identify contributing factors. Here, subjective judgements are a necessary ingredient. This is especially the case when humans have played a central role in the accident sequence. Differences in focus as well as in opinion may occur and need to be resolved.

In Table 5.1, we have applied the accident analysis framework (Figure 5.1) to the analysis of the development of different types of losses. The analytic framework is well suited to the analysis of accidents

Table 5.1 Application of the accident analysis framework to illustrate the development of different types of losses

Type of event	Initial phase (deviations)	Incident	Development of loss
Occupational accidents	Deviations are common.	An incident will always precede the loss.	Person is exposed to sudden and uncontrolled energy flow.
Occupational diseases	Deviations may result in increased exposure.	Low-intensity (and unreported) incidents may contribute.	Person is exposed to low-intensity energy flow ('planned' and unplanned) over a long period.
Material damage accidents	Deviations are common.	An incident will always precede the loss.	Material is exposed to sudden and uncontrolled energy flow.
Poor product quality	Deviations are common.	Not relevant.	Production is outside quality norm.
Poor production regularity	Deviations are most often present.	An incident may precede the stop in production.	Production stops.
Acute environmental pollution	Deviations are common.	An uncontrolled release of toxic substances will always precede the loss.	The environment is exposed to a sudden release of toxic substances.
Long-term environmental pollution	Deviations may occur.	Small but frequent accidental releases may contribute to the loss.	The environment is exposed to a low-intensity release of toxic substances over a long period.

resulting in personal injury, material damage, and acute environmental pollution. By definition, these types of occurrences involve an incident. We also know that 'low-intensity' but frequent incidents may contribute to occupational diseases such as hearing losses, back injuries, and lung diseases.

Deviations are common, but not always present, in the accident sequences. We know that deviations contribute to the occurrence of all different types of losses. The 'control climate' at a company has a significant influence on the risk of accidents as well as on production regularity and quality (Grimaldi 1970). An efficient control of deviations is thus a key strategy for the control of losses in general.

In Sections III through V of this book, we will apply the accident analysis framework in a review of different methods used in the collection and analysis of data of accident risks. Here, we will provide a basis for this review by demonstrating the application of our framework in quantifying the different variables that define the losses, the accident sequence, and the contributing factors and root causes. Our aims will be twofold: first, to be complete (i.e. by presenting all alternative means of measuring and classification), and second, to give specific advice on the preferred method. The reader will find recommended alternatives in tables and checklists.

5.2   Types of data and scales of measurement

An important aspect in designing databases on accident risks (Chapter 14) is how we document and store data. First, we distinguish between two different data types – qualitative and quantitative. By *qualitative data*, we mean free-text descriptions. In an accident investigation, for example, this is similar to establishing a chronicle of the sequence of events. *Quantitative data* are characterised by quantification through the use of a scale. This is done to reduce the complexity of the information in the observations. The results may be presented in a format that is easy to comprehend (e.g. through statistical analysis), but details are lost in the quantification of the data, and this is usually an irreversible process.

We apply four different scales in coding or quantifying data (see e.g. Janicak 2003):

- *Nominal scale*, that is, the discrete classification of an observation by allocating it to a category in a classification scheme. A classification scheme preferably should consist of a complete set of mutually exclusive classes. Sex is an example of a classification scheme with two classes (male/female).
- *Ordinal scale*, that is, placing observations in order of magnitude but without any measurement of differences
- *Interval scale*, where the observation is measured on a scale and the difference between the values of the scale can be quantified. The zero of the scale is arbitrary. The Celsius scale for temperature is an example.
- *Ratio scale*, where the observations are measured on a scale that has an absolute or fixed zero value. This permits the comparison of differences of value. The metre measurement of distance is an example, where it is relevant to say that a distance of 2 km is twice as long as 1 km.

5.3 Consequences of accidents

5.3.1 Types of consequences

The accident analysis framework identifies different types of losses: injury or fatality to people, environmental damage, property damage, and damage to reputation. Information on the immediate consequences to a victim of an accident usually is readily available and is well suited for classification. There are standard schemes for the classification of injury and the body part affected. Table 5.2 shows a simplified version of common injury classification.

Statistics on accident distributions by injury type and body part affected find many practical uses. If the statistics show that there are many eye injuries, for example, an obvious remedy is to introduce requirements for eye protections in those types of work where the injuries occur.

Adverse environmental effects usually are not recorded in a similar way due to the difficulties in identifying the damage, which may be local or regional and acute or delayed. Rather, the type of release is described, such as an oil spill. Insurance companies apply elaborate classification schemes for material damage. In industry, it is common to distinguish between material damage due to fires and explosions from other types of

Table 5.2 Type of injury and body part affected

Type of injury	Part of the body
• Amputations	• Head, including facial area, eyes, ears, and teeth
• Asphyxiation, gassing, drowning	
• Burns (including chemical burns), scalds, frostbite	• Neck
	• Back
• Closed fractures	• Torso and organs, including rib cage, chest area, pelvis, and torso
• Concussions and internal injuries	
• Contusions, bruises	
• Dislocations	• Upper extremities, including shoulder, arm, hand, fingers, and wrist
• Distortions, sprains, torn ligaments	
• Effects of radiation	
• Electrocutions	• Lower extremities, including hip, leg, ankle, foot, and toes
• Open fractures	
• Open wounds, including cuts, lacerations, abrasions, as well as severed tendons, nerves, and blood vessels	• Other parts of the body
	• Whole body and multiple sites
• Poisoning, infections	
• Others	
• Unspecified	

Sources: ILO, *The Sixteenth International Conference of Labour Statisticians*, Geneva, 1998; Eurostat, *European Statistics on Accidents at Work – Summary Methodology*, European Union, Luxembourg, 2013.

material damage. There are no generally accepted schemes for the classification of social and political losses following an accident.

5.3.2   *Measures of loss*

The size of the losses is an important parameter in making priorities on safety measures. Different types of scales are applied. Table 5.3 gives an overview.

Recordable injuries such as the LTIs, when applied widely, were often subject to scrutiny for possible re-classification into non-recordable injuries – the reason being that the incentive scheme in a company (formal and informal) promoted 'zero injury' results. One means of accomplishing re-classification was to ensure that the injured person returned to work on the next day and was assigned an alternative task that was compatible with his/her medical conditions. The U.S. Occupational Health and Safety Administration (OSHA) introduced a definition of recordable injury to counteract this type of 'creativity', see Table 5.4.

Common consequence measures such as LTI and TRI are easy to record but are insensitive to the actual size of the loss. An eye injury resulting in a few days of absence and an amputation of an arm resulting in permanent disability are both recorded as an LTI or TRI. The number of days of absence more accurately reflects the actual losses but is not readily available at the time when the accident is recorded. This especially is a problem in keeping accident records, when the duration of the sick leave extends over several recording periods (months, quarters, and so on).

Consequence measures based on medical evaluations, such as the AIS and the extent of disability, are more valid loss measures than LTI. The AIS measures the probability that an injury may result in a fatality. AIS Category 1 is used for small cut injuries, finger fractures, small burns, and so on. AIS Category 4 is an injury that is life-threatening but where survival is likely, and AIS Category 6 is a fatal injury.

Consequence categories based on assessments of the severity of injuries are applicable immediately after the accident. The problems with delayed recording are thus avoided. They are also well suited for statistical summaries. A consequence matrix applies a common measurement scale for the severity of different types of consequences. Table 5.5 shows an example of a consequence matrix with five severity classes. The matrix in the table has been influenced by the so-called injury potential matrix (Booth 1991).

5.3.3   *Economic consequences of accidents*

The costs of accidents are shared among the individual, the company responsible for the accident, the insurer, and the public sector. Here, we will focus on company costs. These costs normally are not visible in the company's accounting system. A traditional approach in accident research has been to develop models for use in the recording of a company's

Table 5.3 Overview of measures of consequence of loss due to injury or damage

Target	Type of scale[a]	Type of measure
Person	Nominal	LTI = lost-time injury[b] (yes/no)
	Nominal	TRI = total recordable injury[c] (yes/no)
	Nominal	MTI = medical treatment injury[d] (yes/no)
	Nominal	RWI = restricted work injury[e] (yes/no)
	Nominal	Fatality (yes/no)
	Ordinal	Consequence categories (e.g. first aid, temporary disability with a few days of absence, temporary disability over an extended period of time, permanent disability, fatality)
	Ordinal	International Labour Organization's (ILO) consequence categories (temporary incapacity to work, <1 day lost, 1–3 days lost, 4–7 days lost, 8–14 days lost, 15–21 days lost, 22 days to 1-month lost, 1–3 months lost, 3–6 months lost, permanent incapacity to work or more than 183 days lost, fatal injury)
	Ordinal	Classification according to extent of disability (no permanent disability, <5%, 5–10%, 10–20%, 20–50%, >50%, death)
	Interval	AIS = Abbreviated Injury Scale (AAAM 1985)[f]
	Ratio	Number of whole days lost compensated by the insurance company/employer[g]
Environment	Ratio	Amount of release in m³, tons, and so on by substance
Material	Ratio	Losses in monetary unit (e.g. euro or US$)
Production	Ratio	Duration of production stop (e.g. hours)
	Ratio	Lost production (e.g. number of units)
		Losses in monetary unit (e.g. euro or US$)

Sources: ILO, *The Sixteenth International Conference of Labour Statisticians,* Geneva, 1998; Eurostat, *European Statistics on Accidents at Work – Summary Methodology,* European Union, Luxembourg, 2013; OSHA, Recording and Reporting Occupational Injuries and Illness. Regulations, 29 CFR, Part 1904, Occupational Safety & Health Administration, U.S. Department of Labor, Washington, DC, 1971/2015.

[a] For types of scales, see Section 5.2.

[b] Work-related injury (occurring during paid work), resulting from an accident and where the injured person does not return to the next shift.

[c] Work-related injury resulting from an accident that involves one or more of the following: fatality, lost workday(s), loss of consciousness, restriction of work or motion, transfer to another job, and/or medical treatment other than first aid.

[d] Work-related injury other than a lost-time injury, where the injury is serious enough to require treatment that only may be administered by a licensed doctor or nurse. Excluded are diagnostic procedures, such as X-rays and blood tests, including the administration of prescription medications used solely for diagnostic purposes.

[e] Injury at work that does not lead to absence after the day of occurrence because of alternative job assignment.

[f] Measures the probability that an injury may result in a fatality, see No. 7.

[g] Fatality and permanent disability (100%) equal 7500 workdays.

Table 5.4 Definitions according to OSHA record-keeping requirements

Type of injury	Definition
(Total) recordable injury	Death Loss of consciousness Days away from work Restricted work activity or job transfer Medical treatment beyond first aid
Medical treatment	Medical treatment includes managing and caring for a patient for the purpose of combating disease or disorder. The following are not considered medical treatments and thus are not recordable: • Work-related injury, where treatment of the injury does not need to be administered by a licensed doctor or nurse (see first-aid injury) • Visits to a doctor or healthcare professional solely for observation or counselling • Diagnostic procedures, including administering prescription medications that are used solely for diagnostic purposes, and any procedure that can be labelled first aid
Restricted work	Restricted work activity occurs when, as the result of a work-related injury or illness, an employer or healthcare professional keeps, or recommends keeping, an employee from performing the routine functions of his or her job or from working the full work day that the employee would have been scheduled to work before the injury or illness occurred
First-aid injury (FAI)	Work-related injury, where treatment of the injury does not need to be administered by licensed doctor or nurse. First-aid treatment includes: Using non-prescription medications at non-prescription strength; administering tetanus immunisations; cleaning, flushing, or soaking wounds on the skin's surface; using wound coverings, such as bandages; using hot or cold therapy; using any totally non-rigid means of support, such as elastic bandages; using temporary immobilisation devices while transporting an accident victim; drilling a fingernail or toenail to relieve pressure or draining fluids from blisters; using eye patches; using simple irrigation or a cotton swab to remove foreign bodies not embedded in or adhered to the eye; using irrigation, tweezers, cotton swabs, or other simple means to remove splinters or foreign material from areas other than the eye; using finger guards; using massages; drinking fluids to relieve heat stress

Source: OSHA, *Recording and Reporting Occupational Injuries and Illness. Regulations,* 29 CFR, Part 1904, Occupational Safety & Health Administration, U.S. Department of Labor, Washington, DC, 1971/2015.

Table 5.5 Example of a consequence matrix with a common scale for different types of losses

Grade	Personnel	Environment	Assets, production	Delivery	Image
1	First-aid injury	Insignificant damage	<1k euro	Insignificant internal delay	Insignificant impact
2	Lost-time injury	Moderate damage, restitution time < one month	1–10k euro	Internal delay that affects production plans for less than one week	Minor impact
3	Permanent disability	Severe damage, restitution time < one year	10–500k euro	Internal delay that affects production plans for more than one week	Medium impact
4	Fatality, one person	Local irreversible damage, restitution time 1–10 years	500–5000k euro	Delay that affects the client's production plans	Major impact
5	Fatality, two or more persons	Regional irreversible impact, danger of exterminating fauna and flora, restitution time >10 years	>5000k euro	Extensive delay, risk of losing major client	Catastrophic impact

accident costs. Average results are then used in estimations of costs of future accidents. The aim is to make the losses associated with accidents visible to the management of the company. Future reductions in accident costs are also used as input to profitability assessments of safety-related investments (Tappura et al. 2015). It is a concern that the monetary costs of an investments are often certain and well known, whereas the benefits of avoiding monetary losses due to accidents are often uncertain and difficult to calculate. Similar difficulties apply to other benefits such as goodwill and productivity improvements.

Early research on company accident costs applied the *market-pricing model*. Here, the analyst registers the actual losses due to accidents for different factors such as lost working hours, materials, and production. These are assessed in monetary units by applying the market price for each factor. The cost of lost working hours, for example, is set as equal to the hourly wage. Heinrich pioneered this work in the 1920s. He distinguished between direct and indirect costs, where direct costs are those paid by the insurer to the victim (Heinrich 1959). Indirect (or hidden) costs include costs directly carried by the company such as lost working hours, reduced productivity, costs of property damage, and overhead expenses. More recent studies use variations of the terminology introduced by Heinrich such as insured/uninsured costs and costs of accidents that can/cannot be controlled by management (Grimaldi and Simonds 1975; Tappura et al. 2015). Generally, the different models do not include human costs, that is, loss to the affected individuals that is associated with pain and suffering (Lebeau and Duguay 2013).

Heinrich's empirical work showed a 1:4 ratio between direct and indirect or hidden costs (Heinrich 1959). The implication of these figures is that significant economic benefits of safety work are overlooked. However, later studies of accident costs in Great Britain, Denmark, Finland, Israel, Norway, and the United States, applying variations of Heinrich and later accident cost models, have produced different results (Rognstad 1993; Rikhardsson and Impgaard 2004). The ratio between direct/insured costs and indirect/uninsured costs varies from less than 1 to more than 10. Differences in the applied methods and among industries and countries may explain the varying results.

In a study by Rikhardsson and Impgaard (2004), the authors analysed the costs of a selection of 27 accidents at nine Danish companies from various industries. The accident costs that were not visible in the companies' accounting systems amounted to about 35% of the total costs. This figure varied from 2% to 98%, depending on the accident characteristics. In a study of construction companies in Singapore, Feng et al. (2015) found a 2:1 ratio between direct and indirect costs of accidents. The total accident costs accounted for about 0.25% of the total contract sum.

The market-pricing model has been criticised because it is based on unrealistic assumptions regarding perfect market conditions and full capacity utilisation (Rognstad 1993). Companies often apply slack in the manning of the operation and can temporarily reorganise the work to off-set any temporary absence due to an accident, for example (ILO 2012). The so-called frictional cost approach that accounts for only a portion of lost working time results in reduced output. *The accounting model* takes this into account by analysing the effects of accidents on a company's contribution margin (Matson 1988). These effects are made up of changes in revenues and variable costs (including wages) because of the accident. Table 5.6 gives an overview of the different cost elements of the two models.

A series of accident cost studies were performed within the chemical, metallurgical, and mechanical industries in Norway (Sklet and Mostue 1993). The application of the market-pricing model resulted in an average cost to a company per accident of about 1500 euro (adjusted in accordance with the Norwegian Consumer Price Index). Costs were lowest in

Table 5.6 Company's cost elements from an accident

Market-pricing model	Accounting model
• Lost work hours, victim	• Increased costs for personnel (replacement, overtime, and so on)
• Work hours spent on changing routines	
• Work hours spent on the investigation	
• Work hours spent on repair of damaged equipment	
• Lost work hours due to interrupted production or reduced productivity	
• Costs of replacing damaged material	• Costs of replacing damaged material
• Costs for transportation (of victim and so on)	• Costs for transportation (of victim and so on)
• Capital costs (for machinery and so on) during production stop	
• Insurance expenditures	• Insurance expenditures
• Loss of income	• Loss of income
• Costs of safety measures	• Costs of safety measures
• Company's costs for medical treatment	• Company's costs for medical treatment

Source: Sklet, S. and Mostue, B.A., *Kostnader ved arbeidsulykker i prosess- og verkstedsindus-trien* [Costs Associated with Occupational Accidents in the Process Industry]. SINTEF Report STF75 A92032. Trondheim, 1993. In Norwegian.

the chemical industry (1200 euro) and highest in the metallurgic industry (1700 euro). Costs increased with the severity of the accident. For accidents resulting in permanent disabilities and fatalities, company costs varied between 2300 and 15000 euro. Uninsured salaries to injured workers accounted for between 74% and 90% of total costs for temporary disabilities. In-plant accident costs were significantly less than 1% of the salary costs.

When the accounting model was applied, the costs were even lower. For temporary disabilities of up to 10 days of absence, they varied between 10 and 350 euro. For accidents resulting in more than 10 days of absence, companies actually experienced economic benefits of up to 3400 euro. This was due to the companies being able to produce at a normal level with a reduced work force, while the insurer paid the wage costs for the victim.

When these studies were carried out, companies in Norway paid between 20% and 40% of the total costs of accidents. After changes in the legislation, Norwegian companies paid between 60% and 90% of the costs. Even when these legislative changes were accounted for, the costs for HSE work exceeded the in-plant costs of occupational accidents (Rosness 1995).

These results do not take into account the benefits of HSE work that are difficult to assess in monetary terms. This may explain why many large companies have developed HSE policies and management systems based on the business premise that a high safety standard is good for business because it promotes company reputation, stakeholder relations, productivity, product quality, and customer satisfaction (Tappura et al. 2015).

5.3.4   Actual versus potential losses

Both accidents and near accidents may have the potential of resulting in more severe loss than the actual outcome. There are two advantages in evaluating the potential losses of the occurrences:

1. Learning from experience before the occurrence of severe accidents. In practice, this means that certain events are given a more rigorous treatment than would have been the case if only the actual outcome was considered.
2. A higher number of events are evaluated. This gives an increased statistical basis for the identification of trends and the needs for remedial actions.

In assessing the potential losses of accidents and near accidents, the worst consequence that realistically may happen is identified by considering the energies involved. Table 5.7 shows two alternative lists of

Table 5.7 Examples of incident-type classification

Eurostat	ILO
• Contact with electrical voltage, temperature, hazardous substances	• Fall of person
• Drowned, buried, enveloped	• Struck by falling object
• Horizontal or vertical impact with or against a stationary object (the victim is in motion)	• Stepping on, striking against, or struck by object
• Struck by object in motion, collision with object	• Caught in or between
• Contact with sharp, pointed, rough, coarse material agent	• Overexertion or strenuous movement
• Trapped, crushed, and so on	• Exposure to or contact with extreme temperature
• Physical or mental stress	• Exposure to or contact with electric current
• Bite, kick, and so on (animal or human)	• Exposure to or contact with harmful substance or radiation
• Other contacts	• Other types

Sources: ILO, *Recording and Notification of Occupational Accidents and Diseases*, International Labour Office, Geneva, 1996; Eurostat, *European Statistics on Accidents at Work – Summary Methodology*, European Union, Luxembourg, 2013.

incident categories, where each type in both lists is linked to a specific hazard, that is, a source of energy; see further details in Section 12.1. In doing this assessment, existing safety measures and other conditions are considered, but pure luck is disregarded. Consequence categories such as those in Table 5.5 are used in assessing the potential losses. The assessment results in a redistribution of the accidents and near accidents in the direction of a higher severity.

This technique of assessing the potential of accidents can be further developed by assessing the expected frequency of a reoccurrence of the events. By combining expected frequency and potential consequences, a risk score is achieved (low, medium, or high); see Table 21.4. A high-risk score ('red incidents') means that measures must be taken to reduce the risk of an event to a lower level by reducing frequency and/or consequence. A medium-risk score ('yellow incident') means that measures ought to be taken to reduce the risk further, based on the as low as reasonably practicable (ALARP) principle. A low-risk score ('green incident') means that the conditions are acceptable and no measures are necessary.

Experiments have shown that judgements of the potential for accidents made by different safety experts are reasonably consistent. It is important that the person making the judgement does not have a stake in the consequences resulting from his/her judgement. As an example, we use a supervisor responsible for the implementation of remedial actions to prevent recurrence of accidents and incidents. A high-risk score will

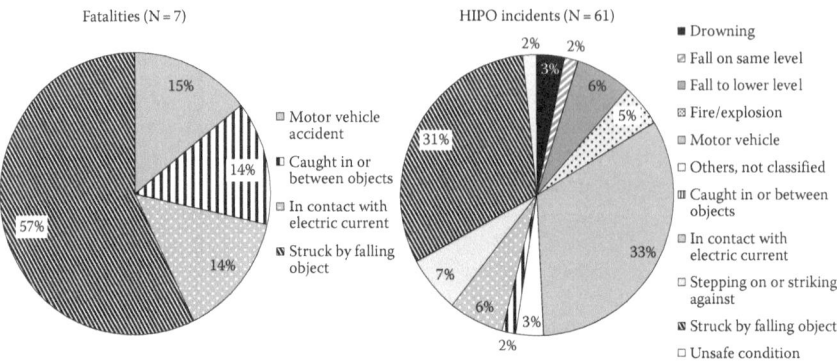

Figure 5.3 Distribution of fatalities and high-potential incidents in four construction projects during a period of three years.

result in more attention from upper management and the authorities and in requirements to implement rigorous safety measures. If the supervisor is given the responsibility to assign risk scores to incidents, he/she likely will tend to arrive at a low or medium score to avoid the 'negative' consequences of increased attention and workload resulting from a high-risk score.

The introduction of the concept of a high-potential (HIPO) incident helps in focusing the accident prevention efforts to those occurrences that represent the potential for fatality or permanent disability. A HIPO incident is defined as one where the most serious probable outcome is a Grade 4 or 5 accident (fatality or serious injury resulting in permanent disability). Figure 5.3 shows the distribution of fatalities and HIPO incidents in four construction projects during three years. The figure illustrates that the four incident categories resulting in fatality account for about three-quarters of the high-potential incidents.

5.4 Incident (uncontrolled energy flow)

An incident involves a sudden and uncontrolled release of energy or of body movements that may come in contact with an energy source (see Figure 4.5). The human operator may be directly involved in this event by losing control of the sources of energy. Failures of technical control systems may also result in an uncontrolled energy release. An uncontrolled movement of a robot arm due to a programming error is one example. Another possibility is that a person loses control over his/her body motions in relation to the energy flow (e.g. falls against a rotating saw blade).

The injury or damage occurs when the energy flow reaches the target (the human body, the environment, or material assets such as buildings, machines, and materials). The severity of the injury or damage is dependent on the type and amount of energy and the way it reaches the target. In the previous section, we have seen fatalities that were associated with a few types of events where large amounts of energies were involved (Figure 5.3). These included movements of vehicles (kinetic energy), being hit by falling objects (potential energy), or being caught in or between objects (mechanical energy, that is, kinetic or potential energy).

An investigator documents the loss of control and subsequent uncontrolled energy transfer in the free-text description of the sequence of events. In addition, it is common for statistical purposes to classify the event by 'incident type'. Table 5.7 shows two different schemes for classification of incident type. The first, by Eurostat (2013), is called contact mode. The second is taken from an ILO (1996) handbook on recording occupational accidents and diseases.

There is a clear relationship between incident type and related hazard (i.e. the source of energy with the potential to harm); for more details, see Section 12.1. Let us take 'fall of person (ILO)/vertical impact with stationary object (Eurostat)' as an example. In this case, the energy source is the potential energy represented by the victim's weight and falling height (from the formula E_p = mgh). When the victim loses control of his/her body movements, the potential energy is transferred to kinetic energy and shared between the body and the stationary object (such as a floor) at the impact. Most of the incident types in the table are related to a source of energy in a similar way. An exception is drowning, where the injury or fatality is the result of disruption of access to the necessary preconditions for life support (in this case, oxygen). 'Physical or mental stress' is another example where the energy model does not apply.

It is also common to classify the object or substance involved in the energy transfer – the so-called injury agency. Table 5.8 shows the classifications of injury agencies used by Eurostat and ILO.

Registrations and coding of accident type and injury agency are based on physical evidence from the accident site and on interviews with witnesses. Such registration is reliable provided that it is based on facts. The coding may give rise to errors because the classes of injury agencies do not represent an exhaustive list of categories that are mutually exclusive.

5.5 Deviations

In this section, we will focus on deviations in the work system and production process. There may also exist deviations from regulations or company standards in the more lasting conditions at the department level and in the higher management systems. We will present an auditing technique to identify and analyse these types of deviations in Chapter 12.

Table 5.8 Examples of injury agency classification

Eurostat	ILO
• Buildings, structures	• Machines
• Systems for the supply and distribution of materials, pipe networks	• Means of transport and lifting equipment
• Motors, systems for energy transmission and storage	• Other equipment
• Hand tools, not powered	• Material, substances, radiation
• Hand-held or hand-guided tools, mechanical	
• Machines and equipment – portable or mobile	
• Machines and equipment – fixed	• Working environment
• Conveying, transport, and storage systems	• Other agencies
• Vehicles	
• Materials, objects, products, machine or vehicle components, debris, dust	
• Chemical, explosive, radioactive, biological substances	
• Safety devices and equipment	
• Office equipment, personal equipment, sports equipment, weapons, domestic appliances	
• Living organisms and human beings	
• Bulk waste	
• Physical phenomena and natural elements	
• Others	

Sources: ILO, *Recording and Notification of Occupational Accidents and Diseases*, International Labour Office, Geneva, 1996; Eurostat, *European Statistics on Accidents at Work – Summary Methodology*, European Union, Luxembourg, 2013.

Nonconformities are defined as the non-fulfillment of specified requirements in the standard ISO 9000 for quality management systems (ISO 2015c). To avoid confusion with this definition, here we will use the term *deviation* instead. It covers those aspects of the accident sequence that represent a mismatch between our norms of a faultless production process and what actually happened. There are different types of norms:

- Formal: the laws, regulations, procedures, standards (refer specified requirement according to ISO 9000)
- What has been planned
- What is normal/usual
- What is acceptable/good enough ('good practice')

Deviations are social constructs whose identification and evaluation are dependent on the types of norms that are in use at any time. The extension of the types of norms included in the definition of deviations beyond specified requirements has been necessary in order to accommodate types of production where the detailed planning and method of work is decided by the individual worker or work team. The focus on deviations helps us in identifying and evaluating the transient and specific circumstances at an accident site. In the next step, we look into how these circumstances can be explained by failures in the prevailing conditions at a workplace and in a company's management systems.

The identification of deviations may be problematic due to differences in opinion – for example, among workers, supervisors, managers, and systems designers – about what is normal. Another problem is the lack of norms in situations that have not been encountered before. These differences of opinion and this lack of norms may in themselves contribute to an increased risk of accidents. One intention of focusing on deviations is to stimulate discussions inside companies, where such differences in opinion are highlighted and preferably resolved. The aim is to arrive at mutually shared norms of what constitutes a faultless production process.

Studies of accident reports show that investigators often fail to document deviations (Kjellén 1982). In order to support the identification and classification of deviations, different checklists have been developed. Table 5.9 shows examples of different systems applied in the classification of deviations.

We recognise the classical man–environment taxonomy according to Heinrich in Table 5.9. It consists of two classes – unsafe acts by persons and unsafe physical conditions. Bird and Germain (1985) have further developed this taxonomy in the International Loss Control Institute (ILCI) model.

Table 5.9 Overview of different systems for the classification of deviations

Basis for classification	Classes of deviations	Source
Man–environment	• Act of person • Mechanical/physical condition	Heinrich, 1959
Man–environment	• Substandard practice/act • Substandard condition	Bird and Germain, 1985
Human-factor systems view	• Personnel • Task • Equipment • Environment	Leplat, 1978
Human errors	• Slips (skill-based) • Mistakes (rule-based) • Mistakes (knowledge-based) • Violations	Reason, 1990
Industrial engineering systems view	• Material • Personnel • Information • Technical • Human act • Intersecting/parallel activities • Environment • Guards	Kjellén and Larsson, 1981; Kjellén, 1984

They start from Heinrich's taxonomy by distinguishing between substandard acts and substandard conditions. Based on practical experience, they have further broken down each class of deviations into subclasses where they distinguish among such substandard acts as operating equipment without authority, failure to warn, failure to secure, operating at improper speed, using equipment improperly, and so on. Among the substandard conditions, we find inadequate guarding, defective tools, congestion, inadequate warning system, and so on. The different classes of deviations are not mutually exclusive, and one act may be classified in two or more different ways. In spite of these shortcomings, the ILCI classification scheme has been in wide use in industry.

The human-factor discipline is concerned with the interactions of man, equipment, and environment and the tasks carried out by man as a component in such a system. A deviation may occur in any of the system's components (man, equipment, and environment) or in the execution of a task.

There are a number of different schemes for the classification of human errors, and Table 5.9 shows one example. There is a distinction between unintentional errors (slips or mistakes) and violations. Slips are execution errors and mistakes are cognitive errors that are either due to incorrect interpretation (rule-based error) or to lack of expertise (knowledge-based).

Industrial engineering is concerned with the design and operation of industrial systems to promote productivity, quality, and safety. Table 5.9 shows one example of the classification of deviations according to this system's view (Kjellén and Larsson 1981). Here, the human actions, the necessary information and instructions, the work material, and the technical equipment in use define the tasks. The environment is split into intersecting/parallel activities (i.e. influences from other work groups) and physical environment (noise, illumination, climate, and so on). Deviations in any of these aspects will affect the risk of accidents. There is a system for control of production linked to each type of deviation.

Table 5.10 shows a checklist of deviations that is compatible with our analytic framework. It represents further development of a checklist presented by Kjellén and Larsson (1981).

5.6　*Contributing factors and root causes*

Incidents and deviations are considered symptoms of underlying deficiencies in human, technical, and organisational systems in a company. The idea is that the identification and amelioration of these deficiencies will have lasting effects on the accident risk level. Different terms are used in the research literature to label these underlying deficiencies – such as latent failures, contributing and root causes, basic causes, lack of control, and organisational influences (Bird and Germain 1985; Cornelison 1989; Reason 1991; DOE 2000; Wiegmann and Shappell 2003). The use of

Table 5.10 Checklist of deviations

Work situation

1. Human error (e.g. wrong action, wrong sequence, omission)
2. Technical failure (e.g. substandard equipment, break down, missing equipment or tools)
3. Disturbance in material flow (e.g. bad quality, delays)
4. Personnel deviations (e.g. absence, not qualified, indisposed)
5. Inadequate information (e.g. job procedure, permit to work, risk assessment, supervision, instructions)
6. Delay in progress

Environment

7. Intersecting or parallel activities (e.g. lack of coordination of work)
8. Bad housekeeping
9. Disturbances from the environment (e.g. excessive noise, high temperature)
10. Substandard building and infrastructure (e.g. roads)

Incident

11. Loss of control of energy or person relative to energy flow
12. Failure in active safety barriers
13. Failure in fixed barriers
14. Failure in personal protective equipment or clothing
15. Persons in danger zone

Development of injury/damage

16. Failure in alarm and mobilisation of emergency response team
17. Failure in limiting injury/damage (e.g. medical treatment and evacuation)
18. Failure in management of information to internal and external stakeholders

---

terminology reflects different conceptions of how and to what extent the combinations of psychological, social, organisational, and technical factors affect the risk of accidents.

Reason (1991) considers *latent failures* as dormant dysfunctions in the design and management of an industrial system. In this sense, they are similar to the medical concept of illness inducing so-called resident pathogens in the human body. Combined with local triggering factors in a work system, latent failures can overcome a system's defences and, as a result, create accidents (Section 4.4.1). Thus, accidents do not arise from single causes but from a combination of latent failures and local triggering factors.

The Management Oversight and Risk Tree (MORT) system uses the term *root causes* in reference to lack of adequate management control that results in substandard practices and conditions and, subsequently, in an accident (Cornelison 1989).

The U.S. Department of Energy (DOE 2000) identifies three types of causal factors – direct causes, contributing causes, and root causes. The direct cause of an accident is the immediate event or condition that caused the accident to occur; these are similar to deviations in the analytic framework presented earlier in this chapter. Contributing causes are events or conditions that collectively increase the likelihood of an incident occurring but that individually do not cause the incident. Root causes are higher-order, fundamental causal factors that address classes of management system deficiencies, rather than single problems or faults. Correcting root causes prevent the analysed accident from recurring and also solve management system deficiencies that could cause or contribute to other accidents.

The Human Factors Analysis and Classification System (HFACS) was presented in Section 4.6.1 and is an example of a method for the analysis of human factors in incident investigations based on experiences in aviation accident reports (Wiegmann and Shappell 2003). The causes of operator error are analysed in relation to three categories of contributing factors: preconditions for unsafe acts (e.g. fatigue, inadequate communication), unsafe supervision (assigning inexperienced operators to a demanding job), and higher organisational influences (e.g. inadequate training due to budget constraints).

In the research literature, we find many different checklists of causal factors. Such checklists are used in accident investigations for the purpose of ensuring that all relevant causal factors are identified.

A review of the accident research literature shows that many of the checklists are based on a hierarchical conception of the relationships between the incident and its causal factors (see Section 4.6.1). This hierarchy of causal factors usually follows the hierarchical levels of a traditional industrial organisation. The analysis starts with the incident and the immediate causes behind it (deviations) and works itself back to the root causes (see Figure 5.4), that is bottom up as opposed to the top-down display of the hierarchical levels of an organisational chart.

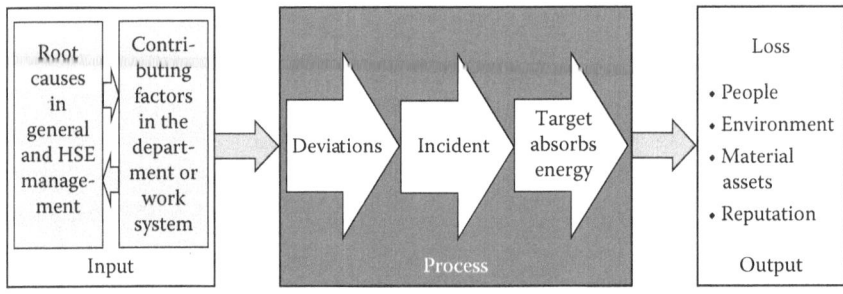

Figure 5.4 Display of relationships between contributing factors and root causes in the analytic framework.

Table 5.11 shows the causal hierarchy characteristics of different accident models. Here, we recognise MORT, which basically is a management model. Also, the ILCI model and Tripod analysis include upper-management elements. Tripod and HFACS analyse the relations between human errors at different hierarchical levels.

We have deliberately limited ourselves to factors inside the company in Table 5.11. This is a natural boundary of the scope of an internal accident investigation. We are concerned with those types of causes that people in the organisation, and management in particular, can do something about. Therefore, we are primarily concerned with the *preventive value* of the causal information. If the *explanatory or predictive values* were in focus, we might well have expanded the analysis to conditions outside the boundary of the company, such as the domestic situations of those involved or the conditions in that branch of industry, and in regulatory bodies and society in general. Under those circumstances, we would have had to define another boundary.

In the subsequent sections, we shall review the contents and underlying theories of some of the checklists. Our intention is to arrive at an understanding of the rationale behind the different checklists and their potentials and limitations. The different approaches are complementary, and there is no single 'true' accident causal model.

5.6.1 Contributing factors at the functional department and work-system levels

An early example of a causal factors checklist that has had significant influence is Swain's (1974) checklist on human-performance-shaping factors. It is human factors–oriented and lists factors that affect the quality of human performance and thus the likelihood of human errors. Table 5.12 shows an extract from this checklist. (A more modern version is presented in Table 10.2.)

This checklist has been used as both a design tool and an accident investigation tool in cases where human errors have played a central role. It draws attention away from blaming the person that made the error. Instead, it focuses on the identification of dysfunctions in the design of the man–machine system from an ergonomics point of view. The intention is to identify accident-prone workplaces rather than accident-prone persons. Behind many of the items in Swain's checklist are objective data on prerequisites for optimal human performance. Human-factor experts are its intended users.

Swain's checklist on performance-shaping factors has had great influence. The ILCI checklist on basic causes, for example, includes many of these items in the part covering personal factors (Bird and Germain 1985). Swain's list is in wide use in industry in accident and incident investigations.

Table 5.11 Examples of accident models applying a causal hierarchy

Level	ILCI model (Bird and Germain 1985)	MORT-based root-cause analysis (Cornelison 1989)	Tripod analysis (Reason 1991)	HFACS (Wiegmann and Shappell 2003; Salmon et al. 2015)
General and HSE management	Lack of control • Program • Inadequate standards • Inadequate compliance	Policy and implementation Risk assessment • Information systems • Hazard analysis • Others Bridge elements • Directives • Budget • Others	Source failure types • Pathological • Incipient – reactive • Worried – reactive • Repair – routine • Conservative – calculating • Incipient – proactive • Generative – proactive	Organisational influences • Resource management • Organisational processes • Organisational climate
Functional department	Immediate causes • Personal factors • Job factors	Specific factors • Maintenance • Supervision • Others	General failure types • Hardware • Design • Maintenance • Procedures • Error enforcing conditions • Housekeeping • Incompatible goals • Organisation • Communication • Training • Defences (barriers)	Unsafe supervision • Inadequate supervision • Failure to correct problems • Failure to enforce • Planned inappropriate operation Preconditions for unsafe acts • Conditions of operators • Personnel factors • Environmental factors

Table 5.12 Extract from Swain's checklist on human-performance-shaping factors

Human-performance-shaping factors

1. *Situational characteristics* such as physical environment (temperature, air quality, noise), manning, work hours, supervision, rewards, organisational structure
2. *Task and equipment* characteristics such as perceptual, anticipatory, decision-making and motor requirements, feedback and knowledge of results, man–machine interface
3. *Job instructions* such as procedures, communication, work methods
4. *Psychological stresses* such as task speed and load, threats in case of failure, monotonous work, conflicts of motives, negative reinforcement
5. *Physiological stresses* such as fatigue, discomfort, hunger and thirst, temperature extremes, vibration, lack of physical exercise
6. *Individual factors* such as knowledge and experience, skills, intelligence, attitudes, physical condition, influence of family, group identification

Source: Swain, A.D., *The Human Element in Systems Safety*, Industrial and Commercial Techniques, Surrey, 1974.

Table 5.13 shows an example of a system for the identification and classification of contributing factors in the work system and in the functional department based on the so-called MTO model. It is rooted in industrial engineering and human factors, and it explains the occurrence of incidents and deviations in production by how the human, technical, and organisational subsystems interact. This checklist and ones similar to it are widely used in accident investigations, such as in the MTO method (Chapter 13).

The SMORT Tier 2 checklist in Appendix B represents a further development of the MTO checklist. It has also been influenced by the MORT diagram (Johnson 1980).

Example: We will illustrate the relationship among an incident, deviations, and contributing factors by an accident during the construction of a pressure tunnel to a hydroelectric power plant. Two wires were attached to a pipe – one at the end of the pipe towards the bottom of the tunnel and one at the other end at the top of the tunnel. In the bottom of the tunnel, the first wire was attached to a winch used to pull the pipe downward. The other wire functioned as a safety wire and was attached to a winch at the top of the tunnel to control the downward movements of the pipe. The pipe skidded on the rails. It moved unevenly, and the operator of the winch at the top of the tunnel decided to remove the safety wire. When the pipe was pulled by the winch at the bottom of the tunnel, it started to skid in an uncontrolled way down the tunnel and was only stopped when

Table 5.13 MTO model for classification of contributing factors

M – Human/ behavioural	T – Technical/physical	O – Organisational/ economic

1. Supervision, instructions
2. Informal information flow
3. Workplace norms
4. Individual norms and attitudes
5. Individual qualifications and experience
6. Special circumstances
	1. Workplace layout
	  • Access to equipment
	  • Walkways, transportation routes
	  • Safe distances between moving equipment
	2. Design of equipment
	  • Physical hazards
	  • Reliability
	  • Man–machine interface
	3. Physical working environment (lighting, inner climate, noise)
	4. Protective equipment, guarding
	5. Work materials, chemicals
	6. Safety equipment and systems
		1. Work organisation, manning, job description
		2. Activity planning
		3. Methods of work, work pace
		4. Instructions, work procedures
		5. Maintenance routines, permit to work
		6. Management of change
		7. Education, training of personnel
		8. Supervision
		9. Systems of remuneration, promotion, sanctioning
		10. Controls of other types (e.g. economic, 'third party')
		11. Systems of shift, work schedule
		12. Routines in safety work
		13. Organisation of on-scene emergency management

Source: Kjellén, U. and Larsson, T.J., *J. Occup. Accid.*, 3, 129–140, 1981.

it hit the winch. The operator managed to move away from the pipe, but the crash resulted in substantial equipment damage. Figure 5.5 illustrates the incident and the different deviations and factors that contributed to it.

This example illustrates how decisions in separate units of an organisation contribute to an accident. Each decision-maker is not likely to see the consequences of his decisions in the total operational context. The example also illustrates how a combination of different decisions has resulted in a specific accident. Even if this particular combination is not likely to repeat itself, each erroneous decision may 'cause' other types of accidents. This type of analysis thus leads to questions about improvement needs in management decision-making routines.

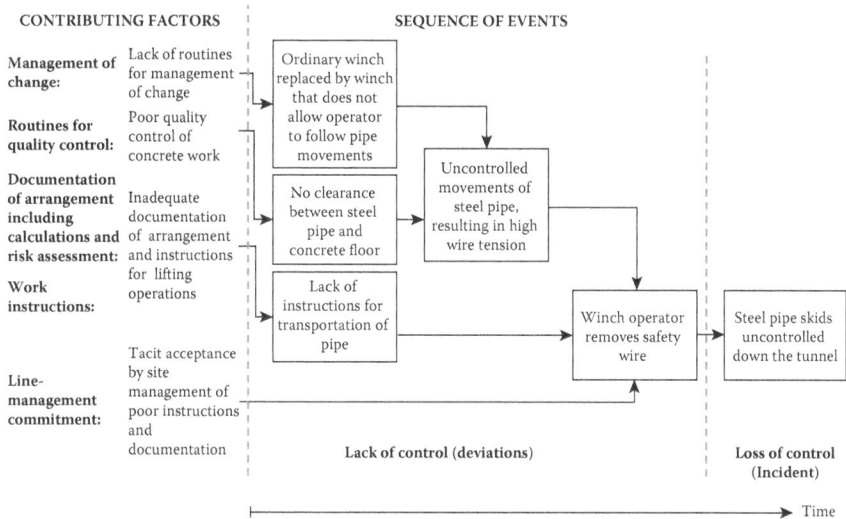

CONTRIBUTING FACTORS   SEQUENCE OF EVENTS

Figure 5.5 Deviations and contributing factors in an accident during the transportation of a 12-ton steel pipe in an inclined tunnel.

In the Tripod model (Section 4.4), the underlying factors behind accidents at the workplace or department level are called latent failures. Based on research into accident causation, 11 different so-called *basic risk factors* have been identified (Groeneweg 1998). They comprise design, tools and equipment, maintenance management, housekeeping, error enforcing conditions, procedures, training, communication, incompatible goals, organisation, and defences. The presence or absence of basic risk factors (BFR) is identified in accident investigations. A failure state profile is developed on the basis of accident statistics.

In conclusion, there is a certain degree of arbitrariness in the selection of elements to be included in the causal factors checklist at the work system and functional levels. Industrial systems are human constructs, and there is no unifying theory on how the different aspects of such systems interact to produce harm. Any checklist must include elements from different accident models.

5.6.2   Root causes at the general and HSE-management-systems levels

In this section, we will take the position that the level of safety in an organisation primarily is decided by general management factors. Adequate HSE management is a necessary supplement but will be inefficient without the support of a well-functioning general management system. This position is supported by evidence from research by Tinmannsvik and Hovden (2003), who distinguish between safety-specific and general management factors (management decisions to change safety-specific factors aimed at promoting safety). These factors include attitudes towards safety, guarding, emergency preparedness, safety work, and experience exchange. Similarly, the objective of changing general management factors is to promote production goals. Examples of such factors are training programs, design of machinery and equipment, maintenance routines, transportation and storage facilities, housekeeping, and supervision. Empirical studies of 14 companies only showed a positive correlation between the injury frequency rate and the general safety factors.

5.6.2.1 Causes derived from quality-assurance principles

The theoretical basis of the different causal models becomes more obvious at the upper management level. MORT was the first comprehensive model to include organisational and individual factors at the top management level. At this level, it draws from quality-assurance management principles similar to those of the plan–do–check–act (PDCA) wheel or cycle discussed in Section 7.3. The SMORT and ILCI models have been influenced by this pioneering work and represent variations on the same theme. The concept of root causes originates from the MORT model. The checklist in Table 5.14 shows the different items in a root-cause analysis.

We recognise the PDCA wheel in the three first main categories of root causes. In the third category, risk assessment, we find the types of themes

Table 5.14 Different elements in a MORT root-cause analysis

Main categories	Subcategories
• Policy	
• Policy implementation	• Line/staff responsibility
	• Accountability
	• Vigour and example
	• Methods and criteria analysis
• Risk assessment	• Safety-information systems
	• Hazard analysis process
	• Safety-program auditing
• Bridge elements	• Management services
	• Directives
	• Budget
	• Information flow

Source: Briscoe, G.J., *MORT-Based Risk Management*, Working Paper No. 28, Systems Safety Development Center, EG & G Idaho, Idaho Falls, ID, 1991.

that will be discussed in Sections III to V of this book. 'Bridge elements' (the fourth category) show how upper-management and HSE-management principles are implemented at the department and work-system levels.

Example: In the accident with the pressure pipe outlined in Figure 5.5, we identified four different contributing factors. Each factor was linked to a decision in a separate part of the organisation. We can proceed with the investigation by analysing deficiencies in the decision-making. Poor management of change was identified as one of the contributing factors. Had adequate criteria been applied in the selection of winches for the work? Was implementation adequate? Had the results been verified and followed up adequately?

SMORT represents further developments of the original MORT concept mainly for the purposes of improving user-friendliness and to add new knowledge. In the application of SMORT, the analysis proceeds from the specific accident occurrence and deficiencies at the workplace and department levels to the general management level in a step-by-step process.

SMORT has similarities to MORT in that it brings up questions concerning deficiencies in the design of industrial systems. The intention here is to learn from accidents in order to improve the design processes and thereby to acquire safer industrial systems in the future. At the top level, we find elements in SMORT similar to those of the quality-assurance management model.

5.6.3   Causes derived from safety culture elements

Here, we will refer to the evolutionary model of HSE culture, also called the HSE culture ladder, presented in Section 4.8. We use this model as a diagnostic tool in the identification and analysis of the influence of safety culture in incident investigation.

5.6.4   Problems in identifying causal factors

We apply a diagnostic process (which will be further outlined in Section 7.3) in identifying causal factors in an accident investigation that is illustrated by Figure 5.6. This starts with the *symptom*, that is, the specific occurrence in the sequence of events (the loss of control of energies in the system and the subsequent development of losses), and the associated deviations in the work system. In the step that follows, contributing factors and root causes (i.e. deficiencies at the department and higher-management levels that may explain these symptoms) are identified. In principle, there are two different methods for establishing such links between accidents and causal factors.

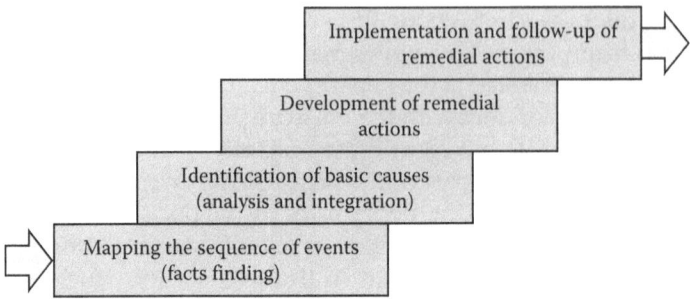

Figure 5.6 Accident investigation stairs.

Analytic methods are used when there are logical relations among the factors at the different levels (i.e. a causal factor is a necessary condition for the occurrence of a deviation in the accident sequence). Analytical methods are fact-based. A typical example is breakdown of equipment. It is often possible to trace such failures back to inadequate design that, in turn, can be traced back to inadequate design standards, inadequate design tools such as calculation programs, and/or to inadequacies in the quality control of design drawings. The following example identifies a factor that was necessary for the occurrence of the described event.

Example: A welder was grinding the welding seam between two pipes when the grinder kicked back and hit his face. In a reconstruction of the accident, it was possible to explain why the kick-back had occurred. The welder had brought the rotating disc in contact with the steel hull close to the spot where he performed the grinding operation. A congested layout was identified as a causal factor.

Subjective or expert judgements are used to identify and interpret causal factors in cases when objective facts about causal relations are lacking. This is a case of establishing causal relations by interpreting facts about the accidents in the light of experiences with the particular workplace and management system in question. The key difference between analytical methods and subjective/expert judgements is that the first is concerned with facts about the accident, while the latter is concerned with interpretations and judgments of causes and causal relations. These judgements normally are made by persons with direct knowledge about the circumstances around the accident to ensure trustworthiness and acceptance. We will come back to this issue and how to ensure that expert judgements are transparent and trustworthy in Chapter 12.

The analytic method is only valid under certain conditions, where physical causal relations are studied. This often is not the case for human or organisational factors. For individual accidents, subjective judgements must play an important part in the evaluation of causal factors. This means of arriving at conclusions is especially relevant when we deal with interpretations of people's behaviour in terms of reasons rather than causes. It is also relevant when we analyse how decisions affect the accidental outcome, and we have to review the distribution of responsibility, authority, and accountability within the organisation. In any collection of data on accident risks, a clear distinction must thus be made between *facts* and *interpretations* based on expert judgement. In making interpretations, it is important to corroborate the basis for these by checking different sources of information as far as possible.

Checklists are applied in accident investigations in order to ensure that all relevant causal factors are considered. In practice, the checklist-supported identification of causal factors is problematic. Items at the top of the checklist are more often addressed than are items further down (Hale *et al.* 1997). Experience also shows that the identification of causal factors is affected more by the items on the checklists and the underlying accident model (see Examples 1 and 2 that follow) than by factual circumstances. A detailed presentation of certain causal aspects will result in an overestimation of the importance of such aspects.

Supervisors have a tendency to choose causal-factor alternatives that are not possible to verify and that involve limited management responsibilities and obligations to implement remedial actions (Kjellén 1993). Typically, causal-factor alternatives related to human factors such as 'improper motivation' are selected. The link between the identified causes and the selection of remedial actions is often weak (see Example 3 that follows).

An interview study of Swedish accident investigators shows the same patterns (Lundberg et al. 2010). Investigators tend to:

- Focus on what they can easily find facts about
- Focus on what is 'politically' correct (individuals not the main cause, do not criticise the top management)
- Stop the investigation when their gut feeling says so, all facts are collected, or when they find issues that can remedied
- Focus on what they know the company is prepared to fix

Example 1: A comparison of the results of investigations of the same dropped-objects accidents on an offshore platform by two different companies, applying different causal models, showed large discrepancies between the results. One of the companies, the operator of the platform, applied the ILCI model on accident causation in the analysis. The three most common basic causes were lack of motivation, insufficient information, and poor design. The drilling company applied a technically oriented perspective, and the three dominant causes were wrong use of equipment, poor maintenance, and poor design.

Example 2: In the same study, 10 safety delegates were independently asked to code a written description of an accident by means of the ILCI model. They produced 10 different results.

Example 3: An evaluation of the accident-reporting system of the operator of the offshore installation showed that remedial actions were identified before and independently of the identification of causal factors. This classification of causal factors was done in order to satisfy formal requirements rather than as a tool in order to come up with an improved quality of safety measures.

The subjective biases of the analysts are also a concern. In *attribution-theory* research, people's judgements in determining accident causes and in selecting remedies are studied along two dimensions: transient–stable and person–environment (DeJoy 1994). Due to the fact that accidents often are causally ambiguous and emotionally charged, subjective biases play an important role in the attribution of causes. Self-protective biases on behalf of the supervisor make him/her likely to deny their own responsibility for an accident. Instead, the accident is often attributed to causes beyond the supervisor's control by blaming the workers involved. They, on the other hand, are likely to favour situational causes in the working environment.

In general, people tend to explain accidents that other people have been involved in by referring to their stable personal characteristics. This corresponds to the third quadrant in Figure 5.7. However, people explain accidents that they themselves have been involved in by referring to transient external causes (i.e. the second quadrant). This so-called fundamental attribution error has great implications for accident investigation routines.

Research has also shown that decision-makers apply shorthand methods or rules derived from earlier experiences in the analysis of causes of accidents. For instance, many decision-makers have developed rules implying that unsafe behaviour on the part of the workers is the important feature of accident causation. A decision-maker may thus stop looking

	Person	Environment
Transient	1. Inattention	2. High wind speed
Stable	3. Poor crane-driving skills	4. Poor sight from the crane cabin

Figure 5.7 Two dimensions in the attribution of causes. They are illustrated by examples from a crane accident.

for more evidence after detecting a human error, even if more important causal factors are at hand.

We also know that incentives and motivational factors play an important role in causal attributions and in the selection of remedies. An illustration of this is when decision-makers attribute accidents to unstable causes rather than stable causes. Inattention is an example of an unstable factor as opposed to poor viewing conditions, which is a stable characteristic. Unstable causes produce less certain predictions about the efficiency of remedial actions. By selecting such causes, the decision-maker may escape from the obligation to decide on binding and resource-demanding solutions to the safety problem (Kjellén 1993).

It is a well-established principle in crisis management to limit the negative public relations effects of a severe accident by focusing on situational factors when dealing with the mass media. These tactics will help to fend off outside forces that demand change. They will, however, when applied within a company, prevent adequate learning. The following example illustrates such communication with mass media and a narrow view of the accident causes.

Example: In a rigging accident at a construction site, two riggers were seriously injured when a pull chain ruptured. The investigation showed that the riggers had violated the safety rules when they stayed in the danger zone while the winches were operated. The investigation also identified serious deficiencies in the planning of the work and in the technical control of lifting equipment. Site management, in their communication with the press, explained that the accident was caused by human error on the part of the operators. Managerial errors were de-emphasised.

Expert judgements are in many cases the only practical way to establish basic causal factors. This condition has some important implications. The person with direct responsibility for safety at the site of the accident (i.e. the supervisor) should not make the judgements alone. A consensus process is recommended, where persons with direct knowledge of the circumstances and persons with different responsibilities and perspectives take part in the investigation into causal factors.

To sum up, rather than 'objectivity', what matters is the organisation's confidence in the persons doing the investigation and in the transparency of the reasoning that leads to the conclusions and recommendations. Examples of factors that will influence the organisation's confidence are:

- The status of the experts in the organisation
- The consensus processes that lead to the results and the extent to which key persons have participated
- The extent to which the expert judgements are transparent and easy to follow

chapter six

The occurrence of accidents over time

We will now change our focus from that of the individual incident or accident to the occurrence of accidents over time. The occurrence of accidents follows basic laws of statistics similar to events such as telephone calls to a home, customers arriving at a department store, or to radioactive decay.

It is a commonly accepted assumption in health, safety, and environment (HSE) management that accident frequency (i.e. the number of accidents occurring at a company in the course of, for example, a year) is a measure of the HSE performance of the company. In this chapter, we will focus how we can use basic statistical theory in evaluating data on the occurrence of accidents. We will apply this theory in Section IV on HSE performance indicators.

Example: A company has a stable number of employees. Over a period of 10 years, the average annual number of accidents was four, as shown in Figure 6.1. In the following year, there were two accidents. Does this number represent a significant improvement compared to the yearly average value or may it be explained by pure chance?

To be able to answer this question, we must look into the statistical theory of *Poisson distributions*. Here, we will merely review the most important characteristics of this statistical distribution. Standard textbooks on statistics should be referenced for further details.

Assuming that accidents occur at random points in time, let c be the *intensity*, which is the average number of accidents per unit of time (e.g. one year). Let x be the number of accidents occurring during t time periods. Therefore, $f(x)$ is the frequency function of the Poisson distribution $(c \times t)$, that is:

$$f(x) = \frac{(c \times t)^x}{x!} \times e^{-(c \times t)}$$

where $f(x)$ is the probability that x accidents will occur during t time periods.

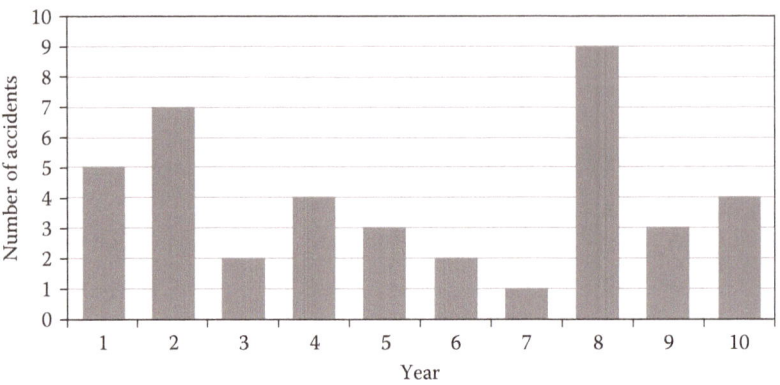

Figure 6.1 Recorded number of accidents per year.

A Poisson distribution has the following characteristics:

- The frequency functions for the number of events (accidents) during time periods that are not overlapping are independent stochastic variables.
- The probability of one event occurring during a short time period Δt is approximately equal to $(c \times \Delta t)$.

    Example: In order to calculate the probability of two or fewer accidents occurring at the company in the aforementioned example:

$$P(x < 3) = \sum_{k=0}^{2} \frac{4^k}{k!} \times e^{-4} = 0.0183 + 0.0733 + 0.1465 = 0.23$$

From this equation, we see that this probability is 0.23, which means that, in almost one out of every four years, we will expect from pure chance that there are two or fewer accidents. Figure 6.2 illustrates this particular Poisson distribution. From Figure 6.2, we can see that the probability of exactly four accidents during one year is only about 0.2. Similarly, the probability of zero accidents during one year is 0.02 (i.e. one every 50 years).

The Poisson distribution has other important characteristics. For 'large' values of $(c \times t)$, the Poisson distribution is approximately represented by a normal distribution with a mean value that equals $(c \times t)$ and a standard deviation that equals $\sqrt{(c \times t)}$.

The aforementioned example is typical. We were interested in knowing whether the number of accidents in one period (e.g. a year)

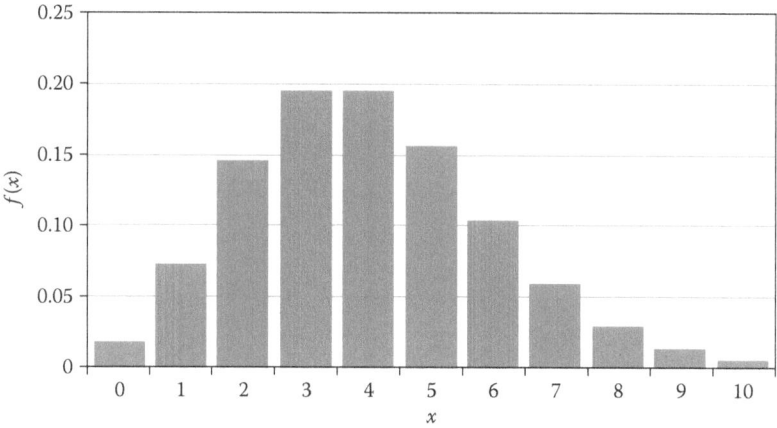

Figure 6.2 Graphical illustration of the frequency function of a Poisson distribution where $(c \times t) = 4$.

significantly differs from the mean number of accidents during previous years. For 'large' numbers of accidents, where we can assume that the number of accidents per period follows the normal distribution, this is easily answered. For most practical circumstances, this means five or more accidents per period. For a normal frequency distribution, the probability of an outcome between plus and minus two standard deviations ($\mu \pm 2\sigma$) is about 95%. This level of confidence is often used when determining whether a change is significant or not.

Continuing with this example, let us say that on an average there were 100 accidents at a particular company each year. Therefore, it follows that, for 19 of 20 years, we will expect a yearly frequency of between $(100 - 2 \times \sqrt{100})$ and $(100 + 2 \times \sqrt{100})$, that is, between 80 and 120 accidents per year. If the number of accidents for one year falls outside this range, we consider this as a significant change in the frequency rate.

The average number of accidents per period is our best *estimate* of the underlying or 'true' value of the accident intensity at a company. The 95% confidence interval of such an estimate can easily be calculated by applying the same method noted earlier.

chapter seven

Management of safety through experience feedback

7.1 What is meant by experience feedback?

In this chapter, we will focus on experience feedback as a process managed at the organisational level and its role in management of safety. We will apply the definition of *experience feedback* according to Wiley's *Encyclopedia of Quantitative Risk Analysis and Assessment* (Melnick and Everitt 2008). It is defined as, 'The process by which information on the results of an activity is fed back to decision-makers as new input to modify and improve subsequent activities'.

First, we need to distinguish between negative and positive feedback. In negative feedback, the input to control a process is a function of the difference between the actual and desired output. Positive feedback, on the other hand, involves system changes through amplification of divergences in output. Whereas negative feedback is essential in maintaining the stability of a system, positive feedback may destabilise a system but is a condition for change (evolution, growth) to reach a new point of equilibrium.

An associated term is *feedforward*. Anticipation is the heart of the feedforward mechanism. Here, the information used as input to control the system is not obtained directly by measuring system performance but indirectly through anticipation. One typical example is when we introduce control measures on the basis of results of risk analyses (see Chapter 20).

Example: Figure 7.1 illustrates the possible effects of experience feedback and learning on the safety performance as measured by the lost-time injuries (LTI) rate at an aluminium plant. The LTI rate measures the number of lost-time injuries per million working hours (see Chapter 16). Based on the LTI rate, we can identify two periods of improvements in safety performance as a result of learning. The first period of 15 years is characterised by improved ability to operate and maintain the plant and, as a consequence, by improved safety results. The LTI rate decreases after the first few years and stabilises around year 25. Fifteen years after start-up, the plant pioneered the use of a formal health, safety, and environment (HSE) management system based on the experiences of the 'internal

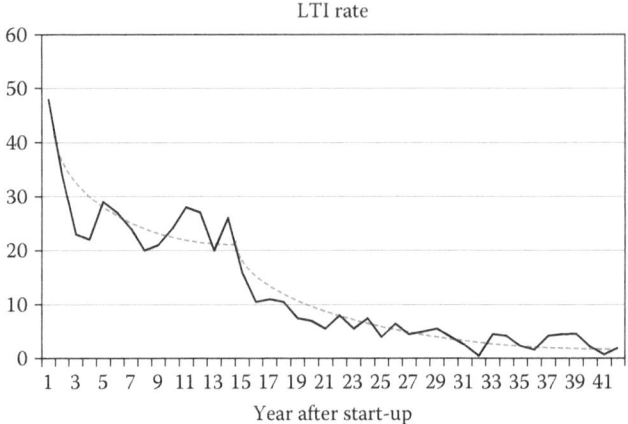

Figure 7.1 Development in the LTI rate at an aluminium plant during 41 years from start-up of production. The stippled line represents two superimposed learning curves.

control regime' of the Norwegian offshore sector. As we will see later in this chapter, such systems utilise a hierarchy of different feedback loops. This resulted in a new learning period, where the LTI rate stabilised around 3 after about ten years.

7.2 Feedback and use of safety-related information in decision-making

We need to design the experience feedback systems in safety management with the users' needs in mind. Figure 7.2 gives examples of the use of information in safety-related decisions in an organisation with two line-management levels and a staff organisation. Top management uses summary data for goal setting and monitoring purposes:

- Input to the establishment of goals or norms based on so-called safety performance indicators
- Measurement and follow-up of results in order to look for trends and to compare the results with pre-established goals
- Input to decisions on action plans when goals are not met

At the workplace, rich data on accidents are used locally for prevention. Here, supervisors and workers solve safety-related problems by evaluating feedback from production in relation to their own experiences.

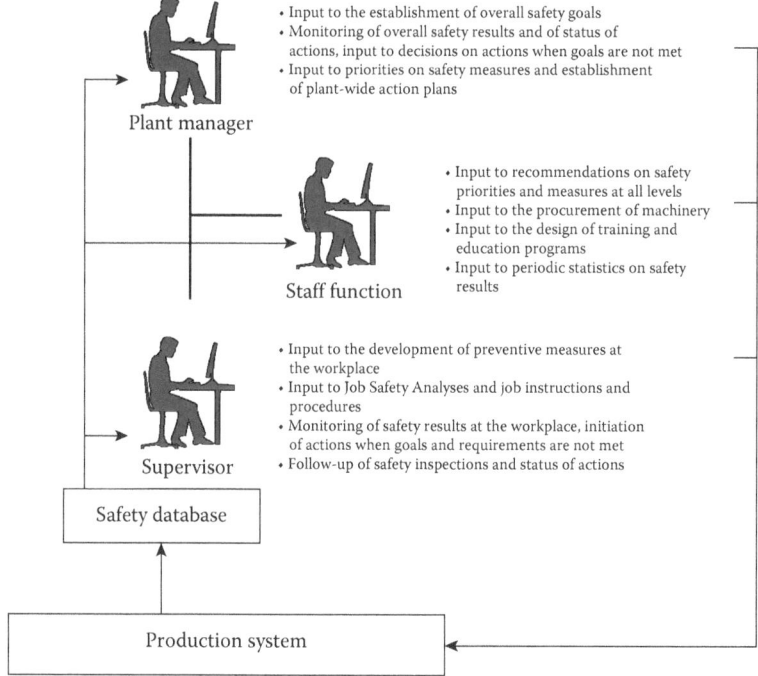

Plant manager
• Input to the establishment of overall safety goals
• Monitoring of overall safety results and of status of actions, input to decisions on actions when goals are not met
• Input to priorities on safety measures and establishment of plant-wide action plans

Staff function
• Input to recommendations on safety priorities and measures at all levels
• Input to the procurement of machinery
• Input to the design of training and education programs
• Input to periodic statistics on safety results

Supervisor
• Input to the development of preventive measures at the workplace
• Input to Job Safety Analyses and job instructions and procedures
• Monitoring of safety results at the workplace, initiation of actions when goals and requirements are not met
• Follow-up of safety inspections and status of actions

Safety database

Production system

Figure 7.2 Examples of feedback and use of safety-related information in decision-making at different levels at a plant.

Following are some examples illustrating different needs of safety-related information at this level:

- Input to priorities, selection, and implementation of safety measures
  Example: The kitchen supervisor of a hotel was committed to a reduction in the number of cut injuries. The safety database was used to analyse accident statistics in a decision about the selection of kitchen tools and equipment and personal protective equipment. The personnel participated in the development of solutions, and they helped in interpreting the accident statistics in relation to their experiences.
- Input to job safety analysis and the development of job instructions, procedures, and technical specifications
  Example: In planning job safety analysis of various activities in a rolling mill, accident statistics were analysed in order to identify hazards during start-up and shut-down, operation, cleaning, main-tenance, and handling of disturbances. The results were used to

document hazards associated with the different activities together with the necessary safety precautions.
- Monitoring of results of safety inspections and close-out of actions
  Example: The results of workplace inspections and the status of actions are fed into the safety database. It is used by the supervisor to check status and follow-up to be sure actions are implemented and closed out.

The HSE staff often are the most frequent users of safety-related experiences. They support line management and worker representatives by providing requested reports. Periodic summary reports on accident statistics, for example, present the necessary information for monitoring and priority-making purposes at the plant and department levels. Risk analyses represent another area requiring input on accidents. Such analyses give overviews of the hazards at the workplace and the associated risk, and they are used as input for decisions on actions to reduce risk. Line management makes the actual decisions. Additionally, the HSE staff play an important role in the systematic collection, quality assurance, and storage of data from incident reports and other safety-related experiences.

We also find practices where line management and staff functions make ad hoc queries, for example, about accident occurrences. Examples are:

- Need for input as to the design or procurement of machinery
  Example: Management at a workshop had decided to buy a new rolling mill. Data from previous accidents with a similar mill were retrieved and analysed, and the safety measures that came out of this review were included as technical requirements in the purchase order.
- Need for input regarding educational programs
  Example: An educational department was planning a first-aid course. They consulted the accident statistics for illustrative examples, showing the benefits of immediate treatment of victims.
- Identification of accident repeaters
  Example: An accident investigation team wanted to consult the safety database for similar occurrences. The results were used to determine the need for further analyses to identify root causes and remedial actions.

7.3   *Feedback mechanisms*

In this section, we will review mechanisms for experience feedback. We distinguish between two main mechanisms for such feedback: the feedback cycle and the diagnostic process. As we will see, the presented feedback mechanisms have their origins in other disciplines other than the science of safety. These two main types of feedback mechanisms are based on key principles of quality management. 'Closing the loop' is central in both of these feedback mechanisms, meaning that decisions and actions for improvement must be based on experience and that follow-up is necessary with regard to the results.

7.3.1 Juran's feedback cycle for the control of anything

Control by negative feedback is a regulating mechanism that identifies deviations and produces corrective action. The system output is monitored continuously or sampled at discrete points in time, and the results are compared to pre-established norms or goals. The measured difference between results and norm (i.e. a deviation) is used as input to control actions. It affects the system in a direction that gives a result closer to the norm.

These principles are applied in quality assurance as illustrated by Juran's (1989) model for the control of anything (see Figure 7.3). Juran's model is applicable to different types of management systems – for example, control of quantity or quality of produced goods and control of accident risk.

With respect to feedback control in the area of safety, we use different types of standards. In workplace inspections, for example, we are concerned with deviations from technical requirements defined in safety regulations or internal procedures. Another example is feedback control of safety performance as measured by one or more safety performance indicators. We will present such indicators in Section IV. We also use Juran's

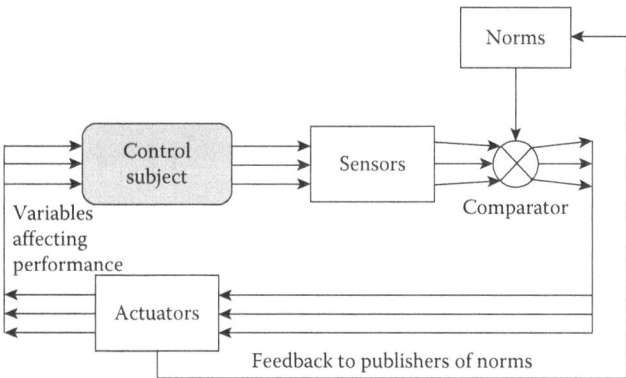

Figure 7.3 Feedback cycle for the control of anything. (Adapted from Juran, J.M., *Juran on Leadership for Quality – An Executive Handbook*, The Free Press, New York, 1989.)

cycle in the feedforward mechanism of risk assessments, when a predicted level of risk is evaluated against a risk acceptance criterion. If the risk is above the criterion, risk-mitigating measures are implemented to reduce the risk to an accepted level.

7.3.2   The diagnostic process

Diagnosis is defined here as the complete decision cycle consisting of: (1) identification of symptoms, (2) determining causes, and (3) prescription of remedy. A *symptom* is a deviation of the system's behaviour from what is considered to be 'normal', a term we are familiar with from the process model of accidents (see Section 4.3). The diagnostic process is based on the scientific method of systematic observation and experiment used in the natural sciences since the seventeenth century (Oxford University Press 2012).

Heinrich introduced a formal decision-making routine similar to the diagnostic process in his pioneering work on industrial safety management (Heinrich 1959). His aim was to shape the behaviour of decision-makers in a direction that gave safety sufficient attention. He introduced five systematic steps that decision-makers had to go through in order to prevent accidents: (1) collection of data about accidents and near-accidents, (2) analysis of these data, (3) selection of remedies, (4) implementation, and (5) evaluation of effects.

A structured approach is also represented by Mintzberg's conceptual model of the decision-making process (see Figure 7.4). The first main phase, identification, involves recognising that there is a problem present – in this case, in the area of safety. Mintzberg sees diagnosis as an inquiry into available information in order to determine the causes of accidents. When these are known, the safety problem to be solved is defined.

In the development phase, the decision-maker establishes goals for problem-solving activities, identifies alternative solutions to solve the problem, and evaluates these in relation to the pre-established goals. Finally, a decision is made on the choice of solutions, these are implemented, and the results are followed up.

Heinrich's and Mintzberg's models are examples of classical rational-choice models of decision-making. They do not describe how decisions actually are made. Instead, they prescribe how organisations should make decisions in order to arrive at satisfactory solutions. There are a number of assumptions behind these models (March and Simon 1958):

- There is a whole set of alternative actions from which to choose.
- Each action alternative is linked to a set of consequences. These may be certain or associated with an uncertainty.
- There exists a utility function that describes the relation between these consequences and the decision-maker's preferences.

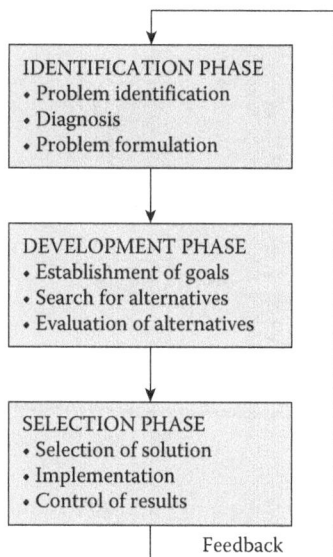

Figure 7.4 Mintzberg's model of the decision-making process. (Adapted from Mintzberg, H., et al., *Admin. Sci. Q.*, 21, 246–275, 1976.)

- The decision-maker is rational in the sense that he or she is fully informed and selects the set of alternatives leading to the preferred set of consequences.

7.3.3   Deming's cycle

Deming's cycle (also called the plan–do–check–act [PDCA] circle or wheel) is named after Edwards Deming, who developed this concept in the 1950s. It is central to quality management (Deming 1993). Deming presented the cycle as a learning process from planning through execution and check to correction (see Figure 7.5).

Deming's four basic processes are further broken down in the following way:

- Plan
  - Where are we?
  - Where do we want to go (goal)?
  - How do we get there?
- Do
  - Communicate and train
  - Secure resources
  - Execute

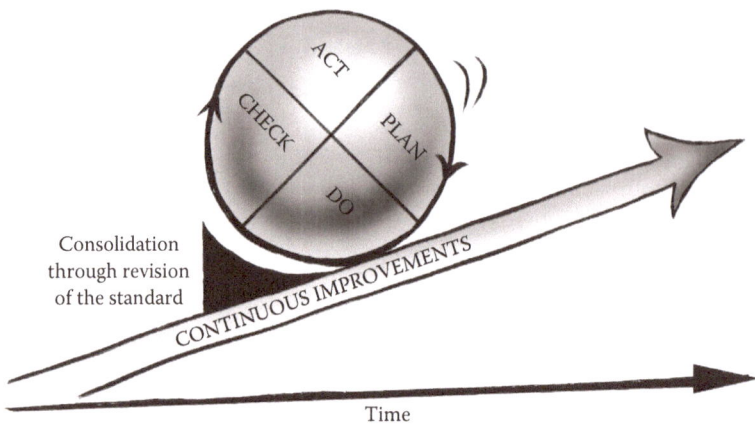

Consolidation
through revision
of the standard

CONTINUOUS IMPROVEMENTS

Time

Figure 7.5 Deming's cycle and the quality standard required to prevent fall-back.

- Check
  - Did we follow the plan?
  - Did we meet our goal?
- Act
  - Implement corrective actions
  - Sum up experiences
  - Standardise

The idea behind Deming's cycle is to ensure continuous improvements through consecutive rotations of the wheel. This concept has been widely adopted by industry and by the International Organization for Standardization (ISO) in the development of the ISO 9000 series of standards for quality management systems and in the ISO 14001 and draft ISO 45001 standards for environmental and occupational health and safety management (ISO 2015/2016). Figure 7.6 illustrates an example of the significance of Deming's cycle in the design of an environmental management system according to ISO 14001.

7.4   *Scope and level of feedback*

7.4.1   *Ashby's law of requisite variety*

When we analyse the types of safety measures that are implemented at a workplace following an accident, we often notice a lack of imagination in the development and selection of measures. We see repetitive use of a few relatively simple types of measures – such as instructions to the

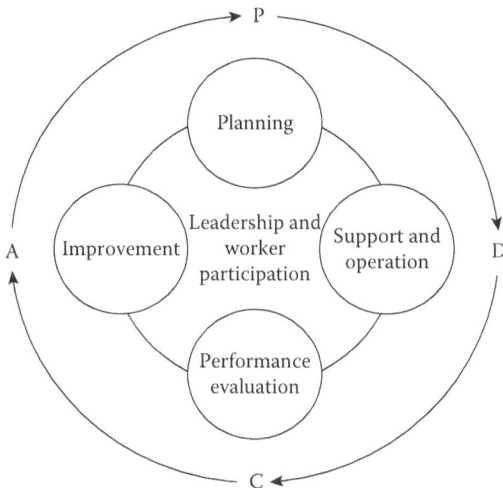

Figure 7.6 Application of Deming's cycle in illustrating the scope of an environmental management system according to the ISO 14001. (From ISO, *Environmental Management Systems – Requirements with Guidance for Use,* International Organization for Standardization, Geneva, 2015. Copyright by the ISO. Reproduced with permission.)

injured person to be more careful, repair of technical faults, erection of guards, and removing of litter. The simplicity and limited scope of the selected measures is remarkable when we consider the complexity of the accident sequence and the conditions at the workplace that have influenced it. *Ashby's law of requisite variety* helps us in analysing the weaknesses in a feedback control system of this type. It tells us about the relationship among the various types of measures that need to be utilised in order to exercise full control and about the complexity of the phenomenon that we want to control. 'For an analyst to gain control over a system, he must be able to take at least as many distinct actions, i.e. as great a variety of countermeasures, as the observed system can exhibit' (Van Court Hare 1967).

In accident control, the requisite variety of countermeasures is dependent on the system level. At the work-system level, the 'analyst' (the worker) faces accident risks that are determined by complex and varying changes in the environment and his/her effect on it. The worker must implement a great variety of measures in order to maintain control. Top management, on the other hand, is concerned with accidents at an aggregate level, where the different short-term variations and details even off. At this level, the performance data are much less complex and, as a consequence, the control actions at this level also may be less complex. Experience shows that a simple high-level decision on changes in HSE policy will influence complex behaviour at lower management levels such that positive effects are achieved.

The successive information filtering from the work-system level through different managerial levels to top management is necessary in order to avoid information overload. There always is a concern, however, that this necessary filtering does not result in distorted or biased information. If this is the case, the feedback processes of the HSE management systems will actually be counterproductive. They may reinforce incorrect accident perceptions and thus result in inefficient countermeasures.

Figure 7.7 shows the results of an analysis of safety measures documented in accident reports at seven shipyards. There were no actions documented in more than a third of the accident reports. Instructions 'to take care', to use prescribed personal protection, or about a method of work were otherwise the most common types of measures. Only very few measures were of a risk-reducing type that changed the more permanent contributing factors in the long term (i.e. changes in method of work, procedure, or organisation). The shipyard results are typical for many workplaces and illustrate that only very few types of measures actually are utilised in accident prevention.

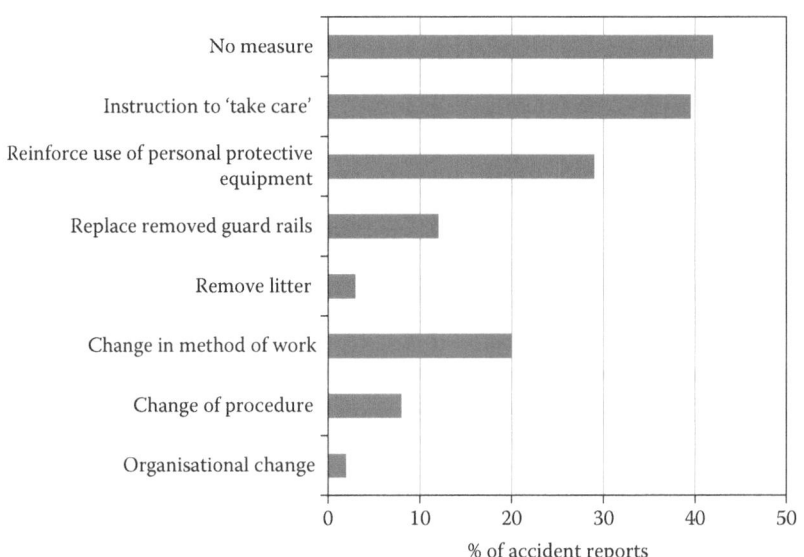

Figure 7.7 Percentage of accident reports at seven shipyards containing information on prevention measures by type. (Adapted from Kjellén, U., In *The Workplace*, Scandinavian Science Publisher, Oslo, 1997.)

There are means for increasing the variety of the measures taken to prevent accidents. One possible way is to train an organisation to conduct more comprehensive accident investigations (see Chapter 13).

The 'defences-in-depth' philosophy (which will be discussed in more detail in Chapter 9) can be interpreted in terms of Ashby's law of requisite variety. Here, the system's designer hopes to prevent major accidents through barrier design. This will reduce the variety exhibited by the system and thus make it less demanding for the operations organisation to manage major accident risks.

Another important aspect of Ashby's law of requisite variety has to do with the rate of change in the system to be controlled. Ashby's law of requisite variety states, 'For an analyst to gain control over a system, he must be able to generate countermeasures at least at a rate corresponding to the rate of variety that the observed system can exhibit'.

We will illustrate this principle with two examples. A construction site is characterised by rapid changes in technology, organisation and manning, methods of work, and so on as a project progresses. A processing industry, such as a fertiliser plant, will look more or less the same from one day to the next. We expect changes with possible negative effects on safety to be introduced at a much higher rate at a construction site than at a processing plant. It follows that requirements will vary as to the frequency of safety inspections in these two cases because of the differences in the ability to detect and process information about changes at each workplace.

To summarise, Ashby's law of requisite variety implies that an analyst must show the following three abilities in order to exercise full control of a system:

1. Ability to take at least as great a variety of actions as can the observed system
2. Ability to take precisely the correct set of action alternatives to counter those changes generated by the system
3. Ability to collect and process information and to decide on and implement measures at a rate at least equal to the system's rate of change.

7.4.2   *Van Court Hare's hierarchy of order of feedback*

Ashby's law on requisite variety is helpful to understand whether sufficient actions have been undertaken to manage a system's safety performance. Another approach to evaluate feedback control efficiency is to consider how far-reaching the changes are that are undertaken. Van Court Hare (1967) distinguishes among different orders of feedback control. Table 7.1 shows these different orders of feedback and gives examples

Table 7.1 Hierarchy of feedback systems arranged by order of feedback

System order	Characteristics	Traditional decision level	Examples from safety management
0	Simple transformation without feedback	Workers	No follow-up of accidents with remedial actions, the loop is not closed.
I	Simple machine with direct feedback but without selective memory	Foremen	Correction of deviations identified by accident investigations or safety inspections.
II	Tactical system with memory organisation, conditional selection of pre-established plans, and predictive feedback	Middle management	Starting a pre-planned eye-protection campaign following an increase in eye injuries.
III	Strategic system that learns from experience and has ability to correct selection of plans and develop new plans	President and staff	Change in routines, instructions, rules, or design on the basis of accident experience.
IV	Goal-changing system that learns and consciously develops, selects, and implements new plans	Board of directors	Changes in HSE policy and goals on the basis of accident experience.

from safety management. The order of feedback is an indicator of the degree of learning from previous experience. At zero and the first order of feedback, there is no such learning. In the area of safety management, this means that the same types of deviations, incidents, and accidents will be able to reoccur. Long-term learning, which manifests itself in actions that prevent recurrence and continuously reduce the risk of accidents at the workplace, is here referred to as third and fourth orders of feedback.

Van Court Hare's hierarchy was developed on the basis of experience in traditional industrial and military organisations. His focus on the type of learning that is involved in feedback control also is, however, of interest for safety in modern industries. In practice, too often we see easy fixes such as replacing a missing guard after a person has fallen from a scaffold or clean-up after a person has tripped on debris. These are examples of first-order feedback. The learning opportunity with more lasting effects has been lost.

Example: An analysis of the accident statistics from a construction site showed that there had been several falls on the same level and others at a lower level during the last year. These types of accidents typically are prevented through adequate gangways and guard rails. Workplace inspections were performed weekly at the site. An analysis of the inspection results showed that corrective action concerning missing guard rails and inadequate gangways dominated the picture. The frequency of these actions (number of actions per inspection) had been relatively stable during the last year; see Figure 7.8. Site management had failed to implement preventive measures in order to avoid these types of deviations reoccurring.

The implementation of measures following an accident may be seen as a means for the organisation to store experience from the event in its collective or organisational memory. Here, we will identify four different types of 'memory' related to accident experience, corresponding roughly to Van Court Hare's hierarchy of order of feedback:

- Correction of deviations, that is, only the 'short-term memory' is utilised. The deviation may reoccur (Order I).
- Long-term storing of experience by means of changes in design, work procedures, and so on at the workplace where the accident occurred. The conclusions drawn from the experience will have lasting effects and may prevent a recurrence but will be context dependent and of limited scope and will not affect accident risks at other workplaces (Order II).

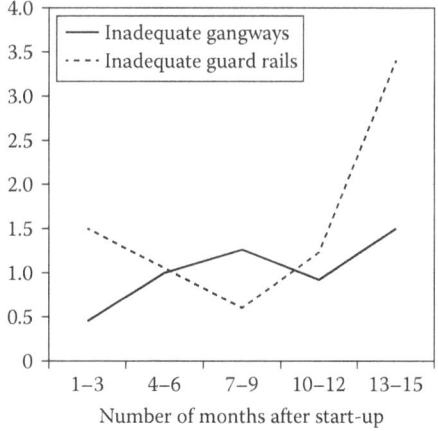

Figure 7.8 Number of deviations per inspection at a construction site by type of deviation. (Adapted from Kjellén, U., *J. Occup. Accid.*, 3, 273–288, 1982.)

- Long-term storing of experience by means of changes in personnel supervision and changes in technical and administrative systems for production control at the functional department. These types of change will also have lasting effects and will affect other workplaces as well (Order III).
- Long-term storing of experience by means of changes in general management and in the HSE management systems and norms (policy, goals, specifications, and so forth). The changes will not only have lasting effects but also have a wide scope and be de-contextualised – thus affecting many workplaces all over the company (Order IV).

The observant reader will note that the four types of accident experience have their counterparts in the different elements of the accident analysis framework shown earlier (see Figure 5.4). When we move upward in Van Court Hare's hierarchy, we can also expect that the measures will involve a higher level in the causal-factor hierarchy.

Example: Table 7.2 shows examples of how measures can be classified according to Van Court Hare's order of feedback. The examples are collected from accident reports. Such classifications can be used to generate an aggregated picture of the quality of learning after incidents (see Chapter 18 for more detail).

7.5   *Safety information systems*

A *safety information system* provides the information needed for decisions and signalling in the management of safety. It is a vital element in a company's safety management system. Such a system provides support to decision-makers in the prevention of accidents that may result in injury to personnel, accidental emissions of pollution to the environment, and in damage to material assets.

In our analysis of the feedback mechanisms in accident prevention, we will use a model of safety information system and its different subsystems according to Figure 7.9.

Our underlying assumption is that accidents are preventable through systematic experience feedback. This feedback goes from the workplaces to the decision-makers at different levels of the organisation. An important part of this feedback takes place in informal settings in the day-to-day contacts among people in the organisation. We will focus on how the experiences are made explicit through categorising the documentation of investigations results into accidents and near-accidents, workplace inspections, safety audits, risk analyses, and so on. Positive effects on safety are only achieved when the *loop is closed*, that is, when the results

Table 7.2 Classification of actions according to Van Court Hare

Accident	Action	Order of feedback
A robot arm squeezed a repairman when he activated the local control by mistake.	Permit to work procedure to ensure that hydraulic equipment is depressurised and isolated before start of repair work.	II
The welder took a step backward and stepped into an opening in the deck. His back was injured.	Opening was protected by guard rail.	I
The operator jumped from 0.6 m. He twisted his foot.	Common accident, no action.	0
The operator hit his head on scaffolding above the gangway. The clearance was only 1.6 m.	Warning sign was put up.	II
The operator's arm got caught in a packing machine while removing a parcel that had gotten stuck.	Implementation of preventive maintenance based on technical failure statistics to reduce the frequency of such failures.	III
The operator was sandblasting in a narrow space. He hit his leg against a support.	Written instructions to use knee protection.	II
Management was unsatisfied with the high accident frequency rate.	Change in HSE policy to signal increased management attention to the prevention of accidents.	IV

of the decisions are implemented in a way that affects the workplaces (existing or future).

A safety information system provides the following:

1. *Reporting and collection of data* on accident risks by means of accident and near-accident reporting and investigations, reporting of unwanted occurrences, workplace inspections, safety audits, and risk analyses. Methods of data collection include observation, interviews, self-reporting, group discussions, electronic registration, and so on.
2. *Storing* of data in a memory (paper file, electronically) and retrieval of data from it.

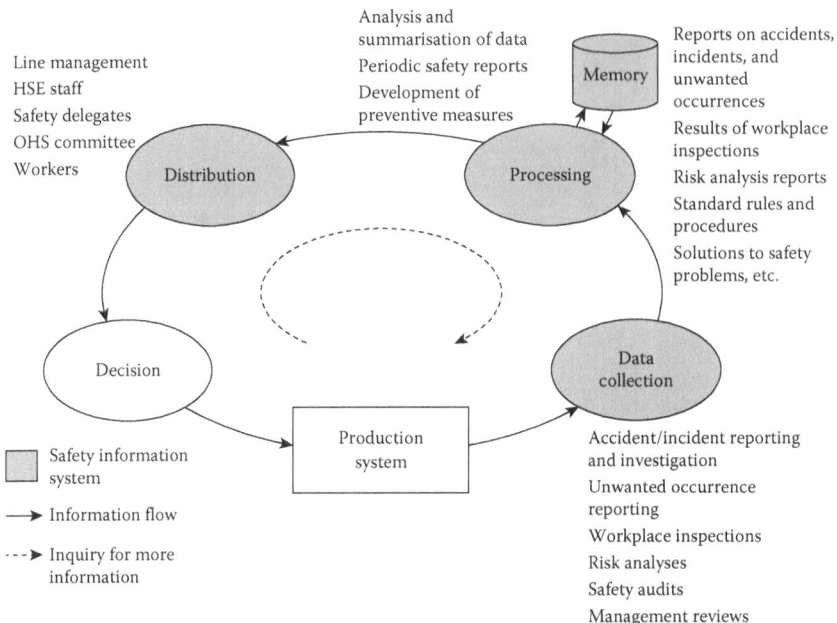

Figure 7.9 Flow of information in a safety information system. Inquiries for more information go in the opposite direction. (Adapted from Kjellén, U., *Analysis and Development of Corporate Practices for Accident Control*, Thesis, Report No. Trita AVE-0001, Royal Institute of Technology, Stockholm, 1983.)

3. *Information processing*, that is, retrieval and analysis of data, compilation into meaningful information, development of remedial actions, and so forth.
4. *Distribution* of the information to decision-makers inside the organisation (e.g. line management, staff officers, working environment committee).

The effectiveness of a safety information system traditionally has been determined by its ability to provide the necessary basis for decisions on remedial actions. In order to close the loop (i.e. to go from experience to action), the efficiency of each subsystem of the safety information system is critical. A weak link, such as an inadequate distribution of information, will break this loop. This is irrespective of the quality of other subsystems. The end result, an improved safety standard, ultimately is determined by the decision-makers' ability to ask for and use the available information.

More specifically, a safety information system must support such different safety management activities as:

• Prioritising safety measures, developing them, and evaluating their effect
• Establishing safety goals and monitoring the development in safety performance for comparison with these goals and other companies' performance
• Providing source data as input to risk analyses
• Evaluating the efficiency of safety program elements and the HSE management system as a whole

Rapid developments in information technology have made it possible to solve many of the traditional problems associated with safety information systems. Computer networks provide efficient support in the collection, processing, and distribution of information as well as 'unlimited' storage space.

7.5.1 Developing the model further

Let us look at the immediate effects or output from a safety information system as shown in Figure 7.10. There are two types of effects (Nonaka and Takeuchi 1995). Effects on the *explicit knowledge* at the organisational level have to do with what we normally associate with results of decision-making in the area of safety, health, and environment. They are expressed in formal documentation of procedures and work methods, changes in design and organisation at the workplaces, safety training programs, and so on.

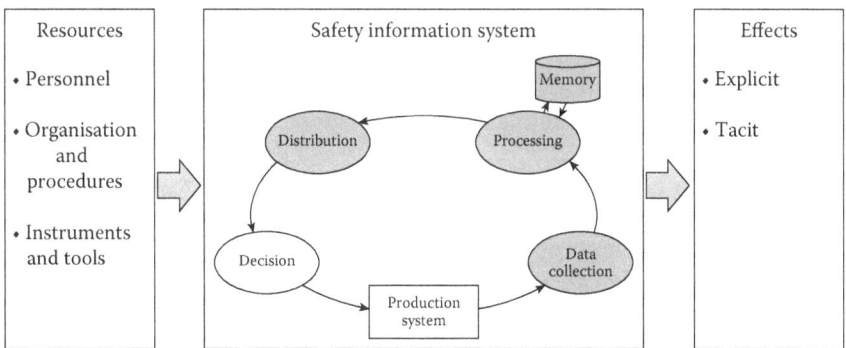

Figure 7.10 A safety information system seen in relation to resources required (input) and its effects (output).

Explicit effects also involve improvements in the personnel's understanding of the safety problem area in a way that they are able to articulate. A safety information system will also affect the unconscious habits, routines, and action patterns in the organisation and the members' unconscious mental models, skills, and value systems. This we call effects on *tacit knowledge*. The aims of a safety information system are to produce explicit and tacit output that will contribute to a reduction in the risk of accidents.

There are certain resources needed to establish and manage a safety information system, including:

- *Personnel* with certain authority and responsibilities and their knowledge, skills, and attitudes.
- *Organisation and administrative procedures* and routines for incident reporting and investigation, workplace inspections, distribution of information including regular reports on safety performance, and so on.
- *Instruments and tools* (i.e. forms, checklists, analytic principles, computer network and software, and so on) for the collection, processing, storing, and distribution of information on accident risks, and so forth.

The different resources needed to establish and maintain a safety information system have to be well-tuned together and adapted to the environment in which the system exists. We will explore different solutions that are dependent on the size of the company, its production, traditions, management philosophy, influences from other companies, and so on. There is no single solution that suits the needs of every company.

7.6   *Organisational learning*

Experience feedback involves learning and the accumulation of qualified experience. This will result in gradual improvements in the organisation's performance as illustrated by the PDCA wheel. Figure 7.5 illustrated how standardisation was used to ensure the capturing of lessons learned and implementation to prevent fall-back. The PDCA wheel does not explain how this learning is accomplished. To do this, we need to look at the areas of organisational learning and knowledge management.

In *organisational learning*, we speak about different learning mechanisms involved in the exchange of tacit and explicit experiences among members of an industrial organisation through a learning spiral (Nonaka and Takeuchi 1995). By making the learning spiral a continuous process, the learning will expand from an individual level to a group level and finally to an organisational level; see Figure 7.11.

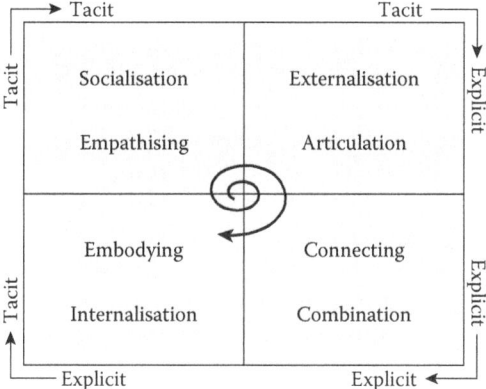

Figure 7.11 Learning mechanisms according to the socialisation–externalisation–combination–internalisation (SECI) process during an exchange of information among members of an industrial organisation. (Reprinted from *Long Range Plann,* 33, Nonaka, I., et al., SECI, Ba and leadership: A unified model of dynamic knowledge creation, 5–34, Copyright 2000, with permission from Elsevier.)

Here, we will use accident investigation as an example to illustrate the so-called SECI (socialisation-externalisation-combination-internalisation) process for knowledge creation through a spiral of four modes of interaction among members of an organisation. In accident investigations, individuals' tacit knowledge and experience are shared with investigators in a process called *socialisation*. This knowledge is subjective and context dependent. It needs to be discussed among the investigators and *externalised* (Nonaka 1994). This involves codification, articulation, and de-contextualisation. The results are made available to the organisation as a whole through documentation in the investigators' report. The organisation will be responsible for documentation and justification of the experience and for *combining* it with existing explicit knowledge. This will result in the development of an improved safety standard for implementation in manuals, procedures, guidelines, specifications, and so on that are made available to the organisation. This represents the consolidation through revision of standards shown in Figure 7.5. Finally, the new knowledge is *internalised* among the members of the organisation and put into use.

The principles illustrated by this example of group problem-solving in connection with accident investigations are also applicable to safety audits and risk analyses. A team structure supports learning through socialisation by exchange of knowledge among the group members. In risk analyses, for example, the team members may internalise the accident models applied in the analysis, and this learning process will affect their unconscious mental models and skills in handling similar situations analysed in the future.

Nonaka considers the *community of practice* within an industrial organisation to be an important element in organisational learning (Nonaka 1994). It is defined by Jubert (1999) as, 'A flexible group of professionals, informally bound by common interests, who interact through interdependent tasks guided by a common purpose thereby embodying a store of common knowledge'. Nonaka identifies certain conditions necessary for experience feedback and learning to take place in such a community:

- The community members need to benefit from participation through joy, improved competence, and increased efficiency in own work.
- The working environment needs to be trustful and caring.
- The members need to represent a moderate degree of variety in types of experience and qualifications.
- There needs to be a balance between accommodation of unexpected opportunities on the one hand, and continuity and time for dialogue on the other.
- To amplify and crystallise experience, the members need to have overlapping competence and areas of responsibility.

The control-system approach and the organisational learning approach are complementary. In our analysis of central principles in safety management, we will be concerned with the handling and feedback of explicit information and the use of this information in decision-making. Feedback control is an essential component in many safety management tools such as workplace inspections and safety audits and in the monitoring of safety performance. We will also be concerned with the significance of organisational learning in the prevention of accidents. Although daily work provides the most important arena for exchange of experiences and learning, the arenas offered by safety management will support similar learning processes. We will also see how certain informal processes may be counterproductive from an organisational perspective by disrupting the exchange of experience and the acquisition of qualified knowledge.

7.7 Obstacles in the management of safety through experience feedback

7.7.1 Limitations in human information processing

In practice, the different assumptions behind the 'ideal' rational-choice models of decision-making outlined in Section 7.3 are rarely present in safety decision-making. One important aspect is the cognitive limitations in the information-processing capacity of the decision-makers. There are costs associated with the search for and processing of information due to the fundamental memory, decisional, response, and attention limitations of human performance. These are treated in many handbooks on decision-making (see e.g. Koeler and Harvey 2004). Here, we will focus on how people handle these limitations by economising on their search for and processing of information. The information search and processing strategies that people apply will result in cognitive biases seen from the perspective of the rational-choice model of decision-making. They are exemplified as follows (see e.g. Haselton et al. 2005):

- Information that is received early in the search is given an unduly high weight.
- Improper generalisations are made from a small sample (perhaps even a single incidence) and refrain from utilising the information sources fully to look for alternative explanations or solutions.
- Have difficulties in taking uncertainty in the information into account.
- Only consider a limited number of (three or four) alternative explanations to the problem at the same time.
- Only able to evaluate a few (two to four) alternative solutions at a time and only able to focus on a few critical characteristics of each solution.
- Overestimates the ability to control events and underestimates the power of others.
- Tend to look for information that confirms the selected solution and avoid information in support of an alternative solution.

In general, people tend to simplify highly complex situations by applying experience-based rules of thumb, known as *bounded rationality* (March and Simon 1958). It also means that decision-makers terminate the search when they have found a satisfying solution rather than searching for more information to find an optimal solution.

People will terminate the search for additional information when they believe that the subjective costs (time, attention, and so on) exceed the benefits (Harrison 1999). In practice, safety-related information has to compete with other types of information to receive the decision-makers' attention. March and Simon (1958) have described bounded rationality as decision-makers having limited time and resources to make optimised decisions. This implies that decisions-makers select the action found first and easiest without further analysis of other options. As a result, safety information systems have to be designed in a way that minimises the cost of searching for and analysing information and maximises the usefulness of the information. It follows that the information has to be relevant to the needs of the decision-maker; it must be comprehensive yet easy to survey. The decision-maker must have easy access to the information when it is needed.

There are decision-making models other than the rational model that better describe how organisations actually arrive at decisions – such as bounded rationality decision-making (March and Simon 1958), muddling through (Lindblom 1959), and naturalistic decision-making (Klein et al. 2003). Organisations that manage according to a *bureaucratic model* of decision-making will apply predefined rules and procedures rather than making trade-offs among different action alternatives. This approach will reduce the demands on the decision-maker's information-processing capabilities. Such an approach mainly is suited for stable systems and for special circumstances, such as crisis handling, when decisions must be made quickly.

We also have to acknowledge that there is a diversity of goals and conflicting interests within an industrial organisation – especially in the area of safety. The *muddling through* decision-making model shows how decisions are made through bargaining and compromises among the different stakeholders (top management, line management, employees, owners, external interest groups, authorities). Data from a safety information system may play an important role in the bargaining process by supplying the different stakeholders with the necessary arguments. Later in this volume, we will see how these motives will also affect the input side of the safety information system through tactical reporting of safety concerns and problems.

7.7.2   *Organisational defences*

Deviations or errors (i.e. mismatches between our plans or intentions and the actual outcome) produce opportunities for learning. Van Court Hare (see Section 7.4.2) distinguishes among four different orders of feedback and the associated results of learning from experience. In practice, we notice that many opportunities for learning are lost, such as when an organisation is only able to accomplish first-order feedback (i.e. no long-term effects are produced as a result of the experience).

Argyris (1992) has analysed the organisational obstacles or defences that prevent efficient learning. He discusses cold and hot variables that distort human and organisational information processing and make feedback inefficient. The 'cold' variables are equal to the limitations in human information processing discussed in Section 7.7.1. By 'hot' variables, Argyris means individual and organisational defence strategies used to avoid embarrassment and threat. At the individual level, we find such psychological defence mechanisms as suppression, denial, and displacement.

In exploring organisational defences, Argyris makes the distinction between single-loop and double-loop learning. Single-loop learning corresponds to Van Court Hare's first order of feedback. It involves measures to correct actions following a mismatch between the achieved and expected results of an action. It is appropriate for routine tasks, where it is necessary to get the job done. Double-loop learning, on the other hand, involves measures resulting from a mismatch that affects the governing variables of an organisation. This type of learning is necessary for long-term effectiveness and, ultimately, for the survival of an organisation. Double-loop learning corresponds to Van Court Hare's higher orders of feedback.

Argyris' studies of organisations show that single-loop learning also is dominant in cases where double-loop learning would have been more appropriate. Typically, when the management of hierarchical organisations faces mismatches between intentions and results, they apply a strategy to achieve unilateral control and self-protection and to discourage inquiry from others. Rationality is emphasised, and feelings – especially negative ones – are suppressed. This creates an atmosphere where problems are not discussed and where errors escalate.

This type of condition will prevail as long as the costs for hiding errors that are difficult to solve or are experienced as embarrassing and threatening are felt to be lower than the costs for remedying them. This typically is the case with accidents. Management often prefers to explain the occurrence of accidents by referring to unique and situational causes rather than to system failures. They thereby reduce the immediate need to change governing variables that serve their purpose well in other instances. Accidents are rare events, and usually it is unlikely that the manager in question will be accountable for a similar type of accident again in the near future. Such a strategy may thus be rational from an individual point of view, even if the company as a whole has missed an opportunity for learning.

Processes to support double-loop learning have to come from the top of an organisation in order to create an atmosphere of trust. These processes involve supporting participation from all involved parties in defining purpose and in making inquiries, minimising unilateral control, creating win–win situations, and in allowing feelings to be expressed.

Argyris' analysis of organisational defences also has implications for the design of safety information systems. A general dilemma in designing information systems has to do with our limited information-processing capacity as human beings. The information provided by such systems should be comprehensive yet manageable (cf. Ashby's law of requisite variety).

At the workplace level, operators and supervisors have direct experience with accidents that are specific, concrete, and rich in details. Typically, this type of information does not allow for generalisation and is unusable by anyone other than those who have generated it. Management at higher levels of the hierarchy, on the other hand, asks for coded summary data on accidents and quantified information on key performance indicators such as the accident frequency rate. This information represents the aggregated result of the complex processes at the local level.

Consequently, operators and first-line supervisors are exposed to direct information about accidents, whereas higher-level management receives filtered and interpreted information from the safety information system. As a result, the different organisational levels hold different conceptions about the causes of accidents and about conditions for effective prevention. The feedback loops to each level will reinforce their respective accident conception.

When top management uses summary data for unilateral control purposes (such as when requiring unrealistic reductions in the accident frequency rate), the local level will respond in ways that may counteract management's intentions (Argyris 1992). They may withhold details or send distorted information through the safety information system. They also may tend to withdraw their commitment and feeling of responsibility. Typical responses from top management in such situations are to require more detailed and tamper-proof information and to increase the orientation towards unilateral control. We thus face a situation where feedback mechanisms among the organisational levels escalate errors and maladaptive behaviour.

To counteract these tendencies, we emphasise the need to develop a spirit of cooperation and shared ownership. The safety information system should not be used as an instrument for unilateral control but as a tool in a problem-solving process where the different levels of the organisation participate. Top management's legitimate need for summary information for monitoring purposes has to be acknowledged. At the same time, the intrinsic limitations of this information have to be understood. It follows that the authority making decisions about remedial actions should be placed at as low a level of the organisation as possible.

7.8 Effects of experience feedback on the risk of accidents

Researchers have studied whether the qualities of the various subsystems of a safety information system affect the accident risk. Three different research methods have been applied. In so-called *ex-post facto studies*, scientists explore whether there are differences in the safety information systems among companies with high and low accident rates. They have found that companies with low accident rates have better injury record-keeping systems (Simonds and Shafai-Sahrai 1977), better formal routines for workplace inspections (Smith et al. 1978), and are more inclined to inquire into minor injuries and near-accidents (Cohen et al. 1975).

In *action research*, scientists participate in the introduction of changes in a company's safety information system. They evaluate how these changes affect the behaviour of the company's organisation and its accident rates. Many studies report reductions in accident rates following the introduction of such measures as improved accident investigation and workplace inspection routines as well as the introduction of near-accident reporting (Komaki et al. 1978; Adams et al. 1981; Menckel 1990; Solem and Kongsvik 2013).

In *evaluation research*, scientists study the effects of program changes made inside companies, without being involved in the changes themselves. Lundberg et al. (2010) conclude from a literature review that there is a general lack of evaluation studies of adequate scientific quality in the area of safety management experience feedback.

Experience feedback is a basic element in the guidelines and standards for occupational health and safety management systems such as draft ISO 45001 and its predecessor from the Occupational Health and Safety Assessment Series, OHSAS 18001 (OHSAS Project Group 2007). We must assume that evaluation research of such systems will produce evidence of the effectiveness of experience feedback in the prevention of accidents. Robson et al. (2007) made a systematic review of the evaluation research literature discussing occupational health and safety management systems. In general, the reviewed studies reported a positive impact. Most of the studies showed methodological limitations, and the authors concluded that evidence was insufficient for generalisation. Gallagher et al. (2003) studied the effectiveness of occupational health systems in Australia and obstacles to their implementation. The ability to produce positive results was dependent on the kind of system used, senior management commitment, integration into general management system, and employee participation.

This conclusion is further supported by the findings from a retrospective case study of an aluminium plant (Kjellén et al. 1997). The plant had introduced principles for HSE management according to the Norwegian requirements for internal control. They included a documented safety information system, involving improved routines for accident and near-accident reporting, workplace inspections, and follow-up of safety results. The plant simultaneously introduced measures to improve product quality and its maintenance standard. An evaluation study comprising a time-series analysis of safety, quality, and financial data over a period of ten years showed that these different measures in combination resulted not only in substantially reduced losses from accidents but also in improved product quality and reduced overall operational expenditures. The effect did not come about as a result of any of the measures seen in isolation but as a result of the total effort. Fernández-Muñiz et al. (2009) came up with similar results in a questionnaire study of 455 Spanish firms. The authors found a positive correlation between safety performance and economic performance and competitiveness.

chapter eight

Criteria for assessing the efficiency of experience feedback

In this chapter, we will summarise the issues dealt with so far in Section II by addressing the focal question: How can we distinguish satisfactory experience feedback systems from those that are less than adequate? To be able to answer this question, we need a set of criteria or requirements that experience feedback systems and their elements must satisfy. We have developed a set of assessment criteria in Section III on learning from incidents and deviations, in Section IV on safety performance monitoring, and in Section V on risk analysis.

When establishing these different criteria, we must acknowledge that one single theory or perspective will not provide a complete answer. Instead, we need to understand the different needs that an experience feedback system will satisfy and hence draw from the different theories and perspectives that which best suits each need. Chapter 7 reviewed different mechanisms for experience feedback. We will rely on the following theories and principles in particular:

- *Feedback control*: We will build on criteria derived from theory on feedback control systems and emphasise efficiency and cost-effectiveness. Safety performance monitoring systems represent the main application of feedback control.
- *Organisational learning*: We will emphasise aspects of incident investigation and risk analysis that promote experience exchange and learning. We will also focus on provisions to minimise dysfunctional effects of organisational defences.
- *Human information processing*: We touched upon this area in Section 7.7.1 in our discussion of the decision-maker's limited information-processing capacity.

Building on the distinction between the feedback cycle and the diagnostic process, in this chapter we will develop different criteria for each of these processes. We will begin by looking at criteria for safety performance monitoring.

8.1 The feedback control cycle

In Juran's feedback control cycle (see Section IV for more detail), safety performance monitoring or measurement is a main factor in safety management. In this book, we define a safety performance indicator as a metric used to measure an organisation's safety performance in terms of its ability to control accident risk in its activities. Feedback control based on safety performance indicators involves establishing safety performance goals and following-up on such goals through measurement of actual performance as well as through corrective action that is undertaken in case the goals are not met. The LTI rate introduced in Section 7.1 is one example of a safety performance indicator.

In this chapter, we will develop two sets of criteria – one for the assessment of safety performance indicators or measures and another for the different elements of the safety performance monitoring system and the system as a whole. The first set of criteria will enable us to assess the quality of safety performance indicators for use in feedback control. We will then shift our focus to the safety performance monitoring system in general. Here, we define criteria for the evaluation of data collection, analysis, and presentation of information and for an evaluation of the system as a whole.

The UK Health and Safety Executive (HSE 2001) recommends the use of 'SMART criteria' for measuring health and safety performance (Doran et al. 1981). SMART is an acronym for specific, measureable, achievable, relevant, and time-related. These criteria were originally developed for general management but have been applied in the field of safety performance monitoring as well (for an overview, see Podgórski 2015). Similarly, Carlucci (2010) has proposed a set of four criteria based on a review of the management literature on various factors considered for the selection of key performance – relevance, reliability, comparability and consistency, and understandable and representational quality.

In the United States, the National Safety Council applied the loss-time injury frequency rate as a main criterion in selecting winners for their excellent safety performance awards. This indicator was criticised because it was so easy to manipulate and also had other weaknesses (see Chapter 16 for further details). This resulted in the development of alternative indicators such as behavioural sampling and the critical incident technique (Rockwell 1959; Tarrants 1963). In his article on behavioural sampling, Rockwell developed a set of criteria that have had a significant impact on future developments (Rockwell 1959; Kjellén 2009):

- Quantifiable and permitting statistical inferential procedures
- Valid or representative of what is to be measured
- Providing minimum variability when measuring the same conditions

- Sensitive to change in environmental or behavioural conditions
- Cost of obtaining and using indicators is consistent with the benefits
- Comprehended by those charged with the responsibility of using the criteria

8.1.1   Requirements for safety performance indicators

In Section IV, we will review different types of safety performance indicators for use in feedback control. Table 8.1 shows the basic requirements that such indicators must satisfy (Tarrants 1980).

The first four criteria are derived from feedback-control theory. A safety performance indicator must be *observable and quantifiable*, that is, it must be possible to observe and measure performance by applying a recognised data collection method and scale of measurement. This corresponds to the criterion 'measurable' according to SMART. The nominal scale is the simplest type. This means that we must be able to tell whether the result represents a deviation from a norm or not. Usually, safety performance indicators are expressed on a ratio scale of measurement. A typical example is the total recordable injury (TRI) rate – for example, the number of recordable injuries per one million hours of work.

The safety performance indicator must also be a *valid indicator of the risk of loss*, corresponding to the SMART criterion 'relevant'. Of special concern is the so-called criterion-related validity. We have to ask whether the safety performance indicator actually measures what we intend to measure – in our case, the risk of losses due to accidents. Because accidents are rare events, we also look for other types of safety performance indicators such as the frequency of unsafe acts and conditions. In these cases, we often have to rely on subjective judgement to assess the validity of the indicators (called 'face validity').

The safety performance indicator must be *sensitive to change*. There is no corresponding criterion in SMART. Sensitivity to change is important in order for the indicator to provide early warning by capturing changes in an industrial system that have significant effects on the risk of losses due to accidents.

Table 8.1 Requirements for safety performance indicators

1. Observable and quantifiable
2. Valid indicator of the risk of loss
3. Sensitive to change
4. Compatible
5. Transparent and easily understood
6. Robust against manipulation

The indicator must also be *compatible* with performance indicators related to overall business objectives in order to deliver on desired performance and prevent decision-makers from receiving contradictory control signals. This is part of the 'relevance' criterion according to SMART.

Example: Lack of compatibility may be illustrated by two indicators from tunnel excavation. One is the progress as measured by the advancement in meters per day, with positive advancements resulting in bonus payments to workers and supervisors. The second is the rate of high-potential incidents in connection with rock fall, where a low rate indicates that sufficient time and resources have been spent on rock support (such as bolting and shotcreting) to ensure an adequate safety standard. This example illustrates the classic dilemma between 'speed and safety'. A poorly designed bonus system may put rock support and the required barriers against falling rock in jeopardy. The safety and progress performance monitoring systems need to be carefully tuned to ensure 'safe speed'.

The safety performance indicator must be *transparent and easily understood* in that its meaning is apparent and compatible with the users' theoretical understanding and unconscious mental models.

It also needs to be *robust against manipulation*. This criterion is not addressed in SMART. It is a variation of the validity criterion. Through safety performance monitoring and feedback, we want to achieve reductions in the risk of accidents. We expect the monitored organisation to change its behaviour in order to achieve improvements. The question is here whether the indicator allows the organisation to 'look good' by, for example, changing reporting behaviour, rather than making the necessary basic changes that reduce the risk of accidents. This may be the case if positive safety performance monitoring results trigger bonus payment or are a necessity for a construction company's contract renewal.

8.1.2   Requirements for the different elements of the safety performance monitoring system and the system as a whole

In this section, we will summarise some important safety performance monitoring system requirements that are necessary to support feedback control (see Table 8.2). The analysis stage is considered to be highly standardised and is not accompanied by any requirements (more than obvious accuracy in calculations).

8.1.2.1   Data collection

First, we will focus on criteria developed from feedback-control theory. Efficient control requires a reliable measurement of performance. *Reliability* is defined for our purposes as the extent to which repeated measurements give the same results. Reliable reporting and calculation of accident and

Table 8.2 Requirements of the safety performance
monitoring system as a whole

Data collection
1. Reliability
2. Accuracy
3. Adequate coverage

Distribution and presentation of information
1. Relevance
2. Comprehensible and easy to survey
3. Timeliness
4. Information available when it is needed

The safety performance monitoring system as a whole
1. Easily understood and acceptable methods
2. Promotion of involvement
3. Cost-efficient

incidents is important, for example, if the frequency of accidents (number of recordable accidents per one million hours of work) is used as an indicator of the safety performance of the company or department in question. This requires that the reporting requirements are uniformly implemented and understood within the organisation. Interobserver reliability is a special case, where agreement among observations by different persons are compared.

The measurement method must be *accurate*. Reliable measurements are not necessarily accurate. We may, for example, have systematic errors due to some misunderstanding of what is to be measured or because of under-reporting. Or, our data may be contaminated by extraneous factors such as rumours or direct manipulation.

Example: The 'number of reports on unwanted occurrences per employee and year' (also known as the RUO rate) is a measure of employee involvement that is based on the employees' self-reporting of their own observations of near-accidents and unsafe acts and conditions. There are examples where construction sites have included observations made by professional safety inspectors in order to 'improve' the results.

The requirement for *adequate coverage* comes from Ashby's law of requisite variety. To exercise efficient control, we must receive data on the different technical, organisational, and human factors that affect the risk of accidents and that are controllable through management decisions. This criterion does not apply to overall indicators of accident risks, such as the recordable injury frequency rate, but to a set of indicators of different variables that affect the risk of accidents such as 'percent of safe behaviour'. There is a conflict between this criterion and the criteria related to information overload and cost-efficiency discussed in the following sections (see also Podgórski 2015).

8.1.2.2   Distribution and presentation of information

We now shift our focus to the limitations of human information processing, and the fact that the decision-maker's attention is a scarce resource. In particular, we are concerned with the decision-maker's experience with the benefits of receiving additional information in relation to the perceived 'costs' (time, attention) of finding the information (Harrison 1999). To avoid overload, the information that is presented to the decision-maker must be *relevant* in relation to the decision-making context. An obvious example is the adaptation of a periodic safety performance report to fit the needs of the individual decision-maker by filtering and adaptation of the report to his or her area of responsibility.

As another way of avoiding overload, the presented information must be *comprehensible and easy to survey*. Managers, especially at top levels, must not be overwhelmed by detailed accident data, where it is impossible to 'see the wood for the trees'.

Example: Aggregate or composite safety performance indicators are based on the weighted average of a number of underlying indicators (Podgórski 2015). An oil company developed a barrier performance indicator, based on the performance of 19 different 'performance standards' (Koren 2006). Examples of performance standards are structural integrity, fire and gas detection, ignition control, deluge system, escape and evacuation, and explosion barriers. Each of the performance standards was regularly inspected, tested, and given a rating. The barrier performance indicator was calculated based on a weighted average of the different scores. Advanced technical knowledge was required to fully comprehend the indicator, and it was not possible to determine what needed to be done in case the indicator showed deteriorating barrier performance without going into the details of the underlying data.

Excessive time delays will jeopardise the possibilities for efficient control. In general, the time it takes to detect and process data on changes and to implement corrective actions must not be larger than the rate of change of the control object (see Ashby's law of requisite variety). *Timeliness* is important in order to avoid hazardous deterioration of a system resulting from the non-detection of hazardous changes over a long period of time. For example, in the case of performance indicators based on safety inspection results, the time lag is determined by the inspection frequency. This means that the inspection frequency must be higher at workplaces with rapid changes (such as construction sites) than at workplaces that remain unchanged during long periods (such as processing plants).

When using the injury frequency rate (i.e. the number of injuries per one million hours of work) as a key performance indicator, timeliness in detecting changes that affect the accident frequency rate is a concern because accidents normally are rare events. Consequently, there is usually a significant time lapse before an increase in the accident frequency rate is detectable with an acceptable confidence for use in feedback control. We will return to this issue in Section IV.

From the decision-maker's perspective, it is important that the *information is available* when it is needed. Information technology has significantly improved the possibilities of accessing performance data for use in decisions.

8.1.2.3   *The safety performance monitoring system as a whole*

At this level, we apply criteria derived mainly from organisational learning theory. To minimise extraneous influences, the methods for data collection, analysis, and distribution of information must be *easily understood and acceptable* to the involved parties. It is important, for example, in the reporting of human errors, that there is a fair or just reporting culture that the employees trust and that reports are not used to punish individuals. The methods should also *promote involvement* of management and employees as well as the development of a shared understanding and ownership as to the safety performance goals and the reporting and measurement system. This is facilitated if there exists a dedicated community of practice inside the company that 'owns' the system and acts as a learning agency for other stakeholders (Drupsteen and Guldenmund 2014).

Finally, we are concerned with *cost-effectiveness* with respect to the resources necessary for the establishment and operation of the safety performance monitoring system, especially as seen in relation to the benefits of the system. Costs are relatively easy to assess by standard methods and include investments and operations. In evaluating the benefits of a system, we have to consider the support it yields in reducing the risk of accidents and other types of unwanted events (production disturbances, reduced quality, and so on) as well as in improvements to the quality of working life.

8.2   Diagnosis

8.2.1   Incident investigations

The diagnostic process is applied in different safety management practices such as incident investigations, risk assessments, and analysis of accumulated data from accidents. In this section, we will focus on incident investigations. The criteria documented in the research literature generally apply to the efficiency of the different steps in an incident investigation process (Lundberg et al 2010; Jacobsson et al. 2012; Salguero-Caparros 2015). Here, we will distinguish among four steps: (1) notify, (2) investigate,

(3) decide, and (4) implement and follow-up (see Chapter 13.2 for a further review).

8.2.1.1   *Notification and initial recording*

The quality criteria for notification coincide with those presented in Section 8.1.2.1 on data collection in safety performance monitoring. Even if *reliability* in the notification and initial recording of an incident is not as critical in diagnosis as in feedback control, opportunities for learning will be lost if the notification and recording of incidents is unreliable.

8.2.1.2   *Investigation*

The next step, investigation, goes beyond the initial recording of the incident (e.g. in the supervisor's first report). In Chapter 13, we will distinguish between an investigation carried out by a team at the workplace and in-depth investigations conducted by an independent team. Lindberg et al. (2010) have identified the following quality criteria for the investigation process, similar to those identified as guidelines for the auditing process outlined in ISO 19011 (ISO 2011):

- Independent investigators not having a stake in the results
- Minimum time lag between the occurrence and the execution of the investigation
- Use of a broad scope of information sources (interviews, documents, visual inspection, technical analysis)
- Use of information from similar investigations in the past
- Validation of results and documentation in a report

The authors also identify the following criteria related to the quality of the investigation report:

- Provide a detailed description of the accident sequence, deviations and barrier failures
- Identify direct causes and underlying (root) causes such as organisational failures and inadequate safety culture

Jacobsson et al. (2012) have identified similar criteria in the development of a method for assessing the effectiveness in learning from incidents. As to the investigation process, the authors include a criterion on the qualifications of the investigators. The report is determined by two criteria – scope and quality. Scope is related to Ashby's law of requisite variety and is assessed by judging the extent to which the report covers relevant aspects of the incident. The quality criterion applies to the depth of the facts presented in the report and whether they allow for a 'thorough analysis'.

This review shows that the process-related criteria are relatively precise and easy to apply, whereas the criteria related to the quality of the report are more vague and subjective.

We will apply the following assessment criteria for incident investigations:

- Investigation process:
  - Qualification of the investigators (management, technical, investigation methodology)
  - Timeliness: Time lag from occurrence of the incident until completion of data collection and report
  - Use of independent sources
  - Validation of results and documentation in a report
- Quality of the investigation report (see Chapter 5)
  - Consistence in the description of the sequence of events
  - Completeness of deviation and barrier analyses
  - Scope and quality of the analyses of contributing factors and root causes

Timeliness is divided into two criteria, one for time lag in the collection of essential data and the second for time lag until completion of the report and submission of feedback to decision-makers. The first criterion is essential in order to secure physical evidence at the scene of the event and information about the experiences of the victims and witnesses before details are distorted or lost. Timeliness in reporting and feedback submission to decision-makers is in accordance with Ashby's law of requisite variety and is also essential (see Section 7.4.1).

8.2.1.3 Decisions

The quality of decisions is closely related to Ashby's law of requisite variety and Van Court Hare's order of feedback (see Section 7.4). We will apply the following criteria:

- *The order of feedback that the measure represents.* We may expect that measures of the fourth order of feedback will have a large range and affect the company's level of risk more generally. Range is one of the quality criteria of decisions that is identified by Jacobsson et al. (2012). Generally, measures beyond the first order of feedback are expected to have more lasting and preventive effects.
- *The scope of the measures as defined* (e.g. by the Safety Management and Organisation Review Technique [SMORT] classification system). An adequate variety of measures with a clear relationship to the analysis of causal factors in the investigation are positive signs of a risk-reducing effect.

8.2.1.4   Implementation and follow-up

In this instance, we are concerned with the timeliness of implementation and whether the measures are subject to any evaluation of risk-reducing effects.

8.2.1.5   Summary

Table 8.3 summarises the criteria identified in this section.

8.2.2   Risk assessment process

In Section IV, we will apply the risk management standard ISO 31000 in our review of different methods of risk analysis (ISO 2009). ISO 31000 identifies five steps in the risk assessment process: (1) establishing the context, (2) identification of hazards and unwanted situations, (3) risk analysis (i.e. establishing a risk picture by determining the frequencies of unwanted incidents and the consequences of the unwanted incidents), (4) risk evaluation, and (5) risk treatment. The standard provides the following criteria for effective risk assessment. The links to the appropriate steps in the risk assessment process appear in parentheses.

- The result of risk assessment helps decision-makers to make informed choices and prioritise actions (risk treatment).
- Risk assessment explicitly takes account of uncertainty, the nature of the uncertainty, and how it can be addressed (risk assessment

Table 8.3 Quality criteria for incident investigations

Step in the investigation	Criteria
Notification and recording	• Reliability • Timeliness

Investigation	• Investigation process
	  • Qualifications of the investigators
	  • Timeliness in data collection
	  • Use of independent sources
	  • Validation of results and documentation in a report
	• Quality of the report
	  • Consistency in the description of the sequence of events
	  • Completeness of deviation and barrier analyses
	  • Scope and quality of the analyses of contributing factors and root causes

Decision	• The order of feedback that the measure represents • The scope of the measures
Implementation and follow-up	• Degree of implementation, timeliness (whole cycle) • Follow-up of results

including identification of hazards and unwanted situations, risk analysis, and risk evaluation).

- The assessment and handling of risk is a systematic, timely, and structured approach that should provide consistent, comparable, and reliable results (all steps).
- Risk assessment is based on the best available information. The inputs to assessment of risk are based on historical data, experience, observations, stakeholder feedback, forecast, and expert judgements (establishing context and risk assessment).
- Decision-makers are informed about limitations of the information (risk treatment).
- The level of detail of the risk analysis is based on the purpose of the analysis and the available resources and information (identification of hazards and unwanted situations and risk analysis).
- The assessment and handling of risk is transparent and inclusive at all levels of the organisation (all steps).
- Continually sense and respond to different types of change: changes of context, changes in knowledge, external or internal incidents, emergence of risks, or other changes (all steps).
- Ensure that controls are effective and efficient (risk treatment).

Regarding identification of hazards and unwanted occurrences, a key challenge is completeness. Are all relevant hazards and unwanted occurrences identified? Risk assessment is different from the diagnostic process of incident investigation because we do not know what incidents may happen. One of the aims of the risk assessment is to state the likelihood of a set of incidents occurring. Some incidents will not be identified due to limitations in knowledge and resources available. Similarly, it is not possible to analyse the frequencies and consequences of all identified incidents. Paltrinieri et al. (2012) distinguish among different incident scenarios – 'known knowns' (what we know will happen), 'known unknowns' (what we know can happen), and 'unknown unknowns' (what we do not know). In risk assessments, there will always be incidents that we cannot imagine will happen.

Furthermore, risk analysis should provide a risk picture based on current and future situations. The analysis should not be solely based on historical data because the future is not necessarily a repetition of the past.

Risk acceptance criteria are a key part of the evaluation of risk. In many cases, risk acceptance criteria should be defined prior to the assessment process by those responsible for the assessment. This is the case for the Norwegian offshore petroleum industry, but it is not unproblematic. Aven and Vinnem (2007) argue that acceptance limits that are too ambitious will make it difficult for companies to obtain the best solutions from both a production and a safety perspective. On the other hand, acceptance

Table 8.4 Quality criteria for the risk assessment process

Step in the risk assessment process	Criteria
Establishing the context	• Balanced risk acceptance criteria • Qualifications of the analysis team • Availability, timeliness, and relevance of background information.
Risk assessment • Identification of hazards and unwanted situations • Risk analysis (analysis of causes, frequency, and consequences)	• Coverage and relevance of identified hazards and unwanted situations • Availability and timeliness of information relevant for hazards and unwanted situations, causes, frequencies, and consequences • Expressed uncertainty of results • Systematic, timely, and structured approach providing consistent, comparable, and reliable results • Transparent and inclusive at all levels of the organisation • Updated in cases of significant changes
Risk evaluation	• The result of the risk evaluation helps decision-makers with informed choices and prioritising actions • Uncertainty of the evaluation must be expressed • Transparent for decision-makers
Risk treatment	• Transparent and inclusive at all levels of the organisation • The order of feedback that the measure represents • The scope of the measures • Degree of implementation and timeliness (whole cycle) • Follow-up of results

criteria that are too weak will be met in most situations without any extra measures but will not provide adequate protection against accident risks.

Table 8.4 summarises the quality criteria for the different steps of the risk assessment process.

chapter nine

Barriers against loss

Barriers play a central role in the prevention of accidents. The most common type of barrier would be some form of physical obstruction that prevents hazardous energy from coming into contact with a vulnerable target such as the human body. Examples of these types of barriers include fencing in cattle to prevent contact with predators, defensive walls, or moats to protect from enemy attack, machine guards to prevent human contact with hazardous machinery, and those barriers that prevent vehicles from veering off a road. Another form of barrier keeps workers out of hazardous danger zones (i.e. safety by distance). Traditional gunpowder production used hydropower to mix potassium nitrate, charcoal, and sulphur. Operators controlled the process from a distance by opening and closing gates, thus remaining outside the danger zone in case of an exploding barrel. A barrier in this sense is either a separate entity or a part of the system needed for normal production that has added characteristics to eliminate or reduce the consequences of unwanted energy flow. In either case, the implementation of barriers entails extra costs that have been determined as acceptable to limit loss in case of an accident.

In this chapter, we will introduce the concept of barrier function; physical separation and separation by distance are two examples of such a function. We will also introduce other types of barrier functions, all of them with the characteristic of being able to intervene in an accident sequence to prevent or limit loss.

The barrier concept experienced a revival through publications on the principles of 'defence in depth' by Perrow (1984, 1999), Rasmussen (1993), and Reason (1997) and through changes in legislation during the last two decades. (See Chapter 4 on the energy model.) European regulations, such as the directive on the control of major accident hazards involving dangerous substances (European Commission 2012) and the machinery directive (European Council 2006), define requirements for the implementation of safety barriers. Examples on the national level are the Norwegian petroleum regulations, requiring oil companies to establish and maintain barriers to reduce the probability of accident situations developing and to limit loss. The Petroleum Safety Authority (PSA 2013) gives an in-depth overview of Norwegian offshore industry regulations on barriers and defences (see Chapter 25).

9.1   Definitions

In a review of safety barrier literature, Sklet (2006) concludes that there is no generally accepted definition of barriers. Sklet defines safety barriers as 'physical and/or non-physical means planned to prevent, control, or mitigate undesired events or accidents'. This definition does not make any clear distinction between barriers and any other safety measure for preventing accidents. The same applies to the definition in ISO 17776 for petroleum and natural gas industries, where a barrier is 'a measure which reduces the probability of realising a hazard's potential harm and which reduces its consequences' (ISO 2000). This standard explicitly mentions procedures, inspections, training, and drills as examples of 'non-physical' barriers. The U.S. Department of Energy defines a barrier as any means used to control, prevent, or impede a hazard from reaching the target (DOE 2000). A distinction is made between physical barriers (such as fences, guard rails, protective clothing, safety devices, and warning device) and management barriers (hazard analyses, knowledge/skills, supervision, training, work planning, and work procedures).

A broad definition undermines the concept of safety barriers and dilutes its specific characteristics applied in the principles of defence in depth through the application of multiple, independent barriers. Here we will use the view of accidents according to the energy model in Section 4.4 as our starting point – a physical phenomenon involving loss due to an uncontrolled transfer of energy to a human or other 'target'. The energy may be mechanical, electrical, involve hazardous substances, and so on. We will define a barrier as *a set of system elements (human, technical, organisational) that as a whole provide a barrier function with the ability to intervene into the energy flow to change the intensity or direction of it.* Haddon's ten accident prevention strategies (detailed in Section 4.4) will serve as the starting point for our definition of barriers (Haddon 1980). We will use the term 'barriers' rather than 'strategies' to denote the physical countermeasures that intervene in the accident process to eliminate or reduce a harmful outcome.

Figure 9.1 shows the accident sequence (i.e. the process part of the accident analysis framework) described in Chapter 5. The figure illustrates that, in principle at least, there are many opportunities to interrupt or change an accident sequence of events before it evolves into a loss. First, there are possibilities of totally changing the preconditions for an accident to occur by eliminating the energy source, modifying the energy characteristics, or by limiting the amount of energy. Second, barriers may interrupt, dilute, or redirect the energy flow during the latter part of the accident process. This could include prevention of the uncontrolled release of energy, dilution of the transferred energy, separation of the victim from the energy flow, or improving the victim's ability to endure the energy flow. Emergency response, being

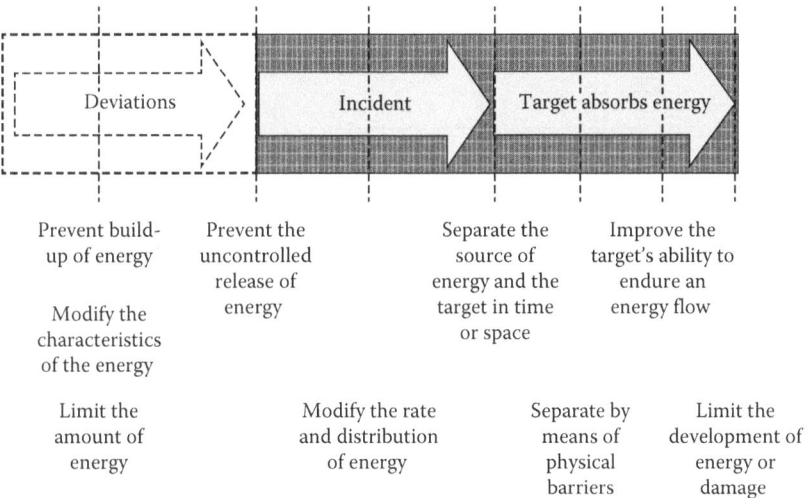

Figure 9.1 Points of intervention by barriers into the accident sequence to eliminate or reduce loss.

part of the accident sequence, aims at limiting the development of loss and is included as a barrier. Finally, stabilisation, repair, and rehabilitation are means of reducing loss after an accident has occurred resulting in injury; but this is not part of the accident sequence and thus not part of our definition of a barrier. The sequence of multiple barriers in Figure 9.1 is recognised as the principles of 'defence in depth' (Rasmussen 1993).

We will distinguish between a barrier function and a barrier system. Here, a barrier function is defined as the ability of a barrier to intervene into an accident sequence to eliminate or reduce loss. A barrier system, on the other hand, is a set of interacting human, technical, and organisational elements that make up the barrier function as an integrated whole. Figure 9.2 illustrates our definition of a barrier as a system of barrier elements and providing a barrier function. The barrier function either prevents a hazard from developing into an incident involving loss of control or limits the loss in connection with the incident.

An example from vehicle safety illustrates the principles of a barrier system. The barrier function of the airbag is to modify the rate and distribution of the energy transfer to the occupant from fixed parts of the vehicle such as the windshield, dashboard, or steering wheel. The airbag inflates quickly at a sudden deceleration above a certain threshold (e.g. due to collision with another vehicle or a fixed object). The airbag system consists of a number of technical elements, where sensors (e.g. accelerometer) give signals to a control unit that triggers the ignition of a propellant that generates gas to inflate the bag.

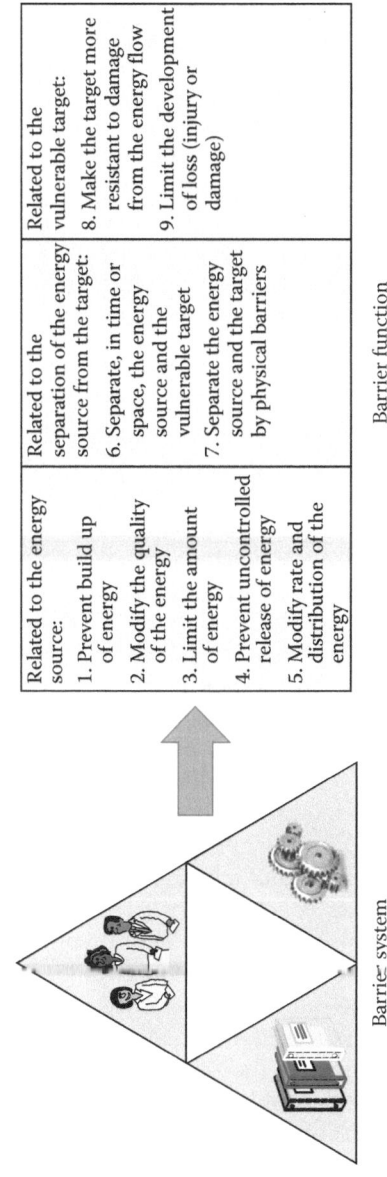

Related to the energy source:	Related to the separation of the energy source from the target:	Related to the vulnerable target:
1. Prevent build up of energy	6. Separate, in time or space, the energy source and the vulnerable target	8. Make the target more resistant to damage from the energy flow
2. Modify the quality of the energy	7. Separate the energy source and the target by physical barriers	9. Limit the development of loss (injury or damage)
3. Limit the amount of energy		
4. Prevent uncontrolled release of energy		
5. Modify rate and distribution of the energy		

Barrier function

Barrier system

Figure 9.2 A barrier is described by its barrier function that is realised by a barrier system consisting of different human, technical, and organisational elements.

It is not always possible to delimit barriers from other parts of a system. The brakes in a car are one example. Their purpose is to control the speed of the car – a normal process to negotiate the movement of the car in the prevailing traffic environment. The same brakes have a second function used in an emergency – to avoid a collision or reduce the consequences of it. In modern cars, brakes are equipped with specific characteristics to serve this barrier function in an optimum way, such as an anti-lock braking system (ABS), electronic brake force distribution (EBD), and a dual circuit hydraulic system. The enclosure of an oil and gas processing system is a second example. The aim is to transport the process media between, for example, separators. A second aim is to prevent loss of containment for flammable gas and liquids (i.e. a barrier function). The piping and flanges are designed to avoid such leakages through increased thickness, corrosion resistance, and so forth. These two examples illustrate that a system may have several functions, and the barrier function is one of them (Rausand and Høyland 2004). In designing systems with dual functions, design must meet the requirements of both functions. For a system that functions as a safety barrier, this often entails additional design margins and/or specific characteristics.

We also make the distinction between preventive and mitigating barriers. The bow-tie diagram puts an incident involving loss of control in the centre (Figure 9.3). Preventive barriers to the left in the bow-tie diagram act on the energy source and will eliminate the risk of loss of control

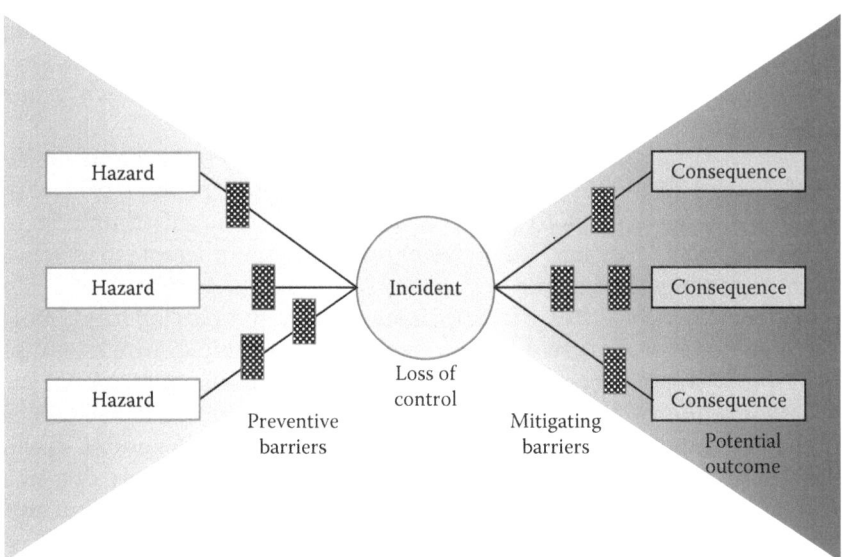

Figure 9.3 The so-called bow-tie diagram, illustrating the difference between preventive and mitigating barriers. (Adapted from Visser, J.P., In *Safety Management – The Challenge of Change*, Pergamon, Bingley, UK, 1998.)

altogether or will reduce the harmfulness of it if released. Mitigating barriers will eliminate or reduce the consequences of the hazardous event through physical barriers, safety by distance, personal protective equipment, or emergency response.

Our definition of barriers, which is based on Haddon's accident prevention strategies, helps us in making a clear distinction among barriers from any safety measure. A dichotomy such as physical and non-physical barriers (Johnson 1980) or the classification of barriers into groups such as physical, functional, symbolic, or immaterial (Hollnagel 2004) is not meaningful according to our definition. A barrier must have at least one physical element in order to intervene in the accident sequence, which is a physical phenomenon. According to Hollnagel, signs are symbolic barriers; however, according to the barrier interpretation presented here, they are an element of a barrier system. In this case, a barrier will include a human operator in the loop with the ability to read a sign and produce the desired barrier function (i.e. limit the amount of energy through complying with a speed limit).

9.2   Passive and active barriers

We distinguish between passive and active barriers, although the demarcation is not always clear. *Passive barriers* are not dependent on any control system or on an action by its barrier elements to realise its barrier function. A typical example is the safety cage surrounding the driver and passengers of a car. Other examples of passive barriers are fixed-mounted machinery guards and fire and explosion walls in a process plant.

Active barriers are dependent on actions by the operators or on a technical control system to function as intended. The carpenter's control of a hammer's movements to keep it a safe distance from his thumb is, in principle, such a barrier. This active barrier is very vulnerable to changes in the carpenter's behaviour.

Figure 9.4 illustrates the principles of how the barrier function is accomplished in an active barrier such as, for example, in our earlier example with an airbag. A sensor detects a specific characteristic of the accident sequence that deviates from the normal and wanted sequence of events in the system. In the case of the airbag, this is a sudden deceleration of the car above a certain threshold level presumably due to a collision. This will initiate an activation of a barrier function – in this example, the filling of the airbag with gas to accomplish a distribution of the energy in time and space when the head of the driver hits the airbag.

In the example with speed limits, the driver will act as the sensor, detecting speed in excess of the limit, and will regulate the velocity of the

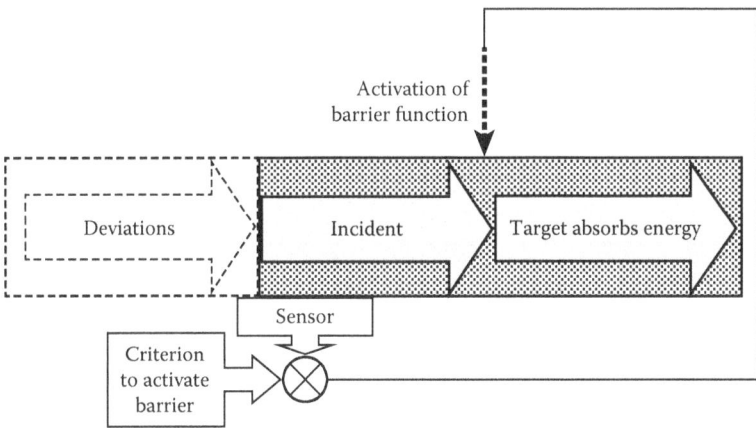

Figure 9.4 The principles of an active barrier.

car accordingly. This example illustrates the case where the active barrier affects the evolution of the initial phase for reducing or eliminating the risk of an accident.

9.3 Defence in depth

Originally, *defence in depth* was the term for a military strategy involving actions to delay and spread out the forces of an attacker rather than preventing an attacker from entering into a territory. Nowadays, the concept also is used in the safety field to analyse the accomplishment of a high level of safety through layers of independent barriers (Perrow 1984/1999; Rasmussen 1993; Reason 1997). Defence in depth primarily is used in safety design where major accident risks are involved, such as in the chemical, oil, and gas, and nuclear industries. By this, we mean major accidents that may result in multiple fatalities inside and outside a plant and extensive material or environmental damage. A major accident may threaten the survival of the company responsible for the operation. Fortunately, the associated hazards are well defined and are possible to identify and evaluate at an early stage. Management of a company will marshal all feasible barriers in order to avoid losses.

The International Atomic Energy Agency (IAEA 1999, p. 17) defines the concept in the following way for nuclear power plants: 'To compensate for potential human and mechanical failures, a defence-in-depth concept is implemented, consisting of several levels of protection including successive barriers preventing the release of radioactive material to the environment. The concept includes protection of the barriers by averting damage to the plant and to the barriers themselves. It includes further measures to protect the public and the environment from harm in

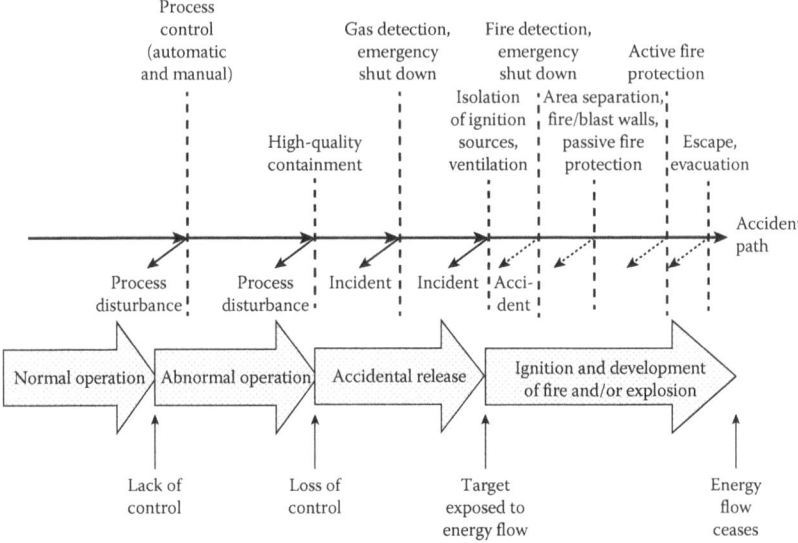

Figure 9.5 Layers of barrier systems used to provide defence in depth in oil and gas processing.

case these barriers are not fully effective'. Defence in depth involves the application of several of the different types of barrier functions shown in Figure 9.1.

Figure 9.5 illustrates the prevention of fires and explosions in a refinery or an offshore installation through the implementation of a number of different barriers, based on the philosophy of defence in depth. A major accident can only occur under those rare circumstances when there is an improbable combination of barrier failures (IOGP 2008).

Let us analyse how these barriers coincide with Haddon's strategies. Oil and gas processing is inherently hazardous due to the presence of flammable hydrocarbons. Thus, it is not possible to prevent the build-up of energy or to modify its characteristics. An important measure is to limit the amount of oil and gas in the process – for example, by dividing the process into separate units – and to isolate these from each other in case of a process upset.

The uncontrolled release of oil and gas is prevented primarily by different means:

- A process control system that keeps process parameters such as pressure and temperature within acceptable limits
- High-quality containment (quality of material/equipment, thickness of piping/vessels)

In the case of an escape of oil and gas from the process containment, gas detection and emergency shut-down prevents harmful consequences. The aim is to modify the rate and concentration of the release of oil and gas. The different parts of the process will be separated from each other through the closure of isolation valves. Hence, only oil and gas from the part of the process where the loss of containment has occurred will be able to escape. Vessels will be depressurised, and the gas will be transferred to the flare tower for flaring.

Normal ventilation – and possibly additional emergency ventilation – will dilute the gas, and potential ignition sources will be shut down to avoid ignition of the hydrocarbons and subsequent fire and explosion. Should a fire or explosion occur, physical barriers such as fire and blast walls will prevent injury or damage outside the immediately affected area.

Finally, there are a number of barriers aimed at limiting the development of injury and damage:

- Passive fire protection of the building and process equipment as well as explosion ventilation will improve the ability of the building and the equipment to endure the explosion and heat from fires.
- Active fire-protection systems such as deluge systems and the fire brigade will have the same effects by combatting the fire.
- Alarms and messages over the personal address system will notify the personnel, who will escape from the affected process areas to safety through marked escape ways. If it is not possible to get the fire or explosion under immediate control, evacuation will follow.

Similar principles apply in the nuclear power industry. The IAEA (1999) identifies five levels of defence in existing nuclear power plants:

1. Prevention of abnormal operation and failures
2. Control of abnormal operation and detection of failures
3. Control of accidents within the design basis
4. Control of severe plant conditions, including prevention of accident progression and mitigation of the consequences of severe accidents
5. Mitigation of radiological consequences of significant releases of radioactive materials

9.4 Limitations of barriers

Accidents occur, even in well-defended systems, with major loss potential. Reason (1997) has developed an illustration of this in the so-called Swiss cheese model (see Figure 4.6). It shows how 'an accident path' escapes through multiple layers of barriers due to an unlikely combination of 'holes in the cheese slice' (i.e. failures in the respective barriers). Some 'holes' or barrier failures occur in the course of the accident (Example 1); others have passed undetected for a long time, the so-called latent conditions (Example 2).

Example 1: In the *Estonia* ferry disaster on September 28, 1994, the Maritime Rescue Coordination Centre in Turku (Finland) failed to acknowledge the mayday from the *Estonia* immediately due to an unclear relay of the message from the ferry. A full-scale emergency was only declared at 2:30 a.m., 40 minutes after the ship had disappeared from the radar screen. This is an example of the vulnerability of barriers; in this case, emergency response – where the initiation of the barrier function is dependent on a human element in the loop.

Example 2: A so-called shotcrete robot was used to convey liquid concrete through a hose and pneumatically project it at a high velocity onto a tunnel surface during excavation to prevent falling rock. The robot was powered by a diesel engine and provided with a 24-voltage DC instrument system, which is intrinsically safe from the risk of electrocution. This was an important safety characteristic due to the wet environment in the tunnel. Management at the construction site had decided to install a floodlight on the robot, powered by the temporary 240-voltage AC power system in the tunnel. Due to a grounding error in the installation of the floodlight, a construction worker received 240-voltage AC through his torso, resulting in electrocution. Both the violation of the intrinsically safe characteristics of the robot and the grounding error represent examples of latent barrier failures, which, in this case, drastically reduced the safety margins of the robot. Latent barrier failures are sometimes also called 'accidents waiting to happen'.

The defence-in-depth principle aims at increasing safety through layers of independent barriers. Perrow (1984) argues that these layers of independent barriers create a more complex system, where the human operator loses overview or becomes negligent due to over-confidence in the safety systems, thus making the system less rather than more safe. Although the system analyst must be aware of this danger, modern industry with its major accident potential would not function without applying the defence-in-depth principle.

Figure 9.6, 'Limitations of barriers', gives an overview of circumstances where barriers may not be present or may fail to perform. Barriers may not be practical or economically unfeasible or may be impossible to implement. Decisions not to implement certain barriers need to be risk-based and made at the proper management level. A second limitation is when the barrier is not in operation. It may not be accessible when needed (such as lacking hearing protection when performing work in a high-noise environment). It may have been deactivated due to maintenance, or it was never reactivated again at start-up of the plant. A permit-to-work system will

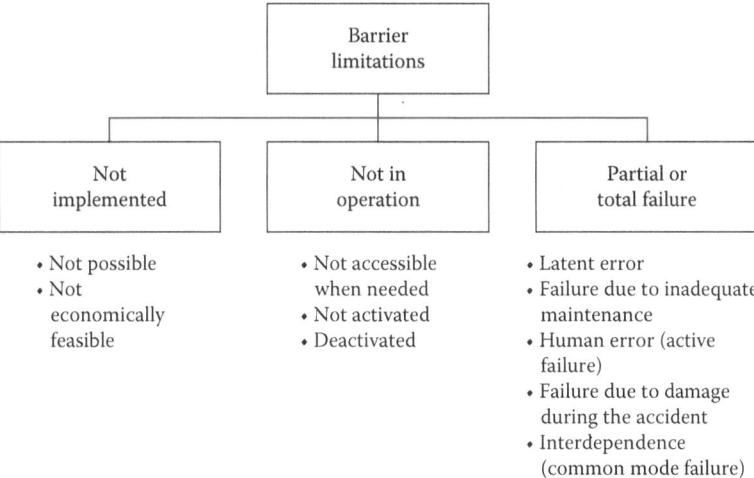

Figure 9.6 Limitations of barriers. (Adapted from Trost, W.A. and Nurtney, R.J., *Barrier Analysis*, Report No. SCIE-DOE-01-TRAC-29-95, Technical Research and Analysis Center, Idaho Falls, ID, 1995.)

implement administrative routines to minimise the risk at such an operation through assessments and checking of the re-institution of barriers before the plant can start up. This will be further discussed in Section 9.6.

Finally, a barrier may fail partly or totally. We have already discussed examples of so-called latent failures. A barrier may need maintenance, and failure to perform such maintenance may result in barrier failure. An example is a car's safety cage that has degraded due to rust. Active barrier systems such as a deluge system are usually more complex and involve several subsystems (fire and smoke detectors, valves and piping, pumps, control system), each of which needs to be tested and maintained regularly to ensure reliable performance.

Failure due to human error is an important subset of barrier failures. The earlier example from the *Estonia* ferry disaster is one such illustration. Due to the risk of human stress reactions putting a barrier in jeopardy, time-critical barriers are often designed to avoid dependence on humans for activation. If the decision is made to rely on human operators, certain criteria need to be met:

- The operators engaged in activating barriers need to be qualified.
- To avoid multiple barrier failure, one operator may only be in the control loop of one barrier, and the actions of operators being responsible for separate barriers must not be too tightly linked.
- Confirmation must be delivered to the operator on the successful initiation of the barrier function.

Barriers may also fail due to the accident itself. This is the case when the barrier has not been adequately dimensioned to absorb the energy flow. Finally, barriers may fail due to what is known as common-mode failure. A classic example is a fire in a culvert for electric and instrument cables putting different barriers out of operation.

An analysis by one of the authors of severe accidents in offshore oil and gas production showed that the barrier failures for process-related accidents with major loss potential followed the principles of Figure 9.5 (Kjellén 2007). Safety was based on the defence-in-depth philosophy, and accidents were the result of a unique combination of barrier failures that included latent errors such as the failure of eroded or deactivated barriers to meet accidental load requirements. There was the potential to improve safety through more thorough analyses of threats against barriers.

Safety against loss due to severe occupational accidents was, on the other hand, characterised by dependence on a few interdependent barriers. An accident frequently was due to 'bad luck', with the operator being at the wrong place at the wrong time. It was concluded that there were improvement potentials in applying a more systematic barrier philosophy based on the implementation of independent barriers.

9.4.1   Quality of barriers

The review in the previous section on limitations of barriers leads us to the question of criteria for barrier quality (also called barrier performance). In this section, we will present a set of criteria similar to those recommended by the Norwegian PSA that includes capacity, reliability, accessibility, efficiency, ability to withstand loads, integrity, and robustness (PSA 2013). Other sets of criteria have been proposed in the literature (Sklet 2006). Here, we will focus on criteria related to the prevention of partial or total failure (see Figure 9.6):

- *Capacity*: The ability of the barrier under given operational conditions to fulfil its purpose through the barrier function (i.e. to intervene in the accident sequence to eliminate or reduce loss)
- *Reliability*: The probability that the barrier performs as specified (capacity, response time) on demand
- *Robustness*: The ability of the barrier to withstand accidental loads (such as fires and explosions) and maintain its function

A critical issue is the barrier's dependence on human actions to fulfil its purpose. There are different types of dependencies, the most obvious being the human operators as a part of the control loop of an active barrier.

Human actions during the testing, inspection, and maintenance of barriers are another source of dependency. Failure to perform will affect the reliability of the barrier.

9.4.2   Inspection, testing and maintenance of barriers

Barriers, including their subsystems and elements, require regular inspection, testing, and maintenance to ensure reliable performance. The requirements in Figure 9.7 may include not only some of the quality criteria from the previous section but also specific design requirements as specified, for example, in the vendor's manual.

Example 1: Figure 9.8 shows the elements of the barrier 'prevent fire development' on an offshore platform. Each of the elements will need regular inspection/testing and maintenance to ensure reliable performance in case of a fire scenario on the platform. Smoke detectors, for example, are tested regularly through the use of non-hazardous smoke to check

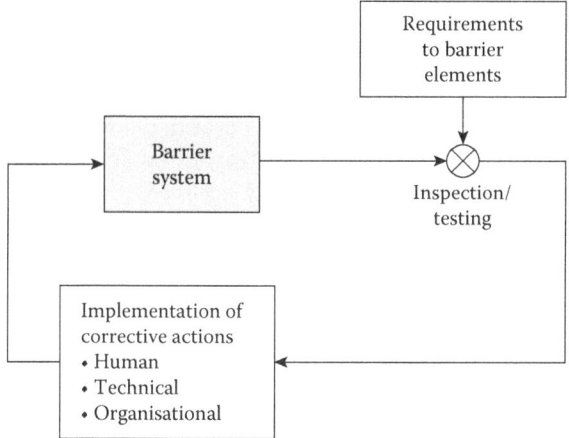

Figure 9.7 Illustration of the maintenance process of barriers as a closed loop.

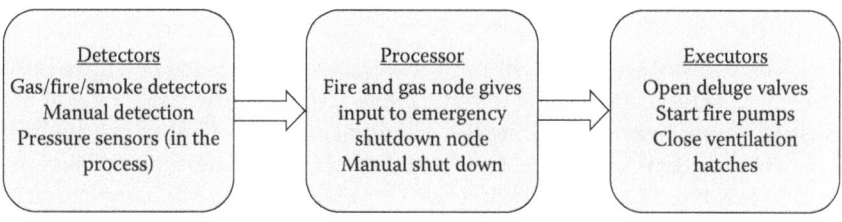

Figure 9.8 The elements of the barrier 'prevent fire development' on an offshore platform.

that the correct signal reaches the control system. Similarly, fire pumps are run regularly to check that they operate according to specifications. The test intervals are determined by the requirements for reliability and by historic experience on failure rates.

Example 2: The management of a hydropower plant has established an emergency response plan and organisation to ensure evacuation of the plant in case of fire or flooding. This is tested through quarterly drills, where the response time and time for evacuation and counting of personnel are recorded.

9.5   Applying barriers in the prevention of occupational accidents

As mentioned in Section 9.4, few barrier opportunities are usually utilised in the prevention of occupational accidents. Looking at the classical example of a carpenter hitting his thumb with a hammer, a moment of lack of concentration may result in a sore thumb. The only barrier against such an accident is the carpenter's ability to control the direction of energy flow as represented by the kinetic energy of the hammer and maintain the safe distance to his thumb. There are many other such examples of hazards in our daily lives that are mitigated by only one barrier. When a human operator is part of the barrier system, we must expect accidents to happen due to inherent variability in human performance. Usually, the consequences are minor and tacitly accepted. Next, we will present examples of more systematic approaches to occupational accident safety using barriers.

9.5.1   Machinery safety

An 'ideal' type of barrier against accidents is one that is passive because it will not be dependent on more or less predictable day-to-day operation to function appropriately. Such barriers have to be implemented during the design phase of industrial systems and equipment. This is the safety philosophy behind the European legislation on machinery safety. We use it here to illustrate the significance of design in the prevention of accidents.

Many machines represent significant potential hazards and have been subject to safety regulations to reduce and eliminate the risks. One of the four freedoms within the European Union is the free float of products. It has thus become necessary to harmonise the machinery safety legislation in Europe. The Council Directives on Machinery is the cornerstone in this harmonisation (European Council 2006). The so-called harmonised European standards detail the requirements. This legislation fits well into the barrier framework presented here.

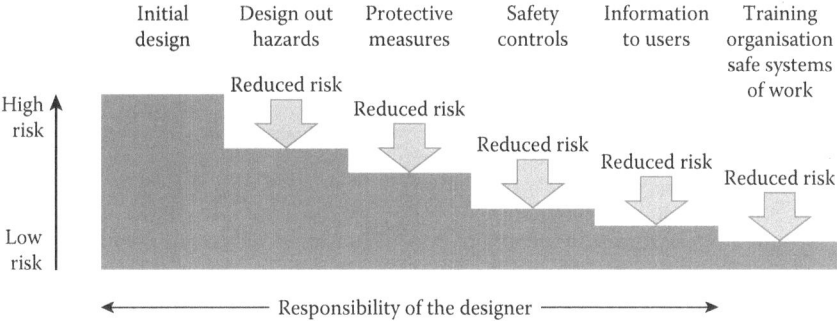

Figure 9.9 Strategies for the selection of safety measures. The priorities go from left to right. (From ISO, *Safety of machinery. General principles for design. Risk assessment and risk reduction*, ISO 12100:2010, International Organization for Standardization, Geneva, 2010. Copyright by the ISO. Reproduced with permission.)

In the harmonised standard ISO 12100 concerning basic concepts and general principles for design of machinery, a strategy for selection of safety measures is described as shown in Figure 9.9 (ISO 2010). This strategy reflects the general agreement among international safety experts that hazards should be prevented at the design stage. Residual hazards may be accepted if design solutions are unfeasible. It is, however, the responsibility of the producer to inform the user about these hazards. The user is responsible for training, organisation, and safe systems at work. This includes providing necessary personal protective equipment.

This standard defines the duties of the designer in order to identify and select the necessary safety measures. A stepwise procedure shall be followed, including:

1. Determination of the intended uses of the machinery, lifespan, space requirements, and so on
2. Identification of the hazards and assessment of the risks
3. Removal or limitation of the hazard
4. Design guards and/or safety devices against any remaining hazards
5. Information and warning for the user
6. Consideration of any necessary additional precaution

The first three steps involve the execution of a risk assessment. An example of how this may be carried out will be presented in Chapter 23. Here, we will focus on the principles behind the legislation regarding the selection of safety measures or barriers. There is a clear relationship among the different types of safety measures according to ISO 12100 and Haddon's strategies for accident prevention (see Table 9.1).

Table 9.1 Examples of safety measures according to ISO 12100 and their relation to Haddon's accident prevention strategies

Examples from ISO 12100	Related accident prevention strategy according to Haddon
Making machines inherently safe:	
• Minimum distances among mechanical components to avoid crushing	To prevent the creation of the hazard in the first place (Haddon No. 1)
• Limitation of forces or velocity of movable elements	To reduce the amount of hazard brought into being (Haddon No. 3)
• Limitation of noise and vibration	
Fully pneumatic or hydraulic control systems and actuators on machines intended for use in explosive atmosphere	To prevent the release of the hazard that already exists (Haddon No. 4)
Selection of safeguards to allow for secure access to the danger zone during normal operation (e.g. interlocking guard, two-hand control device)	To separate, in time or space, the hazard and that which is to be protected (Haddon Nos. 6 and 7)

The first priority is to select measures to reduce or eliminate the hazard. ISO 12100 presents a list of the hazards that machinery is likely to generate and proposes preventive measures. If it is not feasible to eliminate the hazards at the source, guarding should be selected as the second choice. The users should be informed about any residual hazards to be able to implement the necessary additional barriers. These barriers fall outside the scope of the standard.

The *danger zone* is an important concept in machinery safety and in many other areas of safety as well. It means any zone within and/or around an energy source in which a person is subject to a risk to his health or safety by coming into contact with the energy flow from the source (European Council 2006). This definition applies to damage to material assets (buildings, machinery, vehicles, materials, and so on) and to the environment as well, if we replace 'a person' with any of these 'targets'.

The European Council Directive of Machinery applies to machinery manufacturers or their representatives (vendors). Their obligations are summarised in Table 9.2.

The manufacturer (or his representative) draws up a declaration of conformity and affixes a CE mark to the machinery. The letters CE (Conformité Européenne) appear on those products that the manufacturer has declared meets the legal requirements. Before this can be done, the manufacturer has to document that the machinery meets the requirements in the machinery directives. This technical documentation includes:

• Assembly drawing, drawings of the control circuit, and detailed drawings, calculations, and so forth where required to document compliance

Table 9.2 Summary of requirements as to the involvement of an authorised institution in the documentation of the safety of machinery

Not hazardous machines	Hazardous machines that are designed and fabricated in accordance with harmonised standard	Hazardous machines that are not at all or only partly designed and fabricated in accordance with harmonised standard
The manufacturer or vendor works out technical documentation of compliance with the machinery directives	1. Technical documentation is delivered to an authorised institution, or 2. Technical documentation is delivered to an authorised institution, which verifies that the standard has been correctly implemented and draws up a certificate, or 3. Type-examination is carried out by authorised institution	Type-examination is carried out by authorised institution

Source: European Council, *Machinery*, Council Directives 2006/42/EC, Brussels, 2006.

- A list of safety requirements, standards, and specifications applied in the design
- A description of the method applied in eliminating the hazards of the machinery
- If applicable, a technical report or certificate obtained from a competent body or laboratory
- User instructions

To document the method applied in eliminating hazards, a risk assessment has to be carried out. We will return to this topic in Chapter 23.

The directive defines the types of machines that are regarded as hazardous. This list includes, for example, sawing machines, certain woodworking machines, presses, and machinery for working underground. For non-hazardous machinery (i.e. machinery not included in this list), internal documentation by the manufacturer is sufficient. For hazardous machinery, the manufacturer may choose whether to design in accordance with harmonised standards or not. The advantage of applying the harmonised standards lies in the possibility of a simpler procedure for proving compliance with the machinery directives. It may be an advantage to the manufacturer to also receive a certificate of compliance in the case where harmonised standards have been applied. This has to do with the manufacturer's liability in case of an accident with his product.

There are three types of harmonised standards:

- A-type standards define basic concepts and terminology, rules for the writing of subordinated standards, requirements as to risk analysis, and so on. For the machinery directive, there is one A-standard: ISO 12100, *Safety of machinery – General principles for design – Risk assessment and risk reduction.*
- B-type standards define certain specific safety aspects of machinery (including safety of control systems, noise, vibration, safety distances, and so on) or specific types of safeguards (machine guarding, safety mats, and so on).
- C-type standards apply to special types of machines (e.g. cranes, conveyor belts). These standards are all-inclusive and ensure that a piece of machinery that meets the standard also meets A- and B-type standards and thus the machinery directives. For example, the strategy for implementation of safety measures has to be implemented in C-type standards.

9.5.2 Safety against chemical hazards

The European directive on the protection from chemical hazards illustrates a systematic approach in the application of the barrier philosophy to prevent occupational accidents and diseases (European Council 1998). To prevent workers from being exposed to hazardous chemical agents, the employer must take actions in a priority from left to right as shown in Figure 9.10.

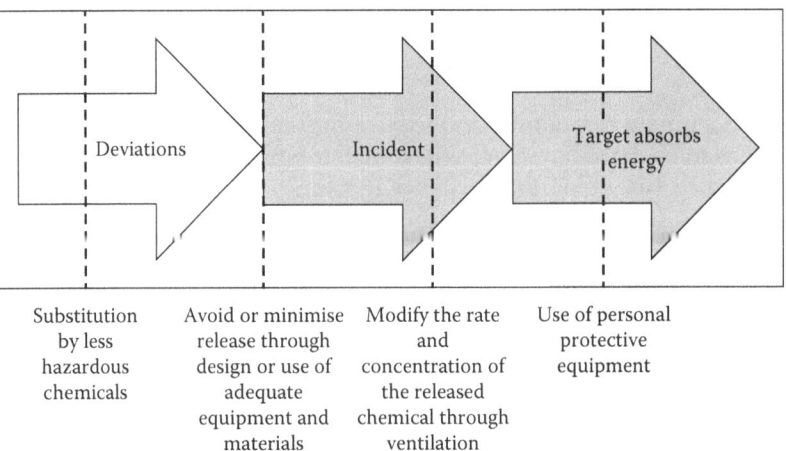

Figure 9.10 Principles for the prevention of accidents or diseases due to chemical agents.

Again, there is a clear order of priority from left to right in Figure 9.10, where substitution has the highest priority and the use of personal protective equipment is considered as the last resort. Second in priority comes containment of the hazardous substance inside the process enclosure, with use of ventilation as third in priority.

9.5.3 Safety in the use of cranes

Example: A serious accident involving a dropped load during lifting occurred during the repair of a dam. Five contracted workers were engaged in repairing the steel gates and hydraulic systems of the dam from a work platform at the dam wall, as shown in Figure 9.11.

Two mobile cranes were located on the dam crest (Cranes A and B in the figure). Crane A was engaged in lifting a 175-litre barrel filled with diesel from the original position to an end position close to Crane B. Due to existing steel structures on the dam crest, Crane A could only lift in a limited sector on the north side of the dam. During movement of the barrel, it slipped out of the sling and fell on one of the workers on the platform at the dam gates. He sustained severe head and back injuries.

This is an example of an accident due to violation of two important barriers during crane handling, as shown in Figure 9.12 (European Council 2001). Had one of these been instituted, the accident would have been avoided.

Table 9.3 shows examples of other types of barriers for the prevention of accidents involving cranes.

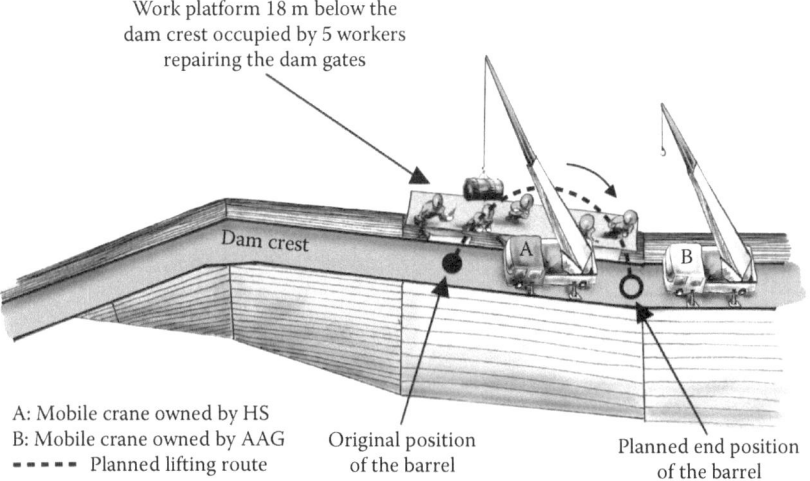

Work platform 18 m below the dam crest occupied by 5 workers repairing the dam gates

Dam crest

A: Mobile crane owned by HS
B: Mobile crane owned by AAG
▬ ▬ ▬ ▬ ▬ Planned lifting route

Original position of the barrel

Planned end position of the barrel

Figure 9.11 Situation at the time of the crane accident.

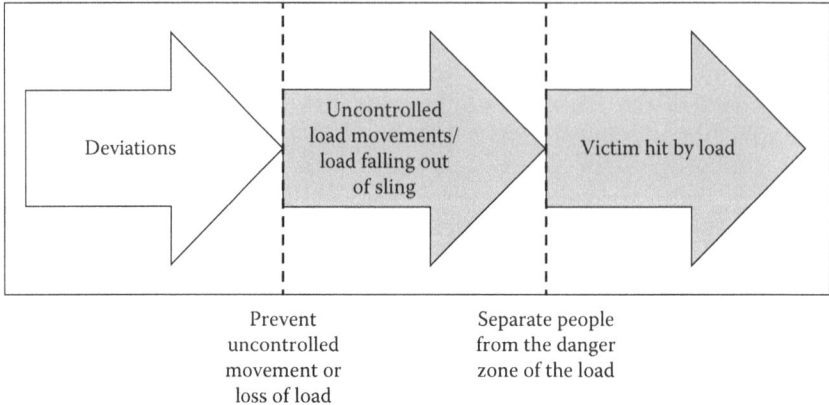

Prevent
uncontrolled
movement or
loss of load

Separate people
from the danger
zone of the load

Figure 9.12 Barriers to prevent accidents due to uncontrolled movements or dropping of load during crane operation.

Table 9.3 Examples of barrier functions to prevent crane accidents

Hazard	Barrier function
Electrocution due to contact between crane and high-voltage transmission line	Maintain safe working clearance, minimum 3 m for 50 kV; 4.5 m for 50–200 kV; 6 m for 200–350 kV; 7.5 m for 350–500 kV
Squeezed by crane tipping over	Prevent crane from tipping by use of outrigger and/or ensure adequate bearing capacity of ground

9.5.4   Safety in the use of heavy mobile equipment

Example: A fatal accident occurred in a tunnel during excavation, where two workers were stationed within the operating area of a muck-loading truck (Häggloader). One of the workers was crushed between the truck and the tunnel wall during forward movement of the machine (see Figure 9.13). The workers attached utility cables to the machine, an operation that was not necessary when the cable handling system of the machine functioned adequately.

This accident illustrates the necessity of maintaining the barrier of keeping personnel out of the operating zone (danger zone) of heavy mobile machinery. It is not possible, as with fixed machinery, to fence off the danger zone. Instead, the barrier must be maintained through a combination of technical and organisational means and through training. Work inside the danger zone of mobile machinery needs to be eliminated through technical solutions such as remotely controlled handling

Figure 9.13 Worker squeezed between a mobile machine and a tunnel wall.

of utility cables. Procedures based on risk assessments must define the operating area as a no-go zone and the workers trained and supervised accordingly.

9.6   Permit-to-work system

The permit-to-work system is an important operational safety measure in industries with severe hazards (HSE 2005; Norwegian Oil and Gas Association 2015). It aims at ensuring that work is planned and carried out under adequate safety precautions in cases where the exposed persons do not control the relevant energies themselves. The permit-to-work system is not a barrier in itself but is a system to ensure that the required barriers against hazardous energies have been instituted and are efficient to ensure safe work execution before work can start. Usually, it is established through administrative procedure, checklists and forms, training of personnel, and auditing to ensure compliance. Examples of hazardous energies often requiring a permit-to-work system are high voltages, liquids under pressure, flammable and other hazardous substances, rotating machinery, and oxygen-depleted or toxic atmosphere (tanks).

The following two examples illustrate the importance of an effective permit-to-work system.

Example: An installation worker needs to remove a flange during the commissioning of a process plant. He mistakenly started to work on a system that was under pressure due to leak testing. When he had removed some of the bolts, the others were ripped off and the flange hit his leg, which he lost.

Example: An instrument operator needs to mount a vibration gauge on a pump. While the man was doing the work, the pump suddenly was started *from* the control room. The rotating shaft caught his arm, and he was severely hurt.

An important safety principle behind the permit-to-work system is to ensure that the relevant parts of the system are de-energised while work is taking place in the danger zone. The challenge lies in designing and implementing a procedure that is simple and robust and that covers all relevant situations.

A padlock system to avoid electrocution is a simple example. An electrician has his own padlock, which he uses to secure the main switch in a safe position and to thereby de-energises the electrical system before he starts to work on it.

In a process plant, work permits are required for:

- Work on systems that contain hazardous substances (e.g. hydrocarbons, radioactive substances) or gas or liquids under high pressure
- Work that requires shut down of safety systems (e.g. a fire alarm)
- Hot work in hazardous areas, where there is the risk of an explosive atmosphere building up
- Work on systems involving moving machinery parts
- Entering tanks
- Work on high-voltage systems, connection and disconnection of electrical equipment

Figure 9.14 shows a typical work flow in the planning, execution, and close out of a permit to work. A permit-to-work system normally is documented in a specific procedure and is based on the following principles:

- All work on process and utility systems, with few and well-defined exceptions, require work permits. The supervisor for the work to be done (e.g. maintenance supervisor) should define the work in the application by type of job and by systems affected.

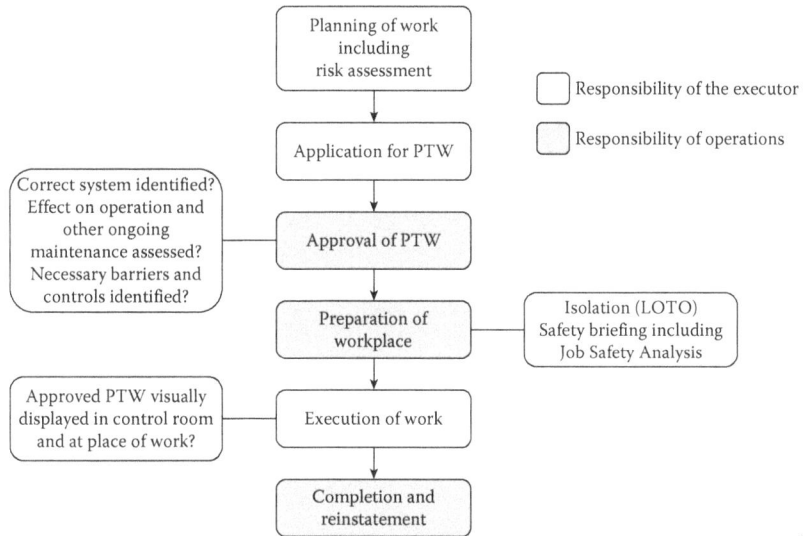

Figure 9.14 Work flow in the planning, execution, and close out of a permit to work.

- The energy sources (hazardous substances such as hydrocarbons, electricity, hydraulic pressure, and so on) are controlled by operations. The responsibility for each energy source is well defined and rests with a supervisor from the organisation such as the shift supervisor or a task safety supervisor (e.g. for work on high-voltage systems). This person shall have the necessary competence to assess the risk associated with the energy source and the barrier requirements. The responsible person shall at any time have the total overview of energised systems and active work permits.
- Responsibilities for the application, approval, preparation for work execution, execution of work, and completion and re-instalment shall be well defined.
- All work requiring a permit shall be subject to risk assessments as a basis for planning and prior to start-up of work.
- The work permits are effective during a well-defined time period such as during daytime (0800 to 1600). There are fixed routines for application and approval of work permits. A special form is used to control and check out all activities related to the planning, execution, and finishing of work requiring a permit.
- The person responsible for operations evaluates the application for work permit and identifies the required safety precautions from a checklist of measures before he approves it. Additional measures may also be stipulated. Table 9.4 shows a typical checklist

Table 9.4 Checklist of measures to be assessed in connection with hot-work permit and their relations to Haddon's strategies

Typical measure	Type of strategy according to Haddon
• Drainage/emptying and flushing (steam, nitrogen) of process systems containing flammable substances	Prevent build-up of energy
• Depressurise process systems	
• Gas detection and shut down	
• Padlock (electrical de-energising)	
• Secure power shafts against accidental start	Prevent the uncontrolled release of energy
• Plugging of hazardous drains	
• Fencing off of affected area	Separate source of energy and target in time or space
• Close and secure valves	Separate by means of physical barriers
• Introduce blind flanges	
• Use of personal protection	Improve the target's ability to endure
• Fireguard, fire alarm, firefighting	Limit the development of injury/damage

of measures. Usually, it is a question of de-energising the affected system and establishing the necessary barriers between the system and the energy source.

• Operations prepares for the work by shutting down the affected system and by introducing the barriers. Checks are also made to ensure that the maintenance operator starts to work on the right system and that he has understood the safety requirements.

• After work has been completed, operations checks the workplace and removes the barriers.

In addition to the general permit-to-work routine, special procedures may be required as part of the general system for work on high-voltage systems, hot work, and for work involving entry into a confined space.

9.7 Emergency response management

This section reviews the principles and practices for the management of emergency response seen from the perspective of an industrial company. The objectives of emergency response are to contain and control unwanted occurrences and to prevent or limit undesired consequences (losses) to human life or health, the environment, company assets or public property, and to the credibility and reputation of the company (Figure 9.15).

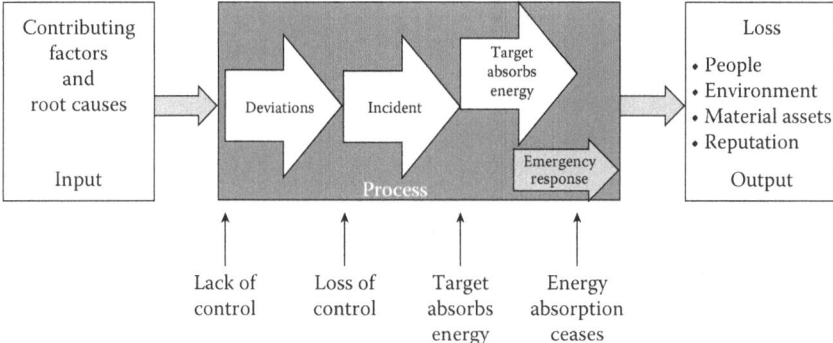

Figure 9.15 The accident analysis framework including emergency response (see Chapter 5).

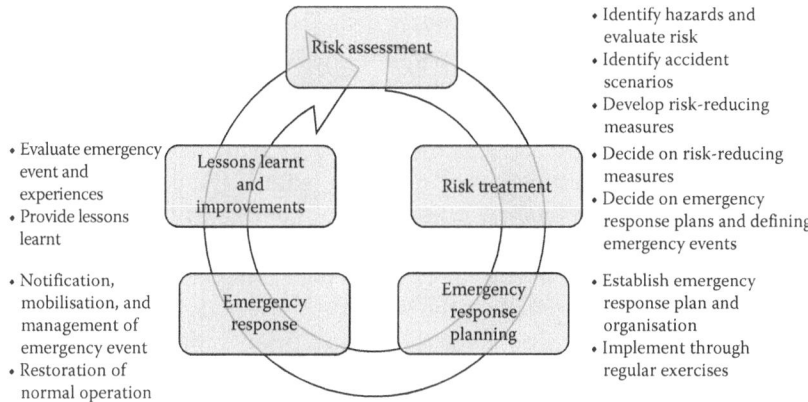

Figure 9.16 The relationship between risk analysis and emergency response.

The planning of emergency response needs to be based on risk assessment in order to ensure adequate direction and balance of resources (see Figure 9.16).

Emergency response is related to Haddon's Strategy no. 9, 'limit the development of loss (injury or damage)', and thus is considered here as an active barrier, although very complex. Risk assessment aims at identifying those risks that, after treatment, still represent events with considerable potentially negative impact and that need to be addressed in the emergency response plan. Examples of such events are:

- An accident during a company activity with severe consequences to that company's own employees, contractors, or to a third party
- Missing person

- Fire or explosion
- Sudden major structural damage
- Major spill of hazardous substances
- Serious security threat including kidnapping, act of terrorism, or serious criminal act
- Natural disaster, earthquake, flooding, landslide, and so on affecting company personnel or assets
- Serious civil unrest, war

The on-scene management of an emergency event goes through different phases as illustrated in Figure 9.17. The first phase involves detection, either by people involved or people witnessing the emergency event or by detectors connected to an alarm system. If people are involved in the detection of the emergency event, they need to notify the emergency response organisation as their first action in order to ensure that people outside the danger zone of the emergency are aware of it. The next action after notification is to evacuate people (i.e. get people out of the danger zone of the emergency event) and to count the number of evacuated people. These actions will be initiated by the people involved and supported by the emergency response team when they arrive at the scene. Search and rescue operations will be initiated by the emergency response team if the personnel count identifies missing persons. The on-scene commander will need to make critical decisions regarding the safety of his/her own personnel when entering the danger zone as compared to the benefit of getting injured people out. Whereas saving people is the first priority, reducing damage to the environment and physical assets is the second priority. Fighting fires and other types of emergencies to reduce material loss will take place if it is possible to do so without putting the emergency personnel at undue risk. When the situation has stabilised, a decision has to be made by the plant or emergency manager with regard to the return to normal.

Emergency management at an operating plant usually consists of two levels, the tactical emergency management and the on-scene (operational) management (see Figure 9.18).

The on-scene commander is responsible for combatting the emergency event and limiting loss. He will organise and coordinate the emergency response activities at the scene of the event, including counting, search and rescue, first-aid and medical evacuation (ambulance, helicopter), firefighting, securing the scene, and guiding the eventual external search and rescue team, ambulance, and fire brigade.

The tactical emergency manager and his team handle organisation and coordination of support to the on-scene team. This includes mobilisation and management of emergency response and transportation resources, organisation of evacuation from the scene of the event, and coordination with the operations manager of the plant's shut down (if relevant).

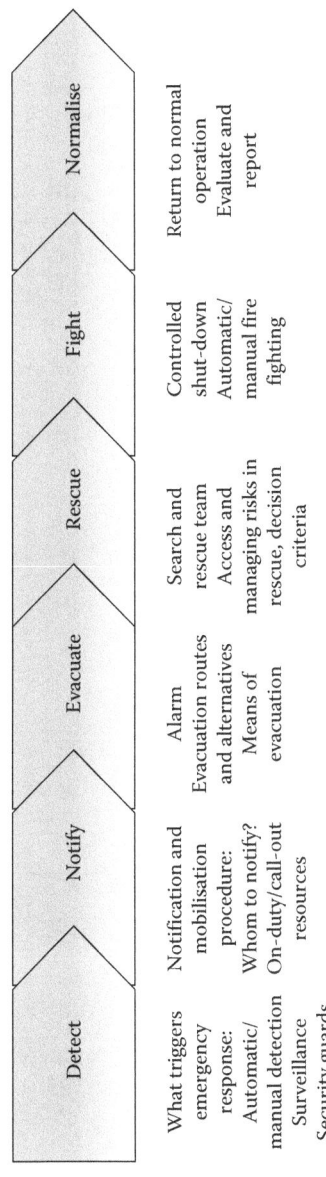

Figure 9.17 The phases of the on-scene management of an emergency event.

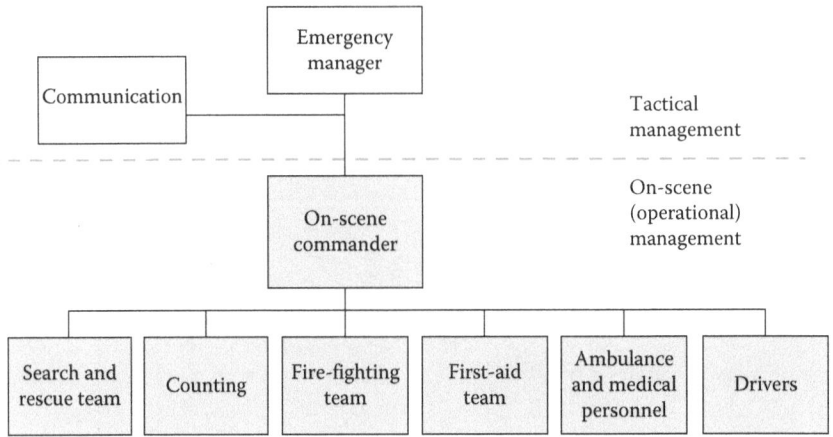

Figure 9.18 Typical emergency response organisation at an operating plant.

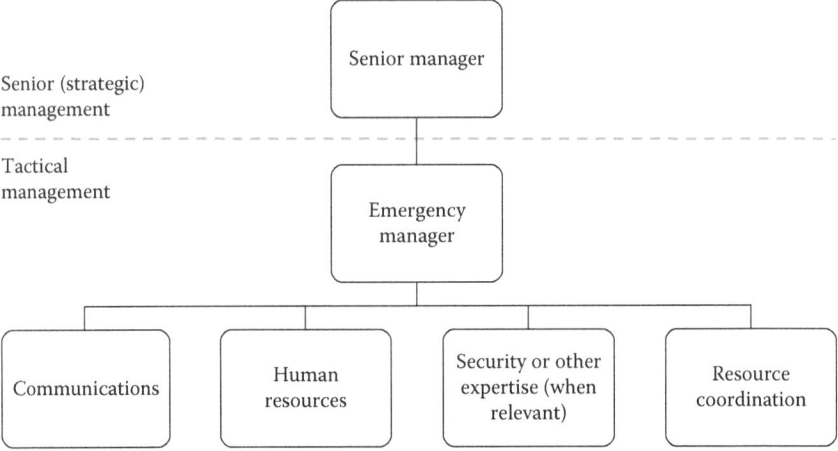

Figure 9.19 Typical emergency response organisation above the plant level in an industrial organisation with tiered response.

Figure 9.19 shows a typical organisation of the tactical and strategic emergency response. The tactical emergency manager will have personnel at his disposition for advice and for delegation of responsibility for specialised areas of emergency management:

- The human resource (HR) manager is responsible for the organisation and management of those emergency responses related to affected company employees, hired personnel and expatriated family members, and to next-of-kin. The company has a duty to do

everything that is reasonable under the circumstances to keep the
employee and his or her family safe from harm.
- The communications manager is responsible for the emergency
  response related to the media, internal communication, and for
  information gathering from external sources.
- The resource coordinator is responsible for the coordination of
  administrative support to the emergency response team and for log
  keeping.
- The senior manager is responsible for monitoring the overall handling of the situation; for considering strategic, long-term consequences and business continuity aspects; for managing contact with
  stakeholders, authorities, and the media; and for providing necessary support to the tactical response.

The principles for emergency response planning outlined in this
chapter follow a set of common guiding principles, although the specific
solutions selected in the individual company need to be flexible and tai-
lored to the specific needs:

1. Regular assessments of safety and security risks to define the
   requirements to emergency response
2. Management of emergency response at the lowest appropriate
   level (parity) and clear criteria for escalation of defined emergency
   response functions to higher levels of the company organisation
3. Responsibility and authority in the emergency response organisation based on the organisation and responsibilities in normal operation (subsidiarity)
4. Reliable notification and mobilisation through:
   - Standardised and reliable paths for notification and mobilisation
   - Utilisation of 24/7 resources, where feasible
   - Redundancy of resources (rosters of trained personnel for each
     function in the emergency response organisation)
5. Well-defined lines and means of communication among emergency
   response teams
6. Training and verification through regular exercises
7. Collaboration with other actors (emergency services, sub-contractors,
   neighbours) in planning and exercises
8. Update of the emergency preparedness plan based on experiences
   from incidents and exercises and on changes in the risk picture

chapter ten

The human element in accident control

The human operator at the sharp end with his/her individual behaviour, knowledge, and attitudes is an important element in the systematic prevention of accidents. In this chapter, we will focus on the role of the human operator in accident causation and on the prevention of accidents through improving the quality of human performance. Employee participation in safety work and utilisation of their knowledge and experience is covered in other parts of this book. In this chapter, we will begin by explaining how individuals – as part of an active barrier system – ensure the execution of the barrier function. We will, in particular, pay attention to macro-cognition, which studies human cognition in real-world settings. Second, we will explain how individuals may generate deviations that may lead to accidents. It is important here to explain how human errors are created by their context. Third, we explain some safety management approaches that can improve the quality of human performance in the execution of hazardous work.

10.1 The human operator as a barrier element

In Chapter 9, we defined barriers as a set of human, technical, and/or organisational elements that – as a whole – provide functions with the ability to intervene in the energy flow to change its intensity or direction. In some barriers, the human operator is one of the important elements for preventing accidents or reducing the consequences of accidents. In this section, we look at the human operator as a barrier element.

An active barrier with a human operator in the control loop is dependent on some kind of action from the individual.

Example 1: A moose jumps into the road in front of a car. The driver detects the moose, hits the brake of the car, and swings the wheel to avoid collision. The driver initiates the barrier function to prevent collision between vehicle and object; this function is then realised by technical barrier elements.

Example 2: A control room operator at an offshore oil and gas installation detects that the pressure in an oil storage tank is increasing and

the automatic shut-down system is not being initiated. The operator intervenes by closing the valve at the tank inlet and thus prevents leakage.

The human barrier element is part of what we defined as an active
barrier in Chapter 9. Figure 9.4 illustrated this principle – in this case, a
deviation or incident is detected by a human operator, who initiates an
action either to stop the sequence of events or to reduce the consequences
from it.

This way of understanding the human barrier element is closely
related to human cognition. Macro-cognition studies cognition in real-
world situations and provides frameworks for understanding the nature
of human performance in naturalistic situations including those that
involve high risk (Klein et al. 2003). Figure 10.1 shows a macro-cognition
model that can be used to understand how humans perform tasks.
Whaley et al. (2016) developed the model based on a thorough literature
review of macro-cognition theories and studies. The five macro-cognitive
functions in the model are not linked linearly but happen in parallel and
are cyclic. The functions may also overlap each other. Furthermore, the
model illustrates that individual cognition is generated by team interaction. *Detection and noticing* are the cognitive processes related to sensation,
perception, and attention to important information in the work environment (i.e. how humans perceive large amounts of information). *Sense-
making and understanding* are processes for comprehending the meaning of
the detected information. *Decision-making* is the process of making decisions or planning action scripts to achieve given task goals. *Action* is used

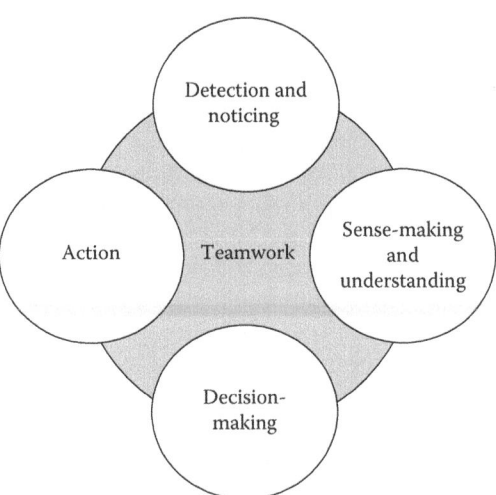

Figure 10.1 The five macro-cognitive functions. (Adapted from Whaley, A.M.,
et al., *Cognitive Basis for Human Reliability Analysis*, U.S. Nuclear Regulatory
Commission Report, NUREG-2114, 2016.)

to implement and achieve a particular goal. *Teamwork* is the interaction among operators that is undertaken to accomplish a task – such as coordination, collaboration, and communication.

The functions of detection, sense-making, decision-making, and action are related to the individual operator. The fifth function, teamwork, is about how a group of individuals interacts and relates to each other in order to accomplish a task. This implies that, to understand human performance, we need to understand the context of individual cognition.

Well-functioning macro-cognitive actions contribute to the realisation of efficient barrier systems and thus to the avoidance of accidents. However, substandard performance by a human operator serving as an element in an active barrier function might lead to accidents. Whaley et al. (2016) present a framework to analyse why macro-cognitive functions fail due to a hierarchy of causes (see Figure 10.2). According to this framework, failures in macro-cognitive functions (detection, sense-making, decision-making, action, and teamwork) can be explained by proximate causes (Table 10.1). The proximate causes are in turn caused by contextual factors – the so-called performance-influencing or performance-shaping factors (PIFs/PSFs). PIFs are commonly used in human reliability analysis, which is part of quantitative/probabilistic risk assessment. There are many classifications of PSIs (Kim and Jung 2003). One example from Groth and Mosleh (2012) is shown in Table 10.2.

These overviews of proximate causes and PIFs support failures in human cognitive functions in analysis of causes. The overview also supports identification of what contextual factors need to be considered in order to ensure adequate human performance. A control room operator, for

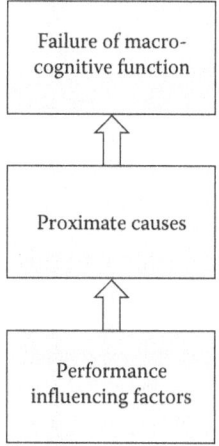

Figure 10.2 Framework for the analysis of cognitive function failures.

Table 10.1 Categories of proximate causes of cognitive failures

Macro-cognitive function	Proximate causes
Detecting and noticing	Cues/information not attended to Cues/information not perceived Cues/information misperceived
Understanding and sense-making	Incorrect data Incorrect mental model used to interpret data Incorrect integration of data and mental model
Decision-making	Incorrect goals or priorities set Incorrect pattern matching Incorrect evaluation of options Cognitive bias
Action	Failure to execute desired action Execute desired action incorrectly
Teamwork	Failure of team communication Failure in leadership/supervision

Source: Whaley, A.M., et al., *Cognitive Basis for Human Reliability Analysis*, U.S. Nuclear Regulatory Commission Report, NUREG-2114, 2016.

example, is dependent on the design of the control room displays and on proper training to correctly identify and react to signals.

10.2 Human error

Early research into accident causes 'showed' that 88% of accidents were caused primarily by dangerous acts on the part of individual workers (Heinrich 1959). Heinrich applied an accident model (the 'domino theory' presented in Chapter 4), where he focussed on the event that went wrong immediately before the occurrence of the 'accident'. These results followed from the central position that the human operator held (and still holds) in the control of industrial production processes. Since Heinrich published his results, the proposition that 80–90% of all accidents are caused by human error has become accepted as an established truth among safety practitioners. Heinrich's model has, however, been replaced by succes- sively more elaborate analyses. It is now understood that accidents have multiple causes and that it is irrational to single out just one of them. It is also necessary to understand why human operators sometimes err. Reason (1997) analyses human error in the tradition of system ergonomics as a consequence of the context (i.e. the system where the operator performs his or her tasks). Similarly, Dekker (2014) looks at human error as symp- toms of deeper trouble in a system. Human errors need to be understood in relationship to the design of tools, tasks, and the operations/organisa- tional environment. By understanding human error in this way, we will

Table 10.2 Categories of performance-influencing factors

Organisation-based	Team-based	Person-based	Situation/stressor-based	Machine-based
Availability and quality of training program	Availability and quality of communication	Attention to task	External environment	Human–system interface
Availability and quality of corrective action program	Direct supervision and leadership	Attention to surroundings	Conditioning events	System response
Availability and quality of other programs	Team coordination	Physical and physiological abilities	Task load	
Safety culture	Team cohesion	Knowledge and experience	Time load	
Management activities (staffing and scheduling)	Risk awareness	Skills	Task complexity	
Workplace adequacy		Bias	Stress	
Availability and quality of procedures		Familiarity with situation	Severity and urgency of perceived situation	
Availability and quality of tools		Morale, motivation, attitude	Responsibility and impact of perceived decision	
Availability and quality of necessary information				

Source: Groth, K.M. and Mosleh, A., *Reliab. Eng. Syst. Saf.*, 108, 154–174, 2012.

be able to understand accidents as being produced by multiple factors; we will understand that safe human performance at the sharp-end depends on cognitive factors that evolve, cascade, and depend on the context of the human performance. In addition, we will avoid blocked learning by a sole focus on human error (Woods et al. 2010).

There are many different definitions of human error. In this case, we will apply a definition similar to that of deviations presented in Chapter 5. Human error is defined here as a subset of human actions that transgresses a norm or limit of what is planned/intended, normal, or acceptable (Kjellén 1987a; Miller and Swain 1987).

There are some important aspects of human error. The norm or limit is defined in relation to the context in which the error occurs and its potential consequence. Some systems represent a benign environment by being tolerant towards large variations in human performance. In such systems, few actions will be defined as human error. Other systems are more demanding, and a small deviation from the ideal path represents a human error that may result in an accident or other unwanted consequence. It also follows that the consequences of errors are not necessarily correlated with the magnitude of the transgression but rather with the energies involved. Furthermore, there is a certain degree of arbitrariness in the definition of human error. In accident investigations, for example, an action may be defined as a human error only in retrospect when the negative outcome is known. The same action may have been carried out successfully many times before and may have been tacitly accepted by operators and management.

There exist several theories and taxonomies describing human error (e.g. Surry 1974; Swain 1974; Norman 1981; Hale and Glendon 1987). However, two models are more dominant than others in the research literature: Rasmussen's (1982) skill–rule–knowledge (SRK) model and Reason's (1990) model of principal types of error, the latter being a development of Rasmussen' model.

Rasmussen's (1982) S–R–K framework distinguishes among three different levels of human cognitive control of the environment:

- *Skill-based behaviour:* At this level, behaviour is automated. Incoming information leads directly to an automatic response without any conscious thought. This control mode is suited for routine situations. Skill-based behaviour is established through training and experience.
- *Rule-based behaviour:* At this level, the operator recognises a known situation and applies a pre-stored rule or action pattern to handle it. The rules have been developed through experience and may be individual or collective. There are, for example, traffic safety rules to avoid collisions when two cars are on an intersecting course.

- *Knowledge-based behaviour:* When the situation is new or the operator is uncertain of what rule to apply, constructive thinking must take place. He or she must interpret the situation and choose among different alternative actions. Here, errors come about due to the human operator's limited information-processing capacity and because of incomplete knowledge. Behaviour in novel situations is knowledge-based. Behaviour evolves from this level to the lower levels through learning and experience.

Based on Rasmussen's taxonomy, Reason (1990) identified the following categories of human errors: slips, lapses, rule-based and knowledge-based mistakes, and violations. Table 10.3 combines Rasmussen and Reason's models.

Table 10.3 Categories and examples of human errors

Control mode (Rasmussen 1982)	Type of human error (Reason 1990)	Examples, car driving (Stanton and Salmon 2009)	Examples, automated production (Döös et al. 2004)
Skill-based	Slips	Driver misreads road signs. Driver presses accelerator instead of brake.	Operator does not notice that his hand is approaching a sharp edge.
	Lapses	Driver fails to recall road just travelled.	–
Rule-based	Rule-based mistakes	–	Operator takes measurement of product without setting the machine to stop, a way of working he knows he should not apply.
Knowledge-based	Knowledge-based mistakes	Driver underestimates the speed of oncoming vehicle.	Operator is blowing away metal shavings with an air gun by stretching in through the ceiling-hatch of a portal. He fails to assess the speed and range of the machine.
Combination	Violations	Driver is exceeding the speed limit intentionally, although he is aware of the speed limit.	–

- *Slips* are attentional failures where one or more tasks were executed wrongly, but the intention was correct. These errors are often observable (e.g. pushing the red button instead of the blue button). We also note that the task could have been planned in such a way that even if it were executed without slips, it would still have failed. This type of human error is related to skill-based behaviour.
- *Lapses* are a result of a failure of the memory, and they differ from slips in that they often cannot be observed. Lapses cause us to forget to carry out an action, to lose our place in a task, or even to forget what we had intended to do. This type of human error is also related to skill-based behaviour.
- *Mistakes* are more subtle and complex than slips and lapses. They are intentional failures in the sense that an operator intentionally performs an action that is wrong. Actions may be performed according to a plan, but the plan is inadequate. We can distinguish between rule-based and skill-based mistakes. Rule-based mistakes involve either misapplication of normally good rules or application of bad rules. Knowledge-based mistakes occur in novel situations where operators fail to work out problem solutions.
- *Violations* are deviations from safety rules and procedures. They can either be deliberate or unintentional acts. Violations can be skill-based (i.e. routine violations). These are what can be called silent deviations, where the same violation is repeated over and over again. Knowledge-based deviations can involve non-compliance with rules because of a desire to achieve personal goals or because non-compliance is necessary in order to get the job done.

Based on this classification of human errors, the Energy Institute (2008b) has developed principles for the management of rule-breaking, applying a combination of measures with a varying mix of coaching and formal discipline. Slips and lapses on the one end are primarily managed through coaching, whereas violations – and reckless violations in particular – require disciplinary action.

10.2.1 The theory of risk homeostasis

Risk homeostasis is a theory that shows how human actions in accident prevention are influenced by macro-cognition (i.e. individual judgements of the risk level). Under certain conditions, the operator is able to choose between becoming exposed to a hazard or avoiding it. The importance of this aspect was recognised first in the area of traffic safety around 1970. It turned out that many measures aimed at facilitating driving did not result in fewer accidents but in the driver adapting to the new situation by, for example, increasing the vehicle's speed. Wilde (1982) has

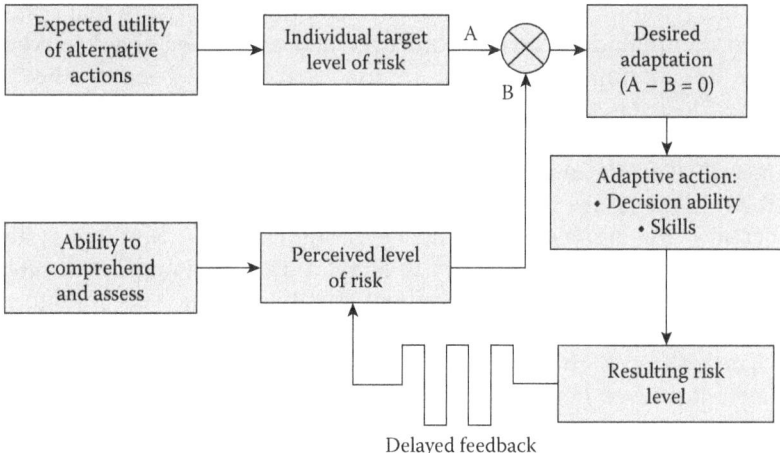

Figure 10.3 Wilde's risk homeostasis theory. (From Wilde, G.J.S., *Accident Anal. Prev.*, 18, 377–401, 1986. With permission).

modelled this situation on risk homeostasis theory (see Figure 10.3). Wilde describes how the operator's caution develops through a process where he/she compares the accepted or target risk with the perceived risk level in a feedback cycle.

The level of accepted risk is assumed to be determined by four classes of expected utility of action alternatives: (1) expected utility/benefits from risky behaviour, (2) costs and inconvenience expected from safe (cautious) behaviour, (3) benefits expected from safe behaviour, and (4) costs and inconvenience expected from risky behaviour. The first two utility factors increase the willingness for risk-taking, while the last two increase the safety level. The only factor that determines the individual's long-term level of risk is his/her individual risk acceptance criteria (Wilde 1986). There are four ways of lowering the individual target level of risk: (1) decrease the expected benefit of risky behaviour, (2) decrease the expected cost of cautious behaviour, (3) increase the expected benefit of cautious behaviour, and (4) increase the expected cost of risky behaviour. The risk homeostasis theory is especially relevant in tasks such as driving. If the behaviour goes unchecked, the operator may use his/her skill and ability to cope in order to balance the level of risk against rewards such as compensation for progress and excitement/arousal.

10.2.2 Successful operations and failure-free organisations

According to the school of resilience engineering, a system successfully copes with complex hazardous processes by adaptation to various situations (Hollnagel et al. 2007). This adaptation will ensure that systems function as intended during both expected and unexpected incidents. People are an important factor in this adaptability because they are required to cope with different kinds of situations (Woods et al. 2010). Hollnagel (2014) focuses on adaptation and coping in his analysis of 'Safety II' (i.e. the understanding of safety as ensuring that things go well/are successful). Hollnagel (2011) claims that successful operations are created by engineering resilience in organisations. Resilient systems are characterised by their ability to adapt to expected and unexpectedly changing situations through resilient capabilities such as monitoring, anticipation, response, and learning.

Theories on high reliability organisations (HROs) provide a similar understanding of how organisations succeed in adapting to expected and unexpected situations (e.g. LaPorte and Consolini 1991; Lekka 2011). Studies of HROs show that some organisations are able to sustain failure-free operation of high-technology systems that are susceptible to major accidents. Important aspects in this research tradition are organisational redundancy and the capacity of organisations to adapt to peak demands and crises (Weick and Sutcliffe 2007).

An HRO uses a strategy of organisational redundancy in avoiding errors and/or recovering from them. Operators receive backup from colleagues who are ready to give advice and to correct errors or to take over if an operator fails to execute an action correctly. Manning cockpits with two pilots is a good example of structural organisational redundancy. Cultural dimensions of organisational redundancy concern the possibility of operators listening to the reasoning of their colleagues, observing their behaviour when performing critical tasks, and understanding the consequences of their actions. Operators must also be able to exchange information and challenge each other's decisions and actions when there are conflicting views as to the correct course of action.

An HRO is able to change its organisational structure in a critical situation. In day-to-day operation, an organisation is hierarchical, and decisions and actions are controlled by well-formulated and tested standard operating procedures. When the workload increases, for example during disturbance handling, operators are protected from interference from the outside and are granted a high degree of autonomy in handling this type of situation. Organisations rely on the technical skills of operators in these cases. Operators must have clear, agreed-upon operational goals and decision criteria to be able to handle the situation efficiently. In an emergency, the organisational structure changes again. There are events clearly specified as emergencies, and the organisation follows carefully assigned and practised operational procedures to handle each type of situation.

Mindfulness is another important part of an HRO. Mindfulness is an organisational capability to handle unexpected situations through the principles of anticipation and containment by (Weick and Sutcliffe 2007):

- Preoccupation with failure. Organisational members look for symptoms and encourage reporting of errors.
- Reluctance to simplify interpretations. Simplifications are avoided because they might create blind spots.
- Sensitivity to operations. Identifying early warnings is emphasised even for normal situations.
- Commitment to resilience. The organisation can withstand various conditions because of flexible organisational members
- Respect for expertise. Decisions migrate to persons with the knowledge to solve the problem.

10.3 Managing the human element in accident prevention

As shown earlier in this chapter, the role of the human element and the role of the context in which the human element is operating are both key to accident prevention and causation. As a result, efforts to manage the human element is essential in accident prevention. Accident prevention measures directed towards individual knowledge and attitudes – such as campaigns, behaviour-based safety programs, and training – have been popular for years. Among other things, they demonstrate management's concern for safety. Personnel-related measures aim to reduce the frequency of human errors and promote safe behaviour. Measures directed at individuals can be categorised as modifications of work conditions, modifications of skill and knowledge (education and training), modifications of attitudes (information campaigns), and selection of personnel (Rundmo 1990; Dyreborg et al. 2015). There are five main objectives of these personnel-related measures:

1. To ensure that the *physical and organisational context, including safety culture*, facilitate safe performance of individuals
2. To ensure adequate *knowledge*, for example, about the technical design of the workplace and about the hazards involved in specific jobs
3. To ensure adequate *skills* as to how to perform the work safely and efficiently

4. To ensure adequate *attitudes* among personnel so that they are able to put their knowledge and skills to use and make a commitment to behave safely
5. To ensure safe *behaviour* among personnel so that they can execute tasks in a manner that prevents incidents from happening

These five objectives are interlinked. Most interventions are an integration of several of these measures. Research has demonstrated that it is most efficient to combine different approaches to the physical and organisational context and modifications of individuals' knowledge, skills, attitudes, and behaviour (Lund and Aarø 2004; Dyreborg et al. 2015). There is a prioritised order of reaching the objectives given in the aforementioned list:

1. Improvement of working conditions (i.e. modifications of the physical and organisational context)
2. Education and training (i.e. modifications of individuals' knowledge and skills)
3. Attitudinal approaches (i.e. modifications of individuals' attitudes)
4. Behavioural approaches (i.e. modifications of individuals' behaviour)

10.3.1 Modification of the physical and organisational context

In this section, we will compare modifications of the physical context (i.e. design of the workplace) and modifications of the organisational context. For the latter, we pay attention to changes in organisational structures but not to those informal organisational parts that are covered in Chapter 4 (safety culture) and in Chapter 10 (high reliability organisations).

Modification of the physical context can happen by engineering control – such as outlined in the discussion of Haddon's strategies in Chapter 4 (elimination and modifications of hazards, machine safeguards, walkways); this will influence the safety of workers but not necessarily their behaviours. Workplace design is an engineering solution that facilitates the use of safe and error-free work practices and promotes error recovery. An advantage of this solution is that design changes are permanent, whereas personnel safety measures such as education and training and campaigns are measures that have to be repeated continuously to be efficient. There are different means of preventing human errors through design. One is to make it impossible to commit the error altogether (e.g. by eliminating the error-prone task). Design may also make errors less likely (e.g. by being compliant with people's ingrained behaviour or so-called population stereotypes). A good working environment will promote safe behaviour in general by reducing the likelihood of fatigue-induced errors.

Modification of the structural organisational context such as procedures and rules and checklists and methods that frame human performance. For this to be efficient, workers need to be compliant. In practice, there can be a difference between 'work as imagined' and 'work as actually done' (e.g. Antonsen et al. 2012; Wold and Laumann 2015). In some cases, non-compliance with rules may be necessary to get the job done safely and efficiently, such as in new situations (Hale and Borys 2012). Hale and Borys (2012) distinguish between two approaches to rule development. The first is rooted in scientific management, where rules are developed by experts and enforced to guard against the errors of fallible operators who lack the time or expertise to devise their own rules. The second approach emphasises that local, situational behaviour based on sharp-end experience of work situations and accident risks represents a necessary input to ensure rules that promote safety. To merge the approaches and thus strengthen the formal safety structures, Hale and Borys (2012) recommend combining the approaches by designing perspicuous and clear rules through participation of users in rule development and implementation.

10.3.2 Personnel-related measures

Education and training aim at modifying knowledge and skills. Knowledge and skills to perform work correctly and safely are acquired through experience, but the learning process may be speeded up and controlled by formal education and training. Government regulations require employers to provide adequate education and training for their personnel. This is especially relevant when operators routinely meet similar hazardous situations and errors, and there is a need to develop adequate routines and ingrained behaviour to handle stress and to recover the system or bring it safely to a halt. Over training may be necessary to make the response automatic. Education and training are also needed in situations where operators routinely meet new hazardous situations that have to be resolved through knowledge and problem-solving skills. When interactions among team members play an important role in error detection and handling, each individual also should be acquainted with the different skills of the other team members. Team training is especially important in this situation.

The selection of personnel for hazardous work involves judgements regarding their physical ability (sight, hearing, motor skills, and so on), competence, motive structure, stress tolerance, and experience. For certain types of hazardous work, such as crane operation and lorry driving, there are formal requirements as to the testing and certification of personnel. Employee selection involves challenges regarding which criteria to use, the validity and reliability of the selection tests, and the availability of an adequate supply of qualified personnel.

Attitude modification involves the use of different means – such as the distribution of posters and pamphlets, safety campaigns, information on the company intranet, showing video films, safety meetings, and direct talks about safety. The aims of such campaigns are to change behavioural patterns (e.g. by defining the 'ideal' safe behaviour), to affect attitudes, to put the focus on safety matters, and to give warnings about negative consequences of unsafe behaviour. The foundation of this approach is that changed attitudes will lead to changed behaviour, which will prevent incidents. Attitudes in themselves do not prevent incidents; it is the actions taken by individuals that will affect change. However, this foundation has been criticised by several authors who have demonstrated that behavioural change is not only a result of changes of attitudes but also of social influences and perceived behavioural control (Dyreborg et al. 2015). Therefore, the effect of different types of attitude modifications on accident control can be weak.

Behavioural modification is based on a theoretical foundation that humans choose various responses based on the perceived consequences of the responses (Skinner 1969). The theory describes links among antecedents, behaviour, and consequences. Antecedents define and signal the behaviour, while the consequences of the behaviour determine whether or not the behaviour will be repeated. Behaviour that is reinforced is likely to be repeated, while behaviour with punishment tends not to be repeated. So-called *behaviour-based safety* (BBS) programs follow these principles: basic training (antecedents) followed by observations and positive feedback (consequences) to enforce safe behaviour (Tuncel et al. 2006). Different types of models and components of BBS programs have been performed, but their effects on safety performance are uncertain (Tuncel et al. 2006). Some common components of BBS programs are safety training, goal-setting, observation and feedback, verbal feedback, data analysis, and problem-solving (Dyreborg et al. 2015). A similar method is *behavioural sampling*, which is also based on behaviour theory. To improve the proportion of safe acts, it establishes a baseline – with mutually agreed targets for improvements in safe behaviour and feedback – for workers that details information on their actual performance. (The behavioural sampling as a safety performance monitoring tool will be discussed in Chapter 17.)

10.3.3 Effectiveness of measures directed at workers

Lund and Aarø (2004) present a conceptual model showing the connections among safety performance (prevention of incidents); three groups of accident prevention measures (modification of attitudes, behaviour, and structural/physical conditions); and personal and organisational factors (see Figure 10.4). The model shows how the prevention of incidents

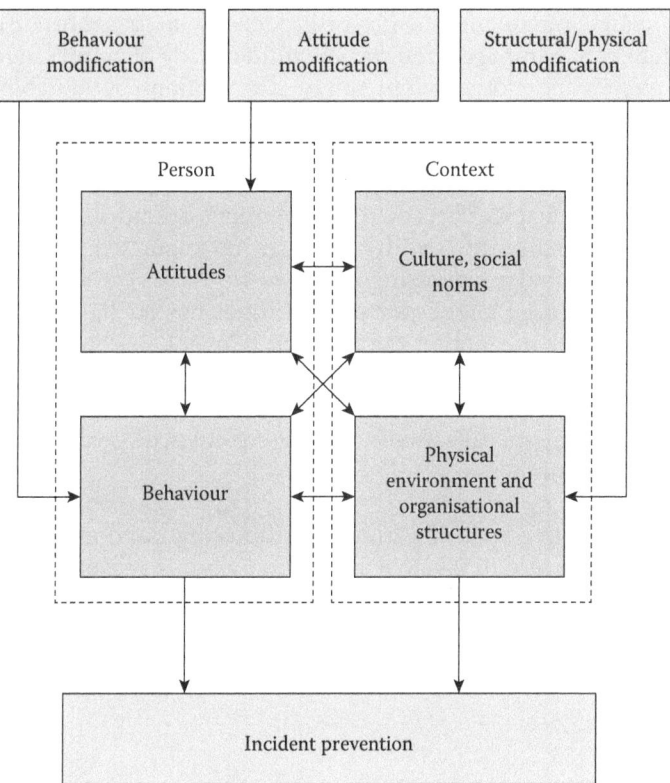

Figure 10.4 An accident prevention model emphasising human, structural, and cultural factors. (From Lund, J. and Aarø, L.E., *Saf. Sci.*, 42, 271–324, 2004. With permission.)

is influenced by individual and contextual factors (culture and physical/organisational environment). An important message in the model is that both personal and contextual factors influence safety performance; this is similar to Reason's (1997) active and latent failures. This model also illustrates that personal and contextual factors are interwoven and influence each other. For example, individual attitudes and behaviour influence safety culture and vice versa. As a result, different types of interventions will influence personal as well as contextual factors.

Based on this model, Lund and Aarø (2004) performed an analysis of empirical studies to investigate the strengths of the various links in the model and the effect on incident prevention. Their analysis shows that the link 'attitude modifications – attitudes/beliefs – behaviour – incident prevention' is weak. But the link 'structural modification – physical and organisational environment – behaviour – incident prevention' shows stronger relationships. The authors conclude that structural modifications are more effective in reducing the risk of accidents than measures directed at individual behaviour and attitudes. A similar study of empirical research on the effectiveness of different safety interventions by Dyreborg et al. (2013) shows the same patterns:

- Modifications of the structural organisational context have moderate to significant effects on the prevention of accidents.
- Safety climate and culture changes have some impact on both behaviour and accidents.
- Modification of knowledge and attitudes has no significant impact on accident prevention but may be relevant in combination with other measures.
- Safety training has no significant impact on accident prevention.
- Behavioural modifications can be effective in combination with other measures.

An important conclusion from the studies by Lund and Aarø (2004) and Dyreborg et al. (2013) is that combining various preventive measures is more effective than single interventions directed towards individuals.

section three

Learning from incidents and deviations

In our model of a safety information system in Chapter 7, we identified four different subsystems (i.e. data collection, data analysis or processing, memory (database), and distribution of information). In this Section, we will first focus on the data-collection subsystem. Chapter 11 presents an overview and Chapters 12 and 13 give details about different methods of data collection on accident risks. Chapter 12 will concentrate on planned health, safety, and environment (HSE) management activities such as hazard identification, workplace inspections, and safety audits. In Chapter 13, we will focus on the reporting and investigation of accidents and near accidents. This will also include employees' self-reporting of near accidents, unsafe acts, and hazardous conditions.

In Chapter 14, we review some basic principles for the establishment and application of a memory (database) on accident risks. This chapter also covers different methods for statistical analysis of incident data.

The principles and methods presented in Section III are reactive in the sense that they are used in an approach where we learn from incidents and deviations after their occurrence.

chapter eleven

Sources of data on accident risks

11.1 The ideal scope of different data-collection methods

In Section III, we focus on the collection and analysis of data on accident risks. We use accident risks as a collective term for any events or conditions that are associated with an increased risk of accidents such as a hazard, incident, deviation, contributing factor, or root cause. There are different means of collecting data on accident risks. Accident and near-accident reporting and workplace inspections represent 'traditional' means, where data are collected 'after the fact' (i.e. after the accident risk has manifested itself). Unwanted occurrence reporting involves an employee's reporting of experienced near accidents and deviations (i.e. unsafe acts and conditions). Safety audits represent a 'temperature measurement' of the present, ongoing management of safety. Hazard recognition identifies potential incidents and is applied both reactively in workplace inspections and proactively as a cornerstone in the different risk analysis methods that will be presented in Section V. Figure 11.1 shows the 'theoretical' coverage of these different means of data collection. Risk analysis offers an opportunity to collect data on accident risks before they have occurred. We will come back to this theme in Section V.

11.2 Filters and barriers in data collection

At the beginning of the 1980s, one of the authors conducted an evaluation of six large Swedish companies' accident and near-accident reporting and workplace-inspection routines (Kjellén 1982). The study was performed before any of the companies had been affected by the then-emerging trends within corporate health, safety, and environment (HSE) management. The data that actually were collected in these activities depart from the ideal scope according to Figure 11.1. There were filters in the data collection at the six companies that prevented certain types of data on accident risks from being documented:

- *Reports on lost-time accidents:* Data on deviations in the initial phase and on contributing factors were sparse or non-existing.
- *Near-accident reporting:* In the two companies that reported near accidents, the information mainly contained data on technical incidents, deviations, and causal factors. The lack of data on human

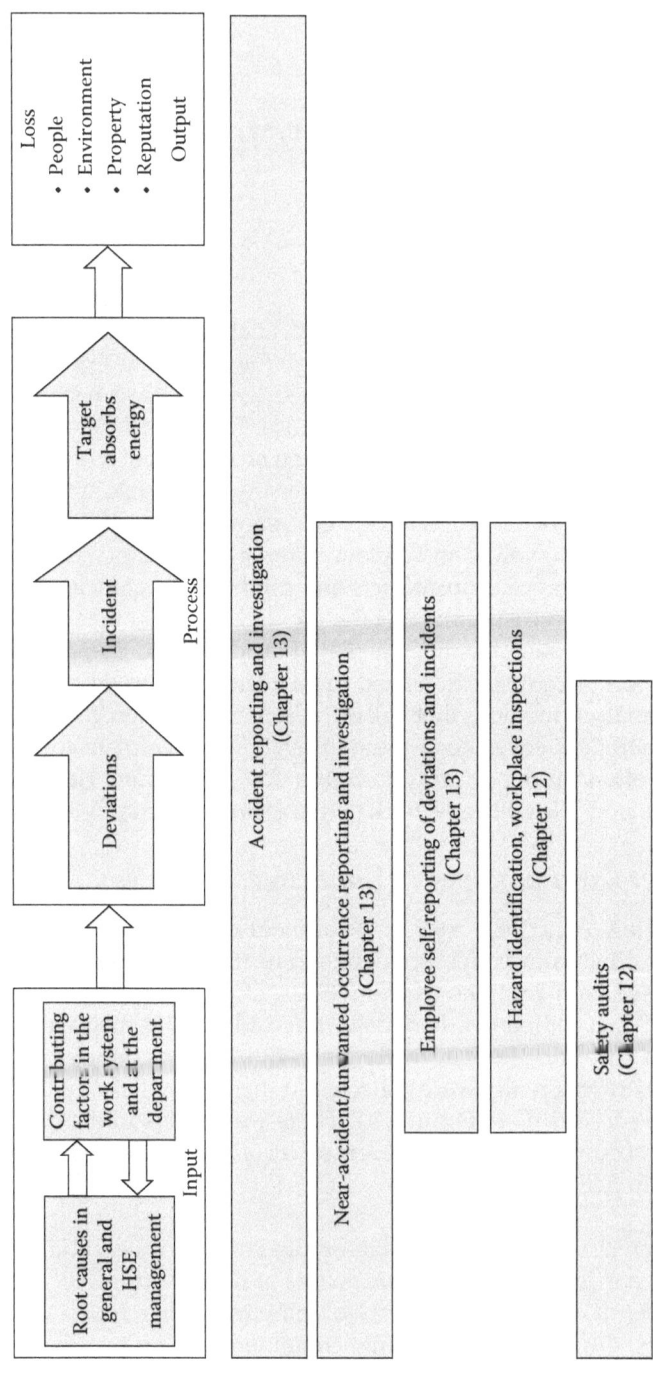

Figure 11.1 Overview of different means of accident risk data collection and their ideal scope in relation to the accident-analysis framework of Chapter 5.

performance was explained by the fact that the reporters' anonymity could not be guaranteed and employee fear of disciplinary action.

- *Workplace inspections:* They functioned as a means of tracing deviations representing unsafe conditions, focussing on technical faults, poor housekeeping, and faulty guards.

These different filters narrowed the scope of the companies' ability to learn from their experiences. The filters limited the types of accident prevention, the remedial actions that were undertaken, and their timing. For example, a focus in accident investigations on the incident rather than on the precursory deviations will direct an investigator towards remedies that stop the energy flow (i.e. protective devices) and not towards improving the 'control climate' (i.e. the management of workplace operation) (Grimaldi 1970). Workplace inspections mainly resulted in corrections of deviations but not in prevention of recurrence, that is, the first order of feedback according to Van Court Hare (Chapter 7).

Let us analyse these findings in relation to Ashby's law of requisite variety (Chapter 7). This law conveys an important message concerning the necessary conditions for us to be able to exercise full control of a system. If the available information on accident risks is limited in coverage, this will restrict the variety of action alternatives taken to control the risk of accidents. The variety of actions may fall well below the critical level, resulting in the risk of accidents remaining at the same level or even increasing. Many opportunities for safety management are lost if the investigator fails to identify how production disturbances affect accident risk.

It further follows from Ashby's law that changes in production affecting accident risk must be detected not only through accident reporting but through other means as well. This is not just because of the need to act before losses occur. Accidents are rare events and will only trigger infrequent data collection. In all practical cases, the rate of data collection and processing will be too low in comparison with the rate of change in production. We are thus dependent on hazard recognition and near-accident and deviation reporting in order to identify these changes quickly enough. It is important that near-accident reporting is efficient in recognising significant incidents and that workplace inspections are carried out with a satisfactory scope and frequency. Changes in production may also be evaluated by risk analysis.

Looking at the 1980s, it is obvious that there was significant improvement potential in many industry safety information systems existing at that time. The author conducted a similar study a decade later (Kjellén 1993). It involved an oil company that had implemented a HSE management system based on International Loss Control Institute (ILCI) principles (Chapter 4). It was interesting to find out whether the systematic application of HSE management principles helped the oil company to overcome some of the filters identified in the earlier study.

In this case, there was an increased focus on the identification of causal factors in accident investigations, and this was accomplished through the support of checklists. The supervisors making the investigations were also responsible for implementation of certain actions, and the oil company had introduced a computerised system for tracking the status of those actions. Interviews and document reviews showed that the supervisors had a tendency to choose causal-factor alternatives that were not possible to verify, resulting in minimum obligations to carry out remedial actions that required a significant effort. A similar lack of variety in the selection of remedial actions was observed in this oil company as was noted in the study of the six Swedish companies a decade earlier. There was no obvious link between the types of causal factors that had been identified and the selection of remedial actions. It follows that the supervisors did not use the rational approach shown in Figure 5.6 in their decisions on accident prevention.

Near-accident reporting, however, had improved dramatically in the oil company, from one near accident being reported per lost-time accident in the two Swedish companies that reported near accidents to 100 near accidents per lost-time accident reported in the oil company. The near-accident reports of the oil company were still dominated by technical events. The reporting of unsafe conditions was a new phenomenon. One explanation for the relatively high frequency of reported near accidents and unsafe conditions was an incentive system that had been established to promote such reporting. The oil company's system for follow-up actions based on the reports was well known among the employees, who used the near-accident and unsafe condition reporting system to resolve working environment problems rather than bringing up these issues with their immediate supervisors.

In designing safety information systems, we must be aware of these filters and biases. The remainder of this section will be concerned with different methods for accident risk data collection and analysis. We will focus on how to accomplish reliable and comprehensive data collection and analysis, taking the knowledge and motive structure of reporters and investigators into account.

chapter twelve

Hazard identification, safety inspections and audits

12.1 Hazard identification

Here, a hazard is defined as a potential source of harm to people, the environment, or material assets. In most accidents, the hazard is a source of energy. The energy model presented in Chapter 4 shows how a transfer of energy to the human body in excess of the body injury threshold causes injury to persons. The science of physics helps us in systemising the different sources of energy (Figure 12.1). The list of types of energy is used in checklists to identify hazards at the workplace and in various risk assessment methods that will be presented in Section V.

A more practical checklist is shown in Table 12.1. It is based on the 'incident type' classification discussed in Chapter 5; each item in the checklist is linked to a source of energy except for loss of a life-supporting environment. An example of the latter is deprivation of oxygen (e.g. through drowning).

Energy is a necessity for everything we do in construction, operation, maintenance, transportation, and so on. It is only when we lose control of the energy (or of the movement of our body in relationship to the source of energy) that incidents occur that can lead to harm. That is why control is so important in preventing accidental loss. In natural catastrophes and in criminal acts, the control of energy is limited or is managed by someone with the intention to harm. It is possible to prevent or reduce loss in these cases as well.

The energy represented by a hazard may be of a different type than that causing loss. In a fall, for example, the hazard is represented by the potential energy of the victim (weight of the victim times distance to ground times gravity), which is transferred into kinetic energy in the fall. This energy is subsequently absorbed by the victim's body and the ground in the impact. The consequences are determined by the amount of energy involved and how this energy is exchanged between the victim and the ground. The severity of the injury will increase with increased distance to the ground and increased concentration of the energy exchange in space and time.

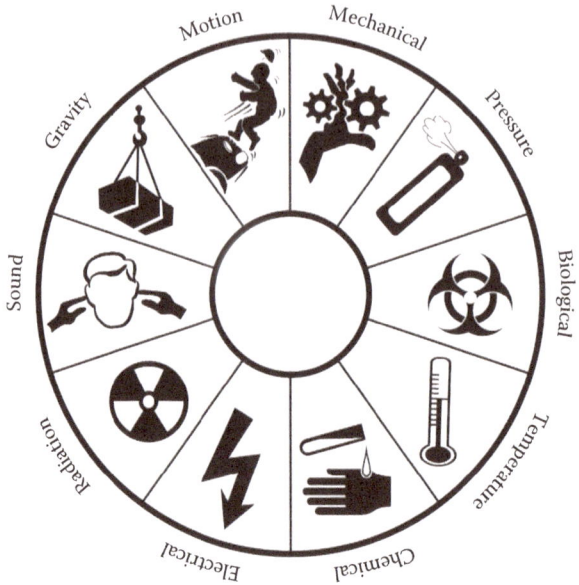

Figure 12.1 Illustration of different sources of energy with the potential to cause harm.

Table 12.1 Practical checklist of incident types used in hazard identification

Contact with vehicle(s) in movement	Fall on same level
Contact with or caught in or between moving parts of machinery	Fall to a lower level
	Hitting against
Release of pressurised gas or fluid	Fire, explosion
Contact with sharp object	Release of/contact with poisonous,
Impact from or squeezed between moving objects	corrosive chemicals
	Contact with electric conductor
Flying object, fragments	Contact with hot or cold surface
Falling object	Loss of a life-supporting environment

The term 'environmental hazard' applies here to acute discharges of harmful substances into the environment. In the hydropower industry, the risk of accidental change in the flow of water in a river system or changes in water level, affecting the biodiversity of the river system, may also be considered as hazards.

Hazard identification (HAZID) is a well-established method in health, safety, and environment (HSE) management. The definition of hazard used in different studies varies from the one used here. It may include not only sources of energy but also inadequate barriers (e.g. lacking guard rails) and other deviations such as unsafe acts, poor housekeeping, and broken equipment (Bahn 2013; Dzeng et al. 2016). Each of these phenomena (barrier failures, deviations) is linked to a source of energy in order to represent a potential for harm. Here, we will apply a stricter definition of hazards in order to avoid including too many different issues and consequently running the risk of defocusing.

Because hazard identification is such a vital activity in HSE management, various observers' abilities to identify hazards in industries such as construction and mining have been subject to research. It has been concluded that length of work experience does not determine the ability of supervisors to identify hazards, for example, in a virtual construction site (Bahn 2013; Perlman et al. 2014; Dzeng et al. 2016). Work experience is not equivalent to the safety-specific experience achieved in safety inspections and training when it comes to the ability to identify and assess hazards. Experienced workers are, however, more confident and use less time to identify hazards (Dzeng et al. 2016). Earlier studies show that hazard identification training has a positive effect on miners' hazard identification skills (Kowalski-Trakofler and Barrett 2003).

Dzeng et al. (2016) make an interesting distinction between obvious and non-obvious hazards in hazard identification. Working at a great height without fall protection is an example of an obvious hazard, and an electric hazard, such as exposing cables to a wet environment, would be a non-obvious hazard example. Results showed that experienced workers used more time to identify non-obvious hazards, but the miss rate was the same as for obvious hazards.

Hazard identification is the first step in a feedback control cycle, which also includes assessment of a risk and design and implementation of mitigating measures. The energy model helps in assessing an identified hazard and its potential to cause harm. It also supports the design of preventive measures. These issues are detailed in Section V. Hazard identification is an important skill among workers and management at the different levels of a company. The skill does not develop through general work experience but must be achieved through training, joint inspections with safety professionals, and through the use of checklists.

12.2   Safety inspections

Inspections are regular activities aimed at maintaining vigilance through adequate control of hazards and deviations at the workplace. The type and timing of inspections are not primarily dictated by the occurrence of events. Rather, such activities are normally integrated as a vital part of the company's scheduled HSE management activities. The type and timing of safety inspections needs to be risk-based. According to Ashby's law of requisite variety (outlined in Chapter 7), workplaces with a significant risk of accidents and those that undergo continuous changes (such as construction sites) need be subject to frequent and systematic inspections.

A main aim of the inspections is to check the technical standards of the workplaces and to identify and correct deviations from regulatory requirements and company standards representing first-order feedback, according to Van Court Hare (Chapter 7).

There are different types of regular inspections. Following are some examples:

- Plant operators inspect and monitor the equipment for which they are responsible on a daily basis. It is, for example, a requirement that crane operators inspect and test certain critical safety functions such as brakes before they start to operate a crane.
- Supervisors and safety representatives co-operate in performing workplace inspections, which represent tangible and repetitive safety practices that stimulate cooperation.
- The technical department or a special plant-inspection department inspects and tests process safety systems such as pressure vessels, piping, safety valves, and gas detectors. The frequency of inspections and tests is determined by the criticality and failure rate of the equipment.
- The fire department inspects buildings and firefighting equipment on a regular basis.
- The scaffolding department performs regular (e.g. weekly) checks on all scaffolding in use.
- The HSE staff inspects each plant department with respect to machine guarding, handling and documentation of chemical substances, and waste handling.
- Management inspections should be performed (e.g. during site visits).

The employer's duty to perform regular workplace inspections is stipulated in the legislation of many countries and in voluntary HSE management standards. According to internal control legislation (in e.g. the Scandinavian countries, the United Kingdom, and the United States), the employer must set up HSE management programs that include routine inspections of the workplaces.

12.2.1   Workplace inspections

We start by looking into the traditional workplace inspections carried out by a supervisor and safety representative, possibly with support from HSE staff. There are documented routines for such inspections in companies that have implemented a formal HSE management system.

A traditional way of defining the *scope of the inspection* is to make a geographical delimitation. The company is partitioned into inspection areas, which usually coincide with the different departments. Each area is inspected at regular intervals by an assigned inspection team.

The theme of the inspection also has to do with its scope. Evaluation research shows that general inspections, where no theme has been pre-defined, only focus on a few types of deviations such as need for repair, missing guards, and poor housekeeping. This narrow scope may be explained by our limited capacity as humans to be attentive to many different items simultaneously. Miller's 'magic number seven plus or minus two' applies here. It means that we cannot expect an inspector to cover more than a maximum of nine items in one inspection. In practice, five items seem to be the limit. Too limited scope, in this respect, violates the criterion of coverage; compare Ashby's law of requisite variety. A common means of circumventing this problem is to select one or more themes for each inspection according to a pre-established plan and to apply theme-specific checklists, as shown in Table 12.2.

Theme-specific checklists are developed in order to secure a reliable and comprehensive mapping of deviations. It is recommended that the checklists be company specific. Typical input to the design of such checklists are regulatory requirements, company standards and experience from accidents, details about near accidents, and previous inspections. An example of a checklist for inspection of fire protection is shown in Table 12.3.

Generally, there are no valid recommendations as to an optimum *inspection frequency*. This will vary depending on the level of risk, the rate of change in production, the age of production, and other circumstances. Weekly inspections may prove necessary, for example, at construction sites where there is a high risk of accidents and a high rate of change. The inspection frequency is thus based on Ashby's law of requisite variety. At the other end of the spectrum are general offices, where yearly inspections may be adequate.

Table 12.2 Example of a checklist of inspection themes

- Housekeeping
- Traffic
- Gangways, escape ways
- Ladders, stairs
- Fire prevention
- Lifting equipment, cranes
- Scaffolds, guard rails, work at height
- Storage and use of chemicals, waste handling
- Machine guarding
- Signs

Table 12.3 Example of a theme-specific checklist (fire protection)

- *Fire-extinguishing equipment* (fire extinguishers, hydrants, hoses, and so on): According to safety plot plan, accessible, signed, fire extinguishers checked at regular intervals and sealed
- *Manual call points*: According to safety plot plan, accessible, signed, functioning
- *Safety showers*: According to safety plot plan, accessible, signed, functioning
- *Emergency exits, escape ways*: According to safety plot plan, accessible, signed
- *Storage of flammable material*: In assigned area, housekeeping, marking of material, fire protection

Different conditions determine the participation of the workplace inspection teams:

- Workplace inspections are a means for the supervisors to follow up on their responsibility for safety.
- One of the participants must have authority to make decisions about remedial actions. Usually the supervisor fills this role.
- The participants must have the necessary knowledge about the specific hazards at the inspected workplaces, about safety requirements, and about available safety measures. Typically, the workplace inspection team is made up of the supervisor, the workers' safety representative, and a safety engineer. Experts may be called in when specific themes are covered such as fire safety, ergonomics, or crane safety.
- The workplace inspections are part of the formal system for cooperation between management and unions on safety issues. The safety representative at the department therefore should participate.

The workplace inspection is *documented* to ensure a timely implementation of measures. At a minimum, this report should include a description of identified deviations and hazards and the measures that have been decided, who is responsible for implementation of measures, and a due date for action. Figure 12.2 shows an example. The reports are distributed to those participating in the inspection and to those responsible for safety measures. *Follow-up* preferably should utilise established management routines. In some companies, it is common to take action on the spot and without documenting the measures. The formal reports from such inspections will not give an accurate picture of the impacts of the inspections.

In workplace inspection systems that function according to the principles of first-order feedback, the same types of deviations will recur. By periodic (e.g. yearly) analysis of the inspection reports, systematic recurrences of deviations will be identified. The group problem-solving

No.	Observation	Evidence	Action	Responsible	Due date	Close out	Evidence
05	Fan guard missing at turbine floor	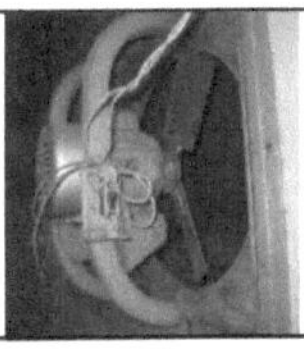	Install fan guard	Maintenance	Sept 13	Sept 10	

Figure 12.2 Example of an inspection item and how it is documented for follow-up and close out.

technique that was presented in Chapter 7 is also applicable to the development of measures to prevent such recurrence.

12.2.2 Management inspections

Management inspection is a tool aimed at demonstrating senior management's concern for safety and reinforcing the safety standard at workplaces. One example is the one-to-one inspection, where a manager makes a safety inspection jointly with the closest subordinate of his/her area of responsibility in a work site. The focus is on how work is carried out, housekeeping, and unsafe acts and conditions. The managers reflect on the current conditions and agree on the present standard and what has to be done.

So-called *one-to-one inspections* aim at strengthening line management's feeling of responsibility for safety. Management at different levels performs inspections of their area of responsibility together with the immediate superior. Here, the focus is on the employees' work practices in order to identify and correct unsafe acts. Deviations related to work at height, use of personal protective equipment, use of equipment and tools, availability and use of safety instructions, and housekeeping are typical findings identified in these inspections. An important aspect is how to communicate and follow up on the findings. A report is written for each inspection and the results are followed up.

Another example of a practical tool with the purpose of reducing the frequency of unsafe acts is the so-called *walk–observe–communicate* (WOC) inspection. Here, managers follow a specific agenda in observing workers and approaching them to discuss their behaviour. The aim is to arrive at a commitment to follow safe work practices. The WOC inspection is announced in advance and it is important to ensure that the inspection is perceived as legitimate by the involved personnel. At the workplace, the manager:

1. Stops and observes people
2. Approaches a worker who has been observed committing unsafe work practices and explains his/her intentions in approaching the worker

3. Asks about the job and how it is performed; safe work practices are praised
4. Asks about possible accidents on the job and the most severe consequences that may follow
5. Asks why the worker used unsafe work practices and about corrective measures
6. Obtains a future commitment from the worker to act safely

12.2.3 Inspection and testing of barrier integrity

Regular inspections and tests are necessary tools to ensure that barriers against fires and explosions are maintained (as discussed in connection with Figure 9.5). Legislation in Europe and the United States make it a responsibility of the employer to ensure that the barriers against major accidents involving chemical hazards are inspected regularly (OSHA 2000; European Commission 2012). The inspections are carried out by experts from the technical department of the plant or from a dedicated plant-inspection department. Table 12.4 lists the typical equipment of an offshore floating drilling and/or production platform that are subject to inspection and testing (Koren 2006). Oil companies have developed criteria for each of these elements shown in the table related to performance, integrity (reliability and availability), vulnerability against accidental loads, and documentation (nonconformity reporting).

The inspection or testing frequency has to be determined. Traditionally, this has been based on the manufacturer's recommendations or on good engineering practice. Reliability analysis is a means of establishing inspection intervals for individual types of equipment. Reliability data serve as necessary input to such analyses. In order to establish a database on barrier reliability, it is necessary to tag each piece of equipment and to record its inspection/test, repair, and maintenance history.

Inspection and testing procedures have to be documented for the different types of equipment and systems. It is important to ensure that overall barrier efficiency is tested. For gas alarms, for example, the complete loop

Table 12.4 Safety systems/barriers on a floating offshore platform

Integrity	Active fire protection
Ventilation	Passive fire protection
Gas and fire detection	Escape and evacuation
Emergency shutdown	Explosion barriers
Open drainage	Well barriers
Ignition control	Ballasting and positioning
Blow down	

has to be tested from the gas detector to the alarm presentation in the central control room. The results of each inspection/test have to be recorded, showing date, name of inspector, equipment tag number, description of inspection/test, results, and actions. In cases where deficiencies are identified, the inspector has to evaluate whether they are of a nature requiring immediate shutdown and replacement of the failing unit. Less critical failure may wait for correction until the next scheduled revision stop.

12.3 Safety audits

Here, an audit is defined as a systematic, independent, and documented process for obtaining audit evidence and evaluating it objectively to determine the extent to which audit criteria are fulfilled (ISO 2011). The focus in this section is on safety audits (i.e. audits of a company's management of safety). This includes, but is not limited to, auditing the company's HSE management system. The aims of an audit may include:

- Determining compliance with regulatory, contractual, or internal safety requirements
- Evaluating the capability of the management system to ensure compliance with applicable safety requirements
- Evaluating the effectiveness of the management system to meet specified safety goals and objectives
- Identifying areas of potential improvements in the management of safety

Audits are a management tool used to create confidence in the audit client that safety is taken care of adequately. The client gives the audit team its legitimacy and authority to make the necessary enquiries. Legislation on internal control and voluntary HSE management standards identify safety audits as an important element in a company's management system.

The auditing of companies' occupational health and safety (OHS) management systems has been subject to evaluation research. The results are inconclusive. Blewett and O'Keeffe (2011) concluded that OHS management system audits are inadequate and lack the nuances necessary to evaluate systematic approaches to HSE management. The authors' conclusions were based on evidence from a literature review that formal OHS management systems did not by themselves make workplaces healthy and safe. Hedlund (2014), on the other hand, found that South African manufacturing companies committed to the NOSA Five Star health and safety auditing program experienced fewer fatalities and permanently disabling injuries and that the number of stars received correlated with lower injury rates.

Blewett and O'Keeffe (2011) and Robson et al. (2012) point at deficiencies in the auditing process including limitations of standardised audit criteria such as those used in the International Safety Rating System (ISRS), varying skills of auditors and poor inter-audit reliability, and discrepancies between actual audit practices and International Organization for Standardization (ISO) 19011.

The approach taken here is holistic, where auditing of a company's HSE management system is only a part of the scope of a typical safety audit. We advocate a risk-based approach, focusing on areas in the management of the operation that are experienced as challenging in relationship to the company's exposure to risks.

The composition of the audit team is also critical. An audit team will typically consist of three to four members with expertise in the following fields: (1) management system auditing, (2) health and safety management, (3) technical expertise in a main focus area of the audit, (4) line management expertise from an organisation similar to that being audited, (5) knowledge about national HSE regulations and authorities, and (6) local language skills. It is preferable that all members have some basic understanding about the type of facility or project being audited.

12.3.1   General guidelines for the auditing of management systems

Our presentation is based on the ISO 19011 guidelines for auditing management systems (ISO 2011). This standard is part of the ISO 9000 family of quality management standards. It gives guidance on the planning, execution, and reporting of audits of all types of systems related to the management of quality, environment, health, and safety.

There are different types of audits with different objectives for the client in each case (Figure 12.3):

1. *First-party (internal) audits* aim at evaluating whether the company's own management of HSE meets regulatory requirements, recognised national or international standards, company requirements, and so on. The client is often top management or the working environment committee.
2. *Second-party (customer) audits* may be part of a plan for prequalification of suppliers, where there is a desire to establish a contractual relationship. Alternatively, they are carried out after the contract has been signed and work has started, when the customer wants to verify that the supplier's management of the work undertaken continues to meet the HSE contractual requirements.
3. *Third-party audits* are conducted by an independent organisation such as a certification body. The aim of inviting a recognised third party is often to verify conformity to a specified standard such as ISO 14000, Occupational Health and Safety Assessment Series (OHSAS) 18001, or American National Standards Institute (ANSI) Z10.

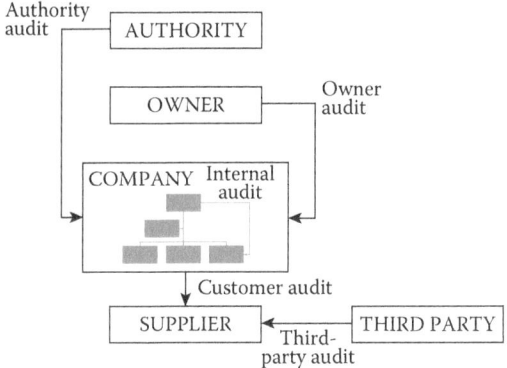

Figure 12.3 Different types of audits.

Certification can be a useful tool to add credibility by demonstrating that a particular product or service meets the expectations of customers. For some industries, certification is a legal or contractual requirement. Following an audit with acceptable results, the third party is a guarantor for the quality of the audited company's HSE management system.

It is important that all participants of an audit recognise the way it is carried out and have trust in the audit team. The ISO 19011 standard is an important tool in standardising the way audits are carried out and in defining the required competence of individuals involved in the auditing process (ISO 2011). Following are some audit terms from ISO 19011 that we will use here:

Audit client	Organisation or person requesting an audit
Auditor	Person with the competence to conduct an audit
Auditee	Organisation being audited
Audit criteria	Set of policies, procedures, or requirements used as a reference against which audit evidence is compared. They include internal requirements as well as external factors such as laws and regulations, ISO standards, and so on
Audit evidence	Records, statements of fact, or other information that are relevant to the audit criteria and are verifiable

Continued

Audit findings (observations)	Results of the evaluation of the collected audit evidence against audit criteria
Nonconformity	Non-fulfilment of a requirement
Corrective action	Action to eliminate the cause of a detected nonconformity or other undesirable situation

The *audit criteria* define the requirements against which the HSE management system is audited. In general, there are three different types of such documents:

1. Laws and regulations
2. Recognised international standards such as ISO 14001, draft ISO 45001, or ANSI Z10 (ANSI 2012; ISO 2015, 2016)
3. The auditee's internal governing documents such as HSE policy and goals, contractual requirements, action plans, job descriptions, methods statements, safety rules, and so on

Safety audits apply the same basic principles as those of quality audits (ISO 2011). The following apply (Figure 12.4):

1. Each individual audit is part of an audit program, usually annual. This is where management defines the priorities and goals for the audit program.
2. In the area of safety, the audit program needs to be risk-based and address the most critical risk according to the company's risk register. A bottom-up approach is recommended, where the individual unit carries out its risk assessments and audit planning as part of the yearly planning and budget process. It is then up to the person managing the audit program in liaison with the main stakeholders to make the necessary priorities and adjustments to reflect the overall risk picture.
3. The implementation of an individual audit starts with the establishment of audit objectives and scope. For safety audits, this often involves a review of incident statistics and other observations on risks at the site as well as documentation on significant nonconformities in the management of HSE. Interviewing key personnel such as the local management, HSE manager, safety representative, and so on

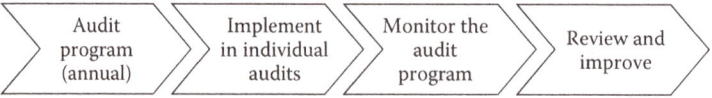

Figure 12.4 Flow in the management of an audit program.

about key concerns is also recommended. It is, however, the duty of the person managing the audit program to establish audit objectives independently and in accordance with the audit plan objectives.

4. A need to review and modify the audit program may come up as a result of the continuous monitoring of its implementation. The audit results may call for changes in the area of safety, but changes in other conditions – such as a severe accident or a change in the exposure to risk – may call for changed priorities.

Figure 12.5 shows the basic steps of an individual audit. The following apply:

1. *Initiation:* Usually, the audit is carried out according to a yearly program approved by the company's board. It can also be carried out on behalf of the board on a short-notice basis due to specific circumstances (e.g. a severe accident). The audit team leader and other members are appointed and an audit plan is established. This includes the organisation to be audited, audit objectives and scope, audit criteria, audit team, and proposed schedule.

2. *Preparing for the audit:* The audit team starts with a review of documentation including relevant parts of the HSE management system, organisation and job descriptions, incident reports, regulatory requirements, other relevant audit criteria, and so on. Based on this and on previous experience, the team prepares checklists. The audit team also develops a schedule for onsite activities, prepares checklists for the collection of information, and sends a notification to the audited organisation (auditee).

3. *Conducting onsite audit activities:* The audit at the site starts with an opening meeting. This meeting is brief, and the main purpose is to confirm the schedule and practicalities during the onsite visit. It is followed by the collection and verification of information through interviews, visual inspections, and document reviews. The audit team summarises its findings and will debrief the audited organisation in a closing meeting. In this meeting, the focus will be on conditions essential to the purpose of the audit and on needs for improvements in the management of HSE. Fault-finding should be avoided.

4. *Reporting:* The audit team prepares a report with their conclusions. Here, any nonconformity is identified together with references to the audit criteria. There is no generally accepted way of classifying audit conclusions, but Figure 12.6 shows an example. The *audit report* conveys the results and recommendations (including nonconformities and other observations) and the audit team's judgement of the audited systems for the management of HSE. This report is

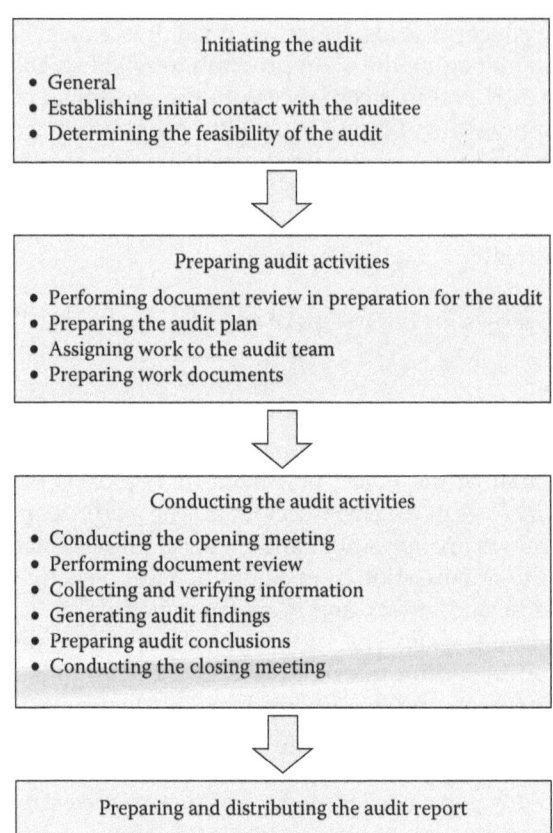

Figure 12.5 Basic steps in an audit. (Reprinted from ISO, *Guidelines for Auditing Management Systems*, ISO 19011:2011, International Organization for Standardization, Geneva, Copyright 2011, with permission from the International Organization for Standardization.)

an important record for experience transfer and shall meet requirements as to thoroughness, accuracy, and objectivity. A draft version of the audit report is sent to the auditee for comments.

5. The final version is approved and distributed to the client (e.g. chairman of the board), the auditee, and the members of the audit team. Further distribution has to be approved by the client.

6. Audit follow-up is the responsibility of line management in the audited unit. They will review all conclusions (classified, for example, as critical, not satisfactory, improvement needed, or satisfactory as shown in Figure 12.6) and plan necessary remedial actions (i.e. how to correct/eliminate the cause, with responsibility and a deadline).

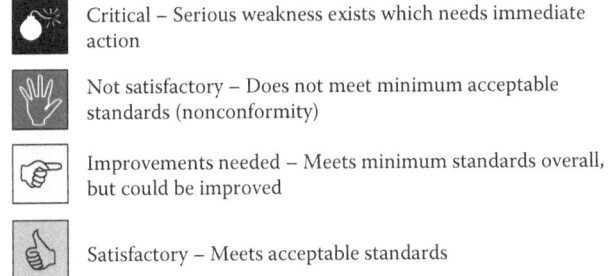

Critical – Serious weakness exists which needs immediate action

Not satisfactory – Does not meet minimum acceptable standards (nonconformity)

Improvements needed – Meets minimum standards overall, but could be improved

Satisfactory – Meets acceptable standards

Figure 12.6 Examples of symbols for classifying audit conclusions.

7. An overview of the status must be documented and available, including updates. The status of actions for handling of deviations may be followed in the company's system until these are closed out.

The individual audit team members need to be independent of the organisation being audited. This excludes, for example, individuals reporting to the operating manager of the auditee. The obvious reason is to avoid possible questions about the impartiality of the team members and the objectivity of the audit conclusions. Assessing independence is always a crucial issue and may be difficult to fully accomplish, especially in small organisations where this criterion has to be balanced against finding individuals with the right qualifications. It is important that the audit team leader is aware of any conflicts of interest and that he/she encourages objectivity.

Another critical issue is the process of arriving at conclusions (see Figure 12.7). Indications of possible nonconformities are noted during the interviews, inspections, and so on and re-reviewed if significant, even if they are not on the checklist for the audit. Information from the interviews needs to be verified by using independent sources – for example, other interviews, inspections, measurements, and registers (triangulation). The audit conclusions (nonconformities and other observations) must be derived from verifiable facts – even if there will, in most cases, be scope for interpretation and judgement by the audit team. This is why the qualifications and the credibility of the audit team are so important.

In the closing meeting, findings (observations) from the audit are presented together with the audit team's general impression. The purpose is to clarify errors, resolve misunderstandings, and to prepare the audited organisation for the audit report. The closing meeting is an important step in creating trust and acceptance of the results.

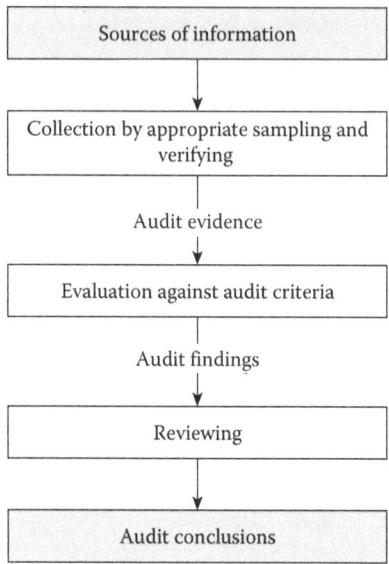

Figure 12.7 Process of arriving at conclusions. (Reprinted from ISO, *Guidelines for Auditing Management Systems*, ISO 19011:2011, International Organization for Standardization, Geneva, Copyright 2011, with permission from the International Organization for Standardization.)

12.3.2   Example: Safety audit of a hydropower plant during refurbishment

A hydropower plant was subject to a major refurbishment of its two 50 MW units. One unit was to be refurbished at a time, while the other continued to produce power to the grid. The plant used a permit-to-work (PTW) system (Chapter 9) to ensure that refurbishment work was executed in a safe way and did not interfere with hazardous energies of the unit in operation.

The owners had initiated an audit to check that the plant complied with regulatory and company safety requirements and that the management of the plant and the refurbishment project were efficient in ensuring safe execution of the project when being conducted in a plant in operation. The owners appointed an audit team consisting of one of the owner's HSE managers, an expert on operations and maintenance of power plants, and an external consultant with competence in HSE regulations and standards, auditing, and local language and customs.

After preparation, notification, and the opening meeting, the audit team conducted interviews with 16 persons from plant and project management as well as individual operators. The audit team also conducted inspections and document reviews.

In the preparatory work, regulatory requirements and the plant's internal regulations were scrutinised to adapt audit checklists from a library of thematic checklists. One of the focal areas of the audit was the plant's PTW system. Following are examples of PTW items in the checklist for information collection:

- Are the routines and responsibilities for work/entry/lock out–tag out permits clearly defined in procedure(s)?
- Is the responsibility for safety at work on systems involving high energies clearly defined?
- Is information about what jobs require a permit clearly defined?
- During work planning, has it been confirmed that certain jobs require permits?
- Are routines established for application and acceptance of work requiring permits? For defining the necessary barriers?
- Has it been confirmed that contractors have the necessary qualifications and instructions to work safely according to the permits?
- Has it been confirmed that the defined measures are implemented before the start of work? At the right place?
- Does the forms for work/entry permit include identification of systems, checklists of hazards, times for start and stop of work, safety measures/barriers?
- Is an overview of active permits available to the shift supervisor at all times? Is there a defined time span during which permits are valid?
- Is the system checked through inspections and audits?

The results of the audit of the PTW system are shown in Table 12.5. The observation was classified as nonconformity because it revealed conditions at the plant that did not meet the requirements. The observation was formulated as a conclusion, followed by seven verifiable findings that substantiated it. This conclusion and the underlying findings were formulated in such a way that they indicated expectations for changes needed in the PTW system to accomplish compliance. This replaced the need for recommendations.

The audit observation points to some serious deficiencies in the way the PTW system has been implemented in the plant in connection with the refurbishment work. One of the key principles in the refurbishment work is that the overall responsibility for the management of safety in the plant including the coordination of all work is uniformly defined and lies with operations. The audit revealed that operations had not taken full responsibility for this task. Operations and maintenance managers could not give a clear account of their responsibilities, the shift supervisor in the control room did not have full overview of active PTWs, and the operations department had not adequately considered the manning needs for administration of the PTW system during the refurbishment work.

Table 12.5 Audit observation of the permit-to-work (PTW) system at the plant

No.	Control	Topic: Management of contractors and sub-contractors
X		The site permit-to-work (PTW) system is not adequately implemented and followed up to ensure proper implementation of the required occupational health and safety standard:

a. Interviewed operations and maintenance managers do not give an unambiguous account of who is responsible for implementation of the PTW system at the plant.
b. There is no overview in the control room of active PTWs in the plant.
c. The PTW No. 168 for refurbishment of the penstock has been signed by the maintenance manager, although the shift supervisor should sign it according to the PTW procedure. It has not been renewed for the day when the audit team inspected work.
d. 'Use complete PPE' is identified as standard 'Job Safety Control' in the JSA [job safety analysis] for rehabilitation of transformers. In this work, paint of the following type (....) was applied. Even if 'fumes' had been identified as a hazard in the JSA, dust masks were used instead of masks efficient to protect against fumes, as specified in the Material Safety Data Sheets for the paint. Interviewed storage foreman could not produce adequate fume masks to the auditors.
e. There is no documented check and sign out in the PTW that identified measures have been implemented before start of work.
f. There has been no assessment of manning needs in operations in connection with refurbishment work to supervise the PTW system.
g. The plant has no routines for regular inspections/auditing of the PTW system.

Audit criteria	Regulatory OHS Standard No. 1083; OHSAS 18001: §4.4.6; Plant OHS procedure No. 20

Another serious issue revealed by the audit was the lack of control of individual task execution (i.e. through the use of JSA). Remedies were filled in in a mechanistic way without any serious consideration for what measures were required to protect against the hazard.

The plant responded by clarifying responsibilities and changing PTW routines. They also implemented new routines to verify the quality of remedial actions initiated by the JSAs. Manning for inspection of jobs governed by PTWs was increased.

12.3.3 Application of SMORT in audits

The Safety Management and Organisation Review Technique (SMORT) system of checklists will be introduced in Chapter 13 in connection with incident investigations. SMORT is also suitable for use as a tool in the planning and execution of safety audits. The steps of a SMORT investigation follow the basic procedures of an audit. The SMORT checklists and questionnaire give support in the development of detailed lists of questions for the audit. The questions must be detailed based primarily on the scope and audit criteria of the actual audit.

Principles for application of SMORT checklists in the planning of safety audits:

- Start by using the SMORT checklist in order to identify and delimit the areas to be covered according to the scope of the audit.
- Identify the audit criteria (regulatory requirements, standards, internal specifications, procedures, and so on).
- Identify the persons to be interviewed and documents to be reviewed.
- Develop detailed questions for each area on the basis of SMORT questions and the audit criteria.

Usually, it is possible to plan the audit in detail in advance and no subdivision into phases is required as in the case of accident investigations. There is a clear difference between use of SMORT in accident investigation and in audits when it comes to the presentation of results. Safety audits are closely linked to requirements as stated in various documents. When presenting *nonconformities*, the reference documentation should always be cited. The audit team also has the freedom to present other types of *observations* where their own judgements go beyond what is clearly specified.

chapter thirteen

Incident reporting and investigation

13.1 Why investigate incidents?

This chapter takes the need for an industrial organisation to report and investigate incidents (i.e. accidents and near accidents) as the starting point. The principles described here also address the needs of other types of organisations conducting incident investigations such as the authorities, insurance companies, and public parties. We will focus on incidents resulting in, or with the potential of resulting in, injury or death to person(s). We will also elaborate on the reporting of unwanted occurrences – a concept that includes near accidents (i.e. incidents without loss) as well as unsafe acts and conditions.

There are different reasons why an industrial organisation establishes and maintains routines for the reporting, investigation, and documentation of incidents:

- Meeting regulatory requirements. Most countries have mandatory requirements for the investigation and record-keeping of occupational accidents and for reporting these occurrences to the authorities (see Section 2.2).
- Compensation to the victim. An occupational accident has to be reported to the national social security agency or to the insurance company in order to ensure that the victim receives compensation for lost wages, medical bills, and for suffering (see Section 2.2). Also, accidents resulting in material or production losses have to be reported to receive compensation from the insurance company.
- Because of the people affected by the incident. Especially in case of accidents with severe loss, the victim and his or her next-of-kin will suffer from the serious trauma. It is important for a company to respect those involved by organising investigations and follow-up to demonstrate that they care and wish to avoid any re-occurrence of the accident in the future.
- To learn from the occurrences in order to continually improve conditions. Accidents and near accidents are unplanned and unwanted events associated with loss. At the same time, they represent

opportunities to learn about workplace hazards and about weaknesses in the systems for the control of such hazards (Kletz 1994;
Drupsteen and Guldenmund 2014). Near accidents may also provide
important information about losses that have been avoided by operators through proper recovery actions. By using such experiences
properly, the organisation will be able to improve its performance
in the area of accident prevention (Hollnagel 2014). In this chapter,
we will focus on learning and improvements.

- Monitoring of safety performance. Accidents and near accidents are
  utilised as input in some commonly used measures of safety performance. We will return to this aspect in Section IV.
- Creation of positive safety attitudes and alertness. Severe accident
  occurrences will shape the accident perceptions of those concerned.
  A well-functioning program for incident notification, investigation,
  and follow-up will support the development of a positive climate for
  safety-related behaviour and awareness.

There are also disciplinary and juridical reasons for the investigation
of accidents. Here, we consider rule violations and criminal negligence. It
is the employees' duty to follow work and safety rules. An accident investigation often will reveal violations of such rules. We also have to consider the legal consequences of the fact that safety is a line-management
responsibility. Supervisors and managers at higher levels may be held
accountable for accidents in cases where an investigation reveals substandard practices or conditions for which management is responsible.

It is the duty of the police to investigate whether criminal negligence
or deliberate violation of the law has taken place. In this chapter, we will
focus on the learning aspects of accidents rather than on the allocation
of guilt or blame. Investigations aimed at identifying persons that have
failed will induce self-protective behaviour on the part of the personnel
involved, such as the hiding and distortion of information. It will also
affect the willingness of employees to report accidents and near accidents in the future. Investigations with such intent should, if possible, be
avoided altogether or conducted separately from investigations focused
on learning, monitoring, and motivation.

13.2   The steps in the investigation

In Chapter 7, we defined an incident investigation as a diagnostic process,
involving: (1) notification and recording of the event, (2) investigation,
(3) decisions about remedial actions, and (4) implementation and follow-
up. The investigation itself can be divided into different steps, including
mapping the sequence of events (fact-finding), identification of causes,
development of remedial actions, and reporting. All investigations have

to go through these four phases in order to close the loop (i.e. to ensure that the experiences are utilised to reduce the risk of accidents).

We find different terms in the literature for the process from the initial notification and recording of an incident to the implementation and follow-up of remedial actions. Examples are operating experience (OE) process (IAEA 2007), operating feedback system (OFS) (Dien et al. 2012), chain of accident investigation (CHAIN) steps (Lindberg et al. 2010), and incident learning system (Jacobsson et al. 2012). Different models of the process in the literature identify a sequence of steps or activities that are required for reporting and investigating incidents and disseminating results; see Table 13.1.

The different models vary in scope. The U.S. Department of Energy (DOE) investigation workbook (2000) is a manual for in-depth investigations and focuses on this step. In contrast, the OE process devised by the

Table 13.1 Overview of the steps in different models of the incident investigation process

Source	Report	Investigate	Decide	Implement and follow-up
Investigation workbook (DOE 2000)		Appoint investigation board, collect data, analyse, develop conclusions, report		
OE process (IAEA 2007)	Identify and report	Review and investigate	Corrective actions	Utilisation and dissemination of OE
Typical steps in eight studied investigation manuals (Lundberg et al. 2009)		Plan and initiate, collect data, analyse, recommend	Adjust	Implement Follow-up
OFS (Dien et al. 2010)	Detect and identify	Collect data, analyse event, define measures	Decide	Implement, assess and validate, store, and disseminate
CHAIN (Lindberg et al. 2010)	Report	Select, investigate, disseminate		Prevent
Incident learning system (Jacobsson et al. 2012)	Report	Analyse	Decide	Implement and follow-up

International Atomic Energy Agency (IAEA 2007) emphasises the value of the collective experience achieved through investigations into incidents and dissemination not only within the plant but also to external organisations. It has been designed for capturing experiences from events in nuclear power plants in order to promote a standardised approach and to facilitate sharing experiences within the nuclear industry.

Lundberg et al. (2009) base their analysis of the investigation process on a review of the incident investigation manuals of two companies and six authorities in Sweden. It shows the typical steps used in these investigation manuals.

Dien et al. (2012) emphasise the follow-up step and the need for assessing and validating the results similar to the principles of Deming's plan–do–check–act (PDCA) cycle (2000). The CHAIN model (Lindberg et al. 2012) and the incident learning system according to Jacobsson et al. (2012) analyse the sequential incident investigation process, which is similar to the model of the incident investigation process used here.

13.3   Investigations at three levels

In safety practice, we will not be prepared to follow the same procedure and use the same amounts of resources on investigations every time an incident occurs. Certain priorities have to be established in order to focus on the vital incidents that offer the most significant opportunities for learning. In this section, we will describe three different levels of goals in learning from incidents based on the actual or potential consequences of the incident. All three levels will apply the same four basic steps in the incident investigation process that were described in the previous section. The steps will be more elaborate depending on the level of investigation. At the highest level, an in-depth investigation by an independent team similar to the approach used in DOE's (2000) investigation workbook will be applied based on the model of typical audit activities according to ISO 19011 (ISO 2011).

A similar approach is applied by the IAEA in their *Best Practices in Identifying, Reporting, and Screening Operating Experience at Nuclear Power Plants* (2007). Events are classified in three categories and, for the most severe category, a team is brought together to perform a root cause analysis.

Figure 13.1 shows the initial steps after an incident and the decisions on the type of investigation. The criteria used in the decisions on type of investigation are shown in Table 13.2. They are described as follows:

1. All reported incidents are investigated immediately at this first level by the supervisor and safety representative.

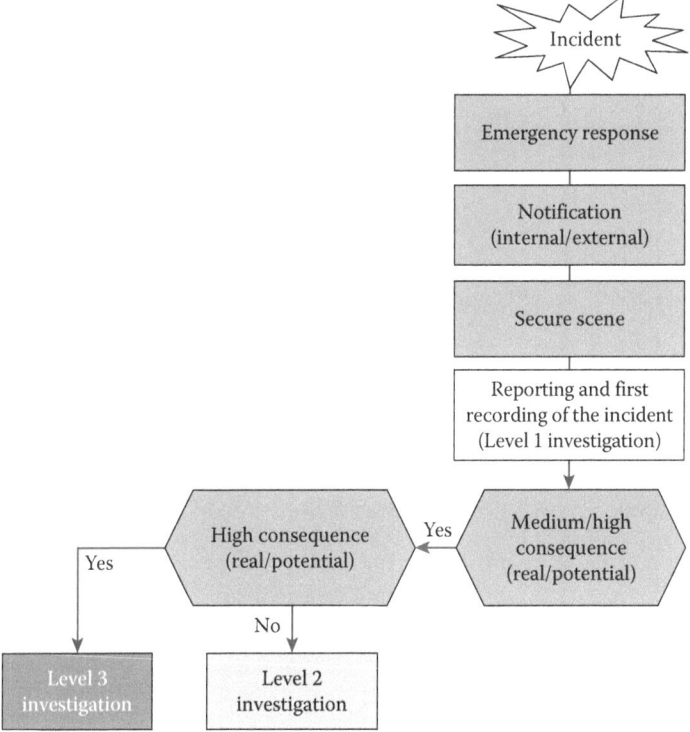

Figure 13.1 Investigation of incidents at three levels.

2. Recordable accidents, frequently recurring incidents, and incidents with moderate loss potential (e.g. sick leave beyond a few days) are subsequently investigated internally by a problem-solving group.
3. On rare occasions, when the actual or potential loss is high, an independent team carries out the investigation. This includes accidents resulting in fatality or permanent disability (>50%). It also includes a selection of so-called high-potential (HIPO) incidents (i.e. incidents with the potential to result in a fatality or serious injury). The selection of HIPO incidents to be subjected to a Level 3 investigation is based on the energy involved and the potential for learning.

The methods used in Levels 2 and 3 will coincide in the first phase of the investigation, which focuses on fact-finding related to the sequence of events and the analysis of deviations and barrier analysis. A Level 3 investigation differs from a Level 2 investigation in that it is carried out by an independent team, and it goes further into the investigation

Table 13.2 Criteria for when to conduct a Level 2 or 3 investigation of accidents resulting in loss

Investigation level	Consequence	Injury to persons	Damage to the environment	Asset damage
Level 1 – Nonconformity management	Low	First-aid injury	Insignificant damage; restitution time < 1 month	Damage with low consequence
Level 2 – Internal investigation	Moderate	Recordable injury	Moderate damage; restitution time 1–12 months	Moderate consequence
Level 3 – Independent investigation	High	Permanent disability (>50%); fatality, including multiple fatalities	Major damage; restitution time > 1 year	High consequence

of root causes in general and the health, safety, and environment (HSE) management at the company. The quality requirements for the report are higher, and this involves the use of a systematic approach in order to arrive at conclusions based on facts. Here, we will apply ISO 19011 quality auditing techniques to ensure that the conclusions are based on transparent interpretations on facts (ISO 2011). A Level 3 investigation is defined similarly to that of quality audits in ISO 9000 as 'a systematic and independent examination to determine whether the accident prevention activities and related results comply with planned arrangements and whether these arrangements are implemented effectively and suitable to achieve objectives (ISO 2015)'.

We will expect different outcomes from the investigations at the three levels, roughly corresponding to the levels of feedback according to Van Court Hare (see Chapter 7). This means that the supervisor's first report (Level 1) corresponds to the first order of feedback (i.e. correction of deviations). The internal investigation (Level 2) will be concerned with the needs of changing contributing factors at the workplace and in the department, roughly corresponding to the second and third orders of feedback. An independent investigation (Level 3) will look at these types of measures as well but will have the duty to evaluate root causes and to come up with recommendations corresponding to the fourth order of feedback (i.e. double-loop learning addressing values, assumptions, and policies, as opposed to single-loop learning, where only the specific situation or process is improved) (see Argyris 1992; Drupsteen and Guldenmund 2014).

The decision to proceed with the investigation at Level 2 or 3 is made by higher management. An alternative approach is to appoint a screening group with representatives of management, employees, and HSE staff, which has the authority to decide about the level of investigation.

The rationale for a three-level approach is:

- Ensuring adequate focus and use of resources on events that have a high learning value and, at the same time, ensuring that all incidents are investigated. The three-level investigation routine meets this dual goal.
- Allowing for local use of incident experience. Accidents and near accidents are rich in information, and it is not possible to record all this information for use by higher-level management and staff officers involved in decision-making. Problem-solving has to take place in a group setting, where there are members with direct experience of the context in which the events have occurred.
- Ensuring an adequate investigation into sensitive issues. There often are circumstances surrounding an incident that are sensitive, especially those related to the distribution of responsibilities and to the safety culture at the workplace and in the company. We cannot expect the involved parties to make a comprehensive and unbiased investigation into such factors. When it is important to delve deeply into these factors, especially in cases of severe consequences, there is a need to ensure that the investigation is carried out by an independent body.

  Example: There was a severe accident at a yard during maintenance of a crane. A worker fell 6 m from the crane boom. This accident triggered setting up a local investigation team and an independent investigation team. The local team focused on the immediate causes of the accident and the absence of adequate fall protection. They proposed changes in the work instructions to ensure that the necessary safety precautions were taken. The independent investigation team looked into these conditions and into root causes as well. It identified an acceptance by management of unsafe behaviour and lack of routines for plan and executing new types of jobs involving hazards with high potential (in this case, the risk of falls). Thus, the recommendations by the independent team were more far-reaching. Top management decided not only to request a follow-up of the independent team's recommendations at the workplace, all other departments with similar types of jobs had to report on how they implemented those recommendations.

13.3.1 *Applying SMORT in investigations at three levels*

The Safety Management and Organisation Review Technique (SMORT) initially was introduced in Chapter 4. It is presented here as a tool for use in in-depth investigations of accidents and near accidents. This tool was originally published by Kjellén et al. (1987) and was based on the Management Oversight and Risk Tree, or MORT (Johnson 1980). It eventually was revised based on more recent experience from Level 3 investigations and on elements from quality and occupational health and safety (OHS) management guidelines according to the ISO 9000 standard family and draft ISO 45001 (ISO 2015/2016). A SMORT accident investigation follows a diagnostic process where the findings at one SMORT tier are used as input to the analysis of causal factors at higher tiers. This is illustrated in Figure 13.2. The different SMORT checklists are demonstrated in an examples presented in Sections 13.6 and 13.7. The complete SMORT questionnaire is presented in Appendix B. These tools are based on many years of experience from accident inspections and from HSE management in general.

The SMORT checklists and questionnaires are open-ended and have the limited purpose of focusing an investigation team's attention on

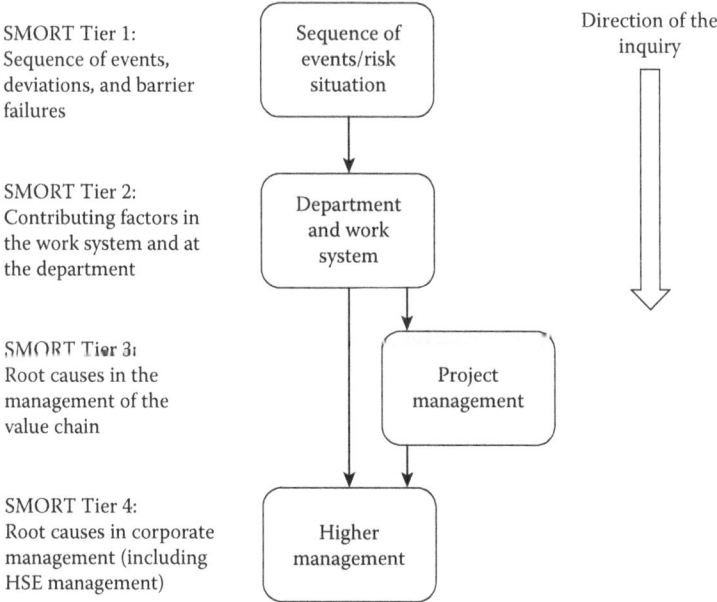

Figure 13.2 The SMORT tiers of investigation and the direction of the inquiry.

various problem areas. Examples of activities and remedial actions at each tier of a SMORT analysis are shown here:

SMORT Tier 1: Initial investigation of the accident sequence or potential accident scenario and identification of deviations and barrier failures at the workplace. This may involve a review of the accident investigation by the first-line supervisor and interviews and observations at the workplace. The immediate use of the findings involves the elimination of deviations at the place of occurrence.

Example: A person inspects roofing tiles. He leans against a guard rail at the end of the roof. The guard rail is not adequately secured. It breaks and the person falls and is injured. The immediate cause of the accident is a faulty guard rail.

SMORT Tier 2: Identification of risk factors at the workplace and department. The results are used to improve the work organisation and production system.

Example (continued): A further analysis of the incident reveals weaknesses in the design of the guard rails and in the routine for workplace inspections.

SMORT Tier 3: Identification of how risk factors have occurred in production through a review of the organisation and the routines for design, construction, and start-up of new production systems. Remedial actions at this tier include improved organisation and routines for project work.

Example (continued): A review of the design activities shows that the requirements for holders for guard rail poles have not been adequately specified in the design of the roof.

SMORT Tier 4: Identification of how risk factors have occurred through a review of supervision and management, including the safety management system. Finally, remedial actions at this tier involve improvements in supervision and safety and in general management organisation and routines.

Example (continued): An overall program that defines the frequency and scope of workplace inspections to identify such deviations as inadequate guard rails is missing.

13.4   Reporting and first recording of incidents

The first reporting of an incident is intended to bring it to the attention of the person(s) in the organisation responsible for the investigation and documentation of such events. Usually, this is done by a person involved in the incident (victim or a witness) or his/her supervisor. The organisation has to determine what types of incidents shall be subject to reporting and documentation (i.e. the reporting criterion). Another important aspect is the reliability of the reporting. These two aspects are interrelated.

The time it takes from when the incident occurs until it is reported also is an important aspect – both with respect to the negative effects of delays on the quality of the investigation and on the needs for quick response with remedial actions (refer to Ashby's law of requisite variety).

The *reporting criterion* defines the types of events to be reported. This criterion may be formalised and also may be informal and based on a shared understanding within the organisation. Formalised reporting criteria for accidents involving loss usually are defined on a consequence scale of measurement such as whether the accident involves lost time after the day of the event or not (yes/no). This particular criterion is relatively simple to communicate and apply because it also has implications for the victim's right to compensation for sick leave. A problem is its limited coverage. Minor incidents with potentially severe consequences will pass undetected, and the organisation will miss opportunities to learn from such experiences. A commonly used reporting and recording criterion includes all accidents resulting in medical treatment and transfer to a different job assignment.

Alternatively, we may base our reporting criteria for incidents on subjective judgements of potential losses. Such criteria involve problems of inter-subjectivity. Different persons may have varying opinions on what constitutes a reportable incident. Because consequence usually is related to the amount of energy involved in the incident, reporters may be trained to make more reliable judgements.

This brings us to the question of *reporting reliability*. This is expressed on a scale from 0% to 100% and is defined as the number of reported incidents in relation to the 'true' number of incidents (as defined by the reporting criterion). The reporting criterion and the reliability of the reporting are closely linked. To achieve a high degree of reliability in the reporting, well-defined and easily understood criteria have to be applied.

The reporting reliability is also affected by the severity of the event. We may expect a high reporting reliability when applying reporting criteria corresponding to a 'high threshold' on the consequence scale. Reporting of fatalities is an obvious example. It is unlikely that a fatal accident at work passes undetected by the organisation, and the reporting reliability is high. There will, on the other hand, be few events that meet such a reporting criterion, and the number of opportunities for learning will be very low. At the other end of the consequence scale, we find such criteria as 'to report all accidents and near accidents'. In this case, it is up to the individual employee to determine whether or not an event is significant enough to be reported. It follows that we would expect a very low and even undefined reporting reliability. When we apply a low reporting threshold, many 'trivial' events with a low learning value will be reported. We are likely to face information overload that ultimately will discredit the system. To counteract this problem, an efficient system for the classification of reports with respect to potential severity is crucial.

Measuring the reliability of accident reporting is not a straightforward process because the dominator is difficult to establish. A commonly used method is to check independent sources for recordable events. Examples of such sources are:

- Interviews of employees about incident experience at the workplace
- Records of treatments at the medical clinic
- Records of absence from work
- Reports on repair of vehicles, machinery, and buildings (e.g. after fire)

13.4.1 Problems of underreporting

Studies in the Scandinavian countries have shown that reporting accidents to the authorities is unreliable and that on average about 50% of reportable accidents are never reported (Kjellén et al. 1986; Swedish Work Environment Authority 2005). It is reasonable to assume that the accidents that are not reported to the authorities are not documented and followed up at the company level either. This means that important experience is lost.

These conclusions are supported by studies from the United States indicating underreporting on the order of magnitude of 50% or more (Weddle 1996; Azaroff et al. 2002). Workers stated reasons for not reporting such as that the injury was too minor to be reported and that they were afraid of being judged as careless or accident-prone. Older workers and those who had worked for the same employer for a long time were less willing to report injuries. Leigh et al. (2004) estimated underreporting to the US Bureau of Labour Statistics to be between 33% and 69% of all recordable injuries.

The UK Health and Safety Executive reports similar figures for underreporting. By comparing records for patients attending a university hospital with national data on reported work-related injuries, 33% of the identified injuries in the hospital records had been reported to the authorities according to Reporting of Injuries, Diseases, and Dangerous Occurrences Regulations (RIDDOR) (HSE 2007). Seven percent of the accidents had been reported within the required timeframe.

The severity of the accident and the size of the company affect the reporting behaviour as well. Senneck (1973) has shown a simple model of the relationship between reporting reliability and the average severity of the reported accidents, as shown in Figure 13.3. A company with a high reporting reliability (Q) is expected to show a low average severity of the reported accidents, compared to a company with a low reporting reliability (P) – everything else assumed equal.

This relationship offers a means of assessing the reliability of a company's reporting system. By analysing the average severity of the accidents in the company's files, we get an indication of the reliability in the reporting.

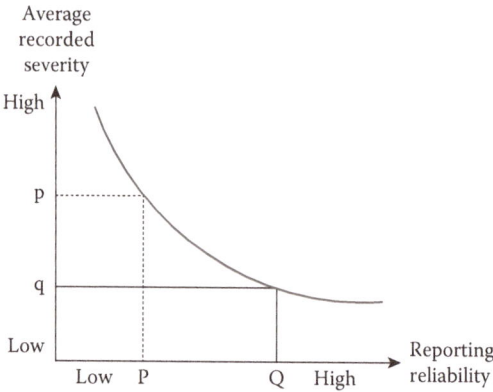

Figure 13.3 Relationship between reporting reliability and the average severity of an accident. (Adapted from Senneck, C.R., *Appl. Ergon.*, 6, 147–153, 1973.)

If this analysis reveals a high average severity, for example, we must expect the company to be de-emphasising accident reporting, with a low reporting reliability of less severe accidents as a consequence.

Research shows that large companies and the government sector have more reliable reporting than small companies (Kjellén et al. 1986). The reporting reliability also varies among different branches of industry. Differences in reporting reliability are to a large extent explained by differences in the incentives to report and consequences in case of failure to report. This applies to employees as well as management.

13.4.2   The significance of reporting near accidents

Early safety-related research concluded that accidents and near accidents were attributed to the same types of causal factors (Heinrich 1959; Tarrants 1980). The consequences of an incident were regarded as largely fortuitous. Pure chance would dictate whether an incident resulted in an accident or not and the degree of severity of the accident. Due to the fact that near accidents are more frequent, these would reveal causal factors earlier and thus make accident prevention possible before the occurrence of loss (Tarrants 1980). These conceptions are rooted in the iceberg theory according to Heinrich (1959). He claimed that out of 300 incidents, one resulted in major lost-time injuries, 27 in minor injuries, and the rest in no injury at all (see Figure 13.4). These figures have been considered to apply as an average across different industries and different types of incidents and causes. It follows from this hypothesis that it is possible to prevent severe accidents by focusing on minor incidents and unsafe acts and conditions.

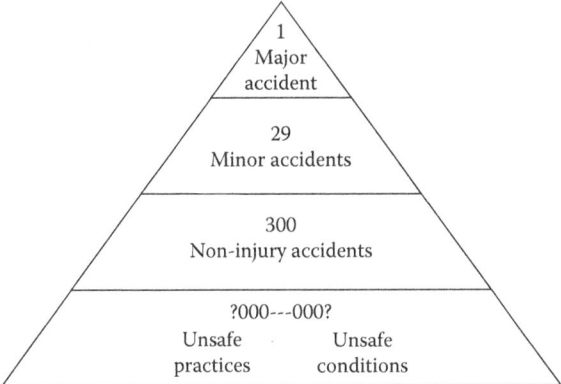

Figure 13.4 Iceberg theory. (Adapted from Heinrich 1959).

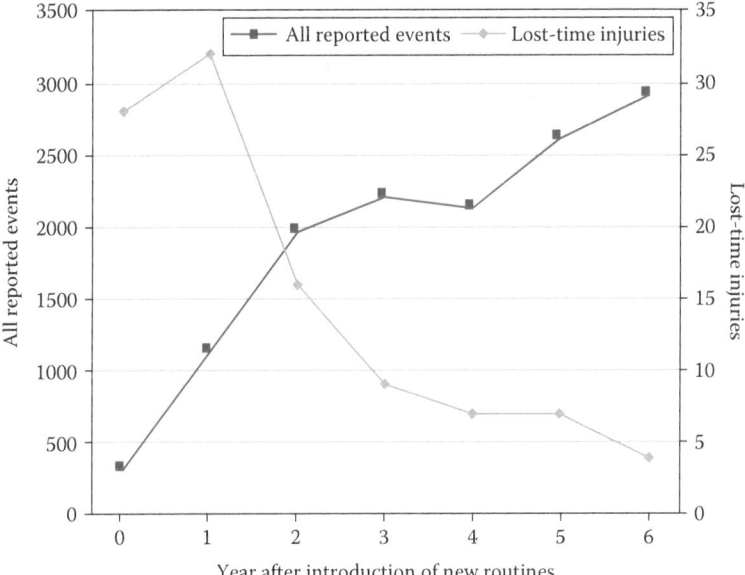

Figure 13.5 Development in the number of reported lost-time injuries and near accidents in a plant during a period of six years after the introduction of new routines for near-accident reporting.

Experiences from industry are used as 'evidence' on the positive effects of near-accident reporting. Figure 13.5 shows a typical example. In the course of a seven-year period, the yearly number of lost-time accidents decreased by almost a factor of 10 at an industrial plant. During the same period, the near-accident reporting frequency increased by about the same factor of 10.

These results have been interpreted as proof of the benefits of near-accident reporting. There are other possible explanations. A change in top management's commitment to safety followed by the implementation of more systematic safety management routines based on the principles of internal control may explain both effects. There is also another concern. The plant did not experience a similar reduction in the frequency of severe accidents, indicating that near-accident reporting will not necessarily direct the resources and attention to remedy the causes of severe accidents.

Figure 13.6 shows the lost-time and fatal accident experience of a large company for a time period of 17 years. For the first 10 years, the lost-time injury (LTI) rate shows a decreasing trend, whereas the fatal accident rate (FAR) remains at approximately the same level. According to the iceberg theory, one should expect a decrease of the FAR as well. Around year 10, there was a debate in the company as to why FAR did not improve in parallel with the LTI rate. It was decided to focus on incidents (accidents and near accidents) with potentially fatal consequences (HIPO incidents). Consequently, management monitored the development of HIPO incidents closely and made them subject to in-depth investigations. From year 12, FAR showed a decreasing trend.

Two catastrophic accidents involving British Petroleum, the Texas City Refinery explosion in 2005 and the Deepwater Horizon blowout in

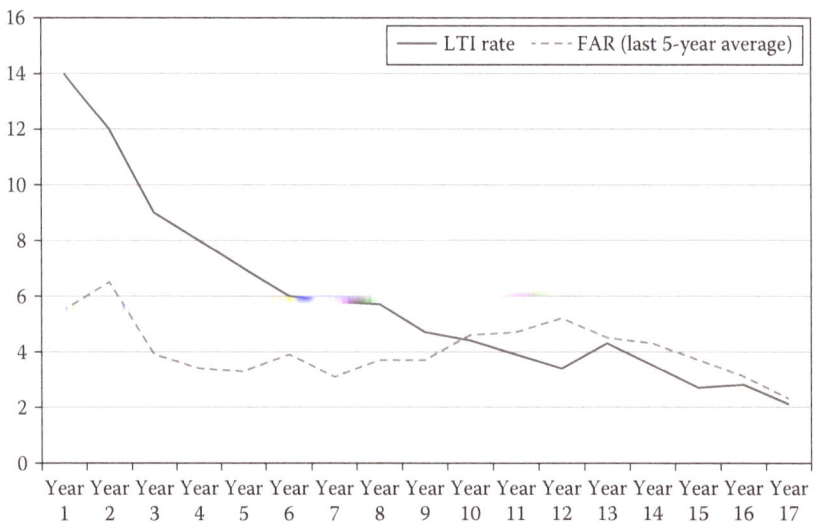

Figure 13.6 Developments in the LTI rate and the FAR at a large company for a time period of 17 years.

the Gulf of Mexico in 2010, point in the same direction. In both cases, the affected industrial system (refinery and drilling rig, respectively) had received awards for excellent safety performance, as measured by the TRI rate, shortly before the catastrophes.

The results of research also contradict the hypothesis. Studies of the consequences of incidents have shown that the severity is dependent on the energy involved (Shannon and Manning 1980; Salminen et al. 1992). Incidents involving falling objects or being hit by a moving vehicle are more likely to result in a fatality than incidents such as a fall on the same level. Generally, there is no valid ratio between the number of recordable accidents and the number of near accidents at a workplace. The distribution of incidents by consequence differs among different types of workplaces, mainly depending on the types of energies involved in production. The conclusion is that each type of accident, as defined by the type of energy involved, has its own severity distribution.

Here, we will take the position that the reporting of unwanted occurrences, including near accidents and unsafe acts and conditions, is an important activity in HSE management. It must be combined with a filter to direct the resources needed to the investigation and follow-up of those near accidents that involve significant energy with the potential to cause major loss. Similarly, unsafe acts and conditions that weaken barriers against major accidents need to be addressed.

13.4.3   Behaviour theory applied to the reporting of incidents

Here, we will analyse workers' and supervisors' inclinations to report incidents based on behavioural theory. This theory primarily has been used in explaining why people deliberately violate safety rules (Saari 1998). Behavioural theory explains how the consequences of behaviour affect some people's judgements and how this results in learned action patterns in recurring situations (Skinner 1969). We can illustrate this theory by a simplified example, where a person has two action alternatives to choose between. One of the alternatives is considered 'safe' and the other 'unsafe' (e.g. to use personal protective equipment or not). Experiences show that people usually choose the 'unsafe' alternative when this has positive foreseeable consequences such as savings in time and effort. The possible negative consequence of unsafe behaviour (i.e. an injury) does not follow each time the unsafe act is conducted. The expected positive consequences of unsafe behaviour are much more likely than the negative consequences, and this feedback mechanism will affect people's behaviour. Deliberate violations will thus become an integral part of a skill.

Workers' and supervisors' inclinations to report incidents can be analysed in terms of action alternatives; the safe act is to report an incident, and the unsafe act is to refrain from reporting the event. For each action alternative, there are positive as well as negative consequences. These are summarised as follows with a focus on the workers (victims).

In order to achieve reliable reporting of recordable accidents, the different incentives and disincentives need to be analysed. The following questions should be asked:

- *Are the incentives that promote the reporting of accidents sufficient?* The compensation to the victim is an important incentive. High reporting reliability will follow if the victim receives economic benefits from it. These benefits may be immediate (higher sick-leave pay) or delayed (injuries emerging at a later time can be attributed to the accident and compensated for). Other examples of such incentives are signs of appreciation and an efficient follow-up of the reporting with remedial actions and feedback of the results to the accident reporter.
- *Have the disadvantages of reporting been reduced to an acceptable level?* Measures need to be taken to reduce the victim's feelings of guilt and blame. The extra time needed for reporting can be reduced through use of a simple reporting form.
- *Are there any negative consequences to the individual if accidents are not reported?* The employees should be informed about the negative consequences of not reporting accidents immediately, for example, the possibility of not receiving compensation in the future.
- *Have the advantages of not reporting accidents been reduced?* Incentive schemes in industry, such as safety prizes for 'zero accidents', may cause underreporting. It is important to evaluate such schemes in relation to the need for reliable accident reporting.

Examples of practical measures to improve reliability in accident reporting:

- Criteria as to which accidents to report should be well defined and easily communicated.
- Simple and well-defined reporting responsibilities and routines. It must be made clear that all employees are responsible for reporting accidents to their immediate supervisor.
- There should be written instructions on accident reporting. It should be available for the supervisors and employees, and the form should be readily available at each supervisor's office. Employees should be informed about the routines when they start work and at regular intervals.
- De-emphasis on blame and guilt and focus on learning and improvement in the subsequent investigations and careful handling of sensitive information.

- Feedback of results of the investigation to those concerned.
- Avoid using incentive schemes that may counteract accident reporting.

13.4.4   *Employees' self-reporting of unwanted occurrences*

We define unwanted occurrences as a common concept for near accidents, unsafe acts, and conditions. Earlier research focused on near-accident reporting, which has a number of advantages (Carter and Menckel 1985; Van der Schaaf et al. 1991):

- Near accidents are less emotionally charged than accidents resulting in loss, and they are less likely to trigger organisational defences against openness and need for change.
- Some near accidents provide important knowledge as to how losses are avoided through recovery actions.
- Near-accident reporting creates an increased knowledge and awareness among the employees about accident risks.
- It also contributes to an improved cooperation between employees and management on safety matters. It represents an alternative channel for communication on safety problems instead of direct contact with line management.

Here, we will assume that the same advantages also apply to the reporting of unsafe acts and conditions.

We have previously discussed research findings that show that near accidents involving human errors are reported to a lesser extent than near accidents with technical causes. The results have been explained by the employees' feeling of guilt and the risk of blame that they may run when reporting their own errors. This leads us to the question of advantages and disadvantages to the employees with regard to the reporting of unwanted occurrences and its relation to the behaviour theory that was presented in the previous section. This may be illustrated by three examples, one from civil aviation and two from the power industry.

Example 1: Figure 13.7 shows an analysis of the incentives that the different participants in the U.S. Aviation Safety Reporting System (ASRS) receive. The ASRS is used for the reporting of incidents and human errors in the prevention of aviation accidents. To promote reporting, the following measures have been taken:

- The research institute responsible for the reporting system (National Aeronautics and Space Administration/NASA) is separate from the enforcement agency (Federal Aviation Administration/FAA).
- Confidentiality and de-identification of accident reporters.

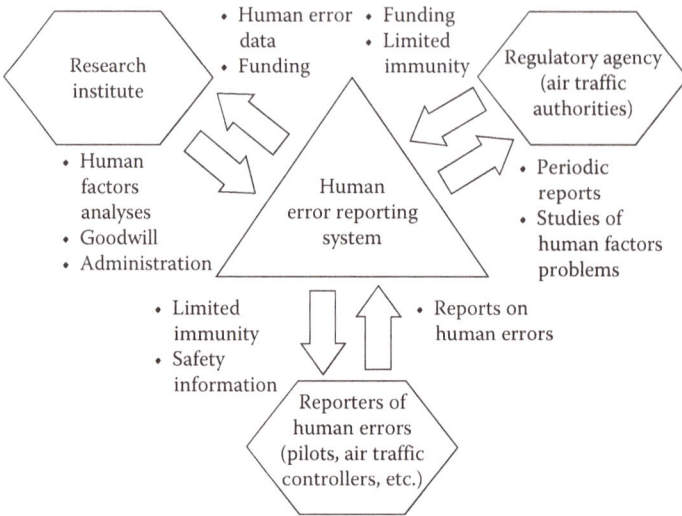

Figure 13.7 What different participants in a human-error reporting system give and receive in return. (Adapted from Reynard, W.D., *The Development of the NASA Aviation Safety Reporting System*, Reference Publication No. 1114, NASA, Mountain View, CA, 1986.)

- Rapid and concrete feedback to the reporting community (pilots and air traffic controllers).
- Easy for reporters to fill in and file reports.

A successful reporting of human errors has been accomplished by giving pilots and air traffic controllers benefits when reporting such events. Most important is the fact that the reporters receive immunity from punitive actions as soon as they have filed a near-accident report. This limited immunity covers negligence but not criminal acts such as drug trafficking. They also receive feedback on the results of their report. The ASRS has been able to maintain a stable reporting frequency since 1990, and the model has been adopted by the aviation safety authorities outside of the United States, including the UK, Canada, and Australia (Connell 2004).

Example 2: A well-functioning near-accident reporting system at a nuclear power plant was jeopardised through changes in the reporting incentives (Chapter 5 by Ives in Van der Schaaf et al. 1991). Use of performance measures became a management focus, and it was decided to measure safety performance by the number of reported near misses, especially those assessed as severe (i.e. Grade 1 on a scale from 1 to 3). A reduction in the reporting frequency by 50% immediately followed the introduction of this new monitoring system. A closer analysis showed that the situation at the plant was unchanged but that the underreporting had increased. Time-consuming and unproductive discussions between the plant and the head office on the classification of events also followed.

Example 3: A hydropower company had implemented requirements for the reporting of unwanted occurrences from its construction projects and operating plants. A performance target had been introduced to achieve a reporting rate of one or more reported unwanted occurrences (RUOs) per employee and year on average. Two subsidiaries meeting the target were approached with questions about the reporting system and initiatives to promote good unwanted occurrence reporting. Similarly, three subsidiaries and one project not meeting the target were approached to explain their performance. The following were the characteristics of the two subsidiaries meeting the target:

- Well-suited tools such as the distribution of a pocket-sized booklet with RUO forms, efficient routines for reporting, and follow-up
- Induction and regular (e.g. yearly) training of all employees, reminders in toolbox meetings
- Target setting and feedback at all levels on RUO rate results (employees, management), quick feedback to the reporter
- Cash/gift incentives and recognition of employees for the 'RUO of the month' with high learning potential

There were different explanations why the lagging organisations failed to achieve the performance target. These included the RUO rate not being introduced as a performance target, RUO reporting not being a priority before more basic safety routines had been implemented, and lack of training, feedback, and incentives. One subsidiary showed declining performance, and this was explained by the fact that HSE staff had recently left the company.

The lessons from these three examples are that the different incentives and disincentives must be carefully analysed when designing RUO routines. The willingness of the workers to report unwanted occurrences is dependent on the extent to which there is openness and trust between workers and supervisors on these issues. The workers must feel a shared ownership of the reporting system and be confident that the results are used for learning and improvements. The opposite – a 'big brother is watching you' attitude, where the purpose is to monitor the organisation in order to allocate blame and punishment – will jeopardise any unwanted occurrence (and accident) reporting system.

A high reporting ratio or frequency is not necessarily a goal in itself. Unwanted occurrence reporting may have counterproductive effects. This is the case when it directs the attention from important safety problems to minor ones and drowns the decision-makers with information that exceeds their capacity. It is important to combine an efficient unwanted occurrence reporting system with tools to help in making priorities and establishing focus on the vital few observations. Commonly used selection criteria include a combination of high loss potential and high frequency of recurrence. The reporting of 'trivial' unwanted occurrences with a low learning value should not necessarily be promoted for the purpose of achieving a high unwanted occurrence rate.

The reliability of unwanted occurrence reporting may be improved by applying the same general principles as those for accident reporting. Measures directed especially at the improvement of unwanted occurrence reporting in industry include:

- Provide possibilities for anonymous reporting.
- Avoid blame and provide immunity against disciplinary action following the reporting of human errors.
- Simple and easily understood forms for self-reporting.
- Efficient follow-up of reports with remedial action and feedback to the reporters with visible results. The employees will use the unwanted occurrence reporting system for the purpose of achieving improvements in their working environment if they know that their reports are taken seriously.
- Communicate realistic goals on the reporting frequency to the employees and provide follow-up and feedback on the results.

It is difficult to measure the reliability in the employees' self-reporting of unwanted occurrence because the definition of what is a reportable unwanted occurrence is not precise. It is mainly left to the judgement of the reporter. Instead, we apply measures of the *reporting propensity* (i.e. the employees' willingness to report).

The most commonly used measures are:

- *Near accident/LTI ratio* = The number of reported near accidents per reported lost-time injury in a period. According to Heinrich, this ratio should be on the order of 300 to 1. In practice, far lower ratios are acceptable.
- *RUO rate* = Number of RUO per employee and year.

13.4.5   *Reporting to the authorities*

Most countries have provisions for the reporting of occupational accidents to the authorities and/or to insurance companies (see Section 2.2). Their aim is to ensure compensation to the victim for sickness leave, medical

Table 13.3 Requirements for reporting accidents to the authorities in Norway

Authority	Occurrence	When?	By whom?
Police	Fatality, serious injury	Immediate notification	Employer
	Serious pollution of the environment	Immediate notification	Principal enterprise
The Labour Inspection Authority	Fatality, serious injury	Notify and report	Employer
Health insurance office	Injuries resulting from occupational accident	As soon as possible, within one year to receive compensation	Employer
Environmental pollution authorities	Accidental contamination of the environment	Immediate notification, report within 24 hours	Principal enterprise
Directorate for Civil Protection	Fire, explosion	Immediate notification, report on request	Employer

treatment, and so on, and to provide the authorities with a basis for regulation of the conditions at workplaces. As an example, Table 13.3 lists the requirements for reporting accidents to the authorities in Norway. Other countries have similar provisions.

13.5   Immediate (Level 1) investigation and follow-up

13.5.1   The quality of the supervisor's first report

Routine investigation into accidents and near accidents usually is carried out by the first-line supervisor and safety representative, or in case of near accidents, by the persons directly involved. These initial reports often are not suited for systematic prevention work. The reports focus on the late phase of the sequence of events. There often is an arbitrary and incomplete documentation of deviations and causal factors.

We must assume that a substandard analysis and documentation of deviations and contributing factors in connection with incidents will be detrimental to the preventive work. Figure 7.7 showed typical results from accident investigations in industry. As illustrated by this diagram, the recommendations on remedial actions are characterised by poor variety and a low degree of learning (compare Ashby's law of requisite variety and Van Court Hare's feedback hierarchy). There is room for large improvements in the quality of accident and near-accident investigations.

A number of factors affect the quality of accident and near-accident investigations in a negative direction (Kjellén et al. 1986). The following factors are to a varying extent present, especially in smaller and medium-sized companies:

- *De-emphasising the use of the report in accident prevention.* The investigations of incidents often are confined to the filling in of forms for insurance purposes and are given minimum attention.
- *The investigators' knowledge about causal factors and investigation techniques.* The investigators usually have fragmentary knowledge about accident-investigation methods. Stereotypical attributions of causes also affect the quality of the investigations.
- *'Punishment' for making a good investigation.* A number of factors often are present at the workplace that discourage the execution of a thorough investigation. One example is the fact that the supervisor has a juridical responsibility for safety at the workplace. An in-depth investigation may reveal violations of regulations and procedures for which the supervisor is responsible. The supervisor will also be responsible for implementation of the results of the investigation. He may want to avoid safety measures that he believes are too 'costly' in terms of working hours and money. Employee re-training and information are examples of less costly measures that are favoured compared to redesign of equipment and changed procedures.
- *Delays the execution of the investigation.* Early statistics from the Swedish Information System on Occupational Accidents and Work-Related Diseases showed that 50% of investigations were carried out only after the workman's compensation board had filed a reminder. This means a delay in the investigation of at least two weeks. Such delays result in loss of information.

Research has explored different possibilities for improving the quality and timing of the investigations. Here, we will focus on the support offered by analytic tools and checklists in improving the quality of the incident reports. In the following section, we will review findings concerning the use of group problem-solving techniques in Level 2 incident investigations.

13.5.2   Well-defined routines for reporting, investigation and follow-up

In order to achieve an adequate immediate (Level 1) reporting and investigation of incidents, the routines and responsibilities must be well-defined and communicated to all concerned. Following are some recommendations:

- *Responsibility for the notification of the incident and to whom.* Normally, the immediate supervisor should be notified without delay, or in case of a more severe accident, in accordance with the emergency notification and response routines.

- *Investigation into the incident.* Normally, this is undertaken by the immediate supervisor jointly with the workers' safety representative. The investigation report form needs to be readily available, in print or on the intranet.
- *Follow-up and close-out of the investigation.* Normally, the immediate supervisor is responsible for the local follow-up and close-out of actions. The HSE department (if any) is responsible for quality control of the Level 1 reports. They should keep a register of all reported and investigated incidents and the status of corrective and preventive actions similar to the principles according to Figure 12.2.

Regarding the timing of the different events, the *'one–one–one'* principle is recommended:

- One hour (or in practice without delay) after the occurrence, management shall be notified about the incident.
- One day after the occurrence, the Level 1 report shall be submitted.
- One week after the occurrence, the Level 2 report shall be submitted as applicable.

The incentives or disincentives of the supervisors in producing adequate Level 1 investigation reports and identifying remedial actions need to be evaluated when designing the incident reporting and investigation routines.

13.5.3   Use of checklists and reporting forms

In Chapter 5, we looked into some of the checklists that have been developed for use in incident investigations and in the coding of data from such investigations. Two alternative principles lie behind the design of checklists on deviations and contributing factors. One principle is that the investigator makes an 'open' query into the immediate and underlying causes of the event. The checklist is used as support to ensure that no important issues have been left out and that the investigator is considering all different circumstances of the event. The results are documented in free-text descriptions. Later, the checklist may be used for coding purposes. An alternative principle is to design a checklist with a multiple-choice format that is used during the interview to catch all the essential information in its detailed categories of deviations and contributing factors. Therefore, the circumstances of the event are documented by checking the correct alternative in the checklist.

The checklists on deviations and contributing factors in Chapter 5 have been designed according to the first principle. Different studies show that the use of tools designed according to this principle improves the reliability and coverage of data collection in accident and near-accident investigations (Kjellén 1983).

The International Loss Control Institute (ILCI) model has been applied in the design of accident investigation checklists according to the second principle. Such checklists are relatively easy to apply in accident investigations and in the subsequent analysis and coding of the results for statistical purposes. They suffer from a poor reliability (different investigators will come up with different results) and from an inadequate coverage of the detailed circumstances of the accidents and near accidents.

Experiences of the British Airway Safety Services' (BASIS) near-accident reporting system support the use of a form for self-reporting with open-ended questions according to the first principle (Reason 1997). BASIS first tried a form with questions concerning types of human errors and contributing factors, where the answers were given in a multiple-choice format. The resulting data suffered from poor validity and reliability. This was because the flight crew personnel did not have a good understanding of the underlying concepts. The introduction of a new form with open-ended questions resulted in a dramatic improvement in the reliability of the data. It was also possible to collect data on a variety of issues that had not been covered before. The introduction of the new form was accompanied by a change in the organisation, where the causal-analysis work was moved from the flight crew personnel to a team of human-factor analysts.

The use of detailed checklists is no panacea for improving the quality in accident and near-accident investigations. Such checklists may even be counterproductive if they replace thoughtful reflection based on experience and documentation of this in free-text descriptions.

Figure 13.8 shows an example of an accident-investigation form that can be used in combination with the checklists in Chapter 5 for documenting the sequence of events and immediate causes. The accident in the example is the same accident described in Section 4.3.

Following are some guidelines for the immediate investigation (Level 1):

1. The reporting routines should ensure that the supervisor gets an immediate notice about the occurrence.
2. The safety representative should participate in the investigation.
3. The site for the occurrence should remain unchanged until the immediate investigation has been carried out.
4. In interviewing the victim (if possible) and witnesses, explain that the purpose of the investigation is to learn and that the question of guilt is irrelevant.
   a. Let the interviewee tell his story by describing the sequence of events in his own words.

INCIDENT REPORT FORM

Type of incident	✓ Personnel injury ☐ Work related illness ☐ Material damage	☐ Security breach ☐ Accidental pollution ☐ Motor vehicle accident	☐ Fire/explosion ☐ Community relations incident ☐ Near miss	☐ Unsafe condition or act ☐ Other:
Time and place	Location *External gangway, house 4*			
	Date: *20.12.2016* Time: *09.20*	✓ During working hours ☐ During business travel ☐ During leisure time	Organisational unit	
Employee Info.	Employee no. *12345*	Name *Ola Anderson*		
	Date of birth *22.03.1984*	Occupation/position *Concrete worker*		
	Company/Department/Employer *Oslo Construction AS*			
Description of incident/ event/ illness/ unsafe condition or act	Main activity at the time of the event? *Assembly of concrete slabs*			
	Description (sequence of event etc.) *While Ola Anderson and Bjarne Boe were assembling concrete slabs on top of concrete beams, a slab got into the wrong position. Anderson went out on the beam to adjust the slab. The beam was icy; he slipped and fell to the floor below.*			
	Machinery, tools, chemicals or materials involved? *Spit; Crane no. 22B*			
Injury or damage info.	Describe injury/damage/type of illness *Broken ribs, punctured lung*	✓ Lost-time injury ☐ Medical treatment injury ☐ Restricted work case injury ☐ Other incl. first aid injury		✓ Emergency preparedness organisation mobilised
	☐ Eyes ☐ Head, face ☐ Back ☐ Trunk ☐ Arm	☐ Hand, wrist ☐ Leg ☐ Feet, ankles ✓ Internal ☐ Other	☐ Amputation ☐ Burn, scald ☐ Concussion ☐ Crushing ✓ Cut, puncture	✓ Fracture ☐ Hernia ☐ Bruise ☐ Sprain, strain ☐ Other
Deviations	Describe deviations from regulations, procedures, instructions, common practice *No fall protection*			
Causal analysis	What were the causes? *Time pressure, crane needed elsewhere. Fall protection not in place.*			
Actions taken	Immediate actions? *Temporary guard-rails*			
	Actions to prevent recurrence *Review planning of guard rails in new jobs. Evaluate needs for more cranes.*		Responsible *Site manager*	Due date *15.01.2017*
Signatures	Date	Employee	Date *20.12.2016*	Supervisor *Lars Larsen*

Figure 13.8 Example of a form for use in Level 1 incident investigations. (Illustrated by an accident at a construction site.)

b. Rehearse the explanation by putting control questions such as: 'What activities were the victim and his work fellows involved in immediately prior to the event?' 'Was everything normal or were there any irregularities or deviations?' 'What triggered the incident?' 'What were the consequences and why?'

c. Keep the checklist on deviations in mind when making sure jointly with the interviewee that the important circumstances have been identified.

d. Try to establish a mental picture of the sequence of events. Is it logical and complete?

e. Ask for the interviewee's opinion about the causes of the accident.

f. Ask for suggestions for remedial actions.

5. The actions from this first investigation usually are of limited scope and involve correction of deviations. Consult the checklist on contributing factors when developing more lasting measures.

6. Ensure to close the loop by prompt follow-up of the investigation with remedial actions.

13.6 Internal (Level 2) investigation

13.6.1 Case: Incident involving flow of water into tunnel during repair work

Here, we will introduce a case involving an incident with accidental flooding of a tunnel with water during repair work. The case will be used to illustrate different methods for Level 2 and Level 3 investigations.

Figure 13.9 shows a simplified cross-section of a hydropower plant. The power plant has been stopped for repair work in a pressure tunnel and a headrace tunnel (transfer tunnel for water) as indicated in the figure. One team, consisting of personnel from the contractor Geo-injection (GI), performs inspection of the pressure tunnel lining in order to repair cracks through injection of cement. To get to the place for inspection and injection, the GI personnel enter through a manhole into the enclosed waterway close to the turbine and follow this until they reach the pressure tunnel. A second contractor, Promac, is performing repair work in the headrace tunnel. This includes removing water from the headrace tunnel and improving rock support to avoid collapse of parts of the tunnel. Both contractors reported to a project manager for the tunnel repair work.

On the day of the incident, the Promac supervisor for the tunnel repair work decided to start removing water that had accumulated between the raised support structures in the headrace tunnel floor. The work started at about 19:00. The work took place about 10 km from the pressure shaft and was carried out by using a tractor with a front-loader (bucket).

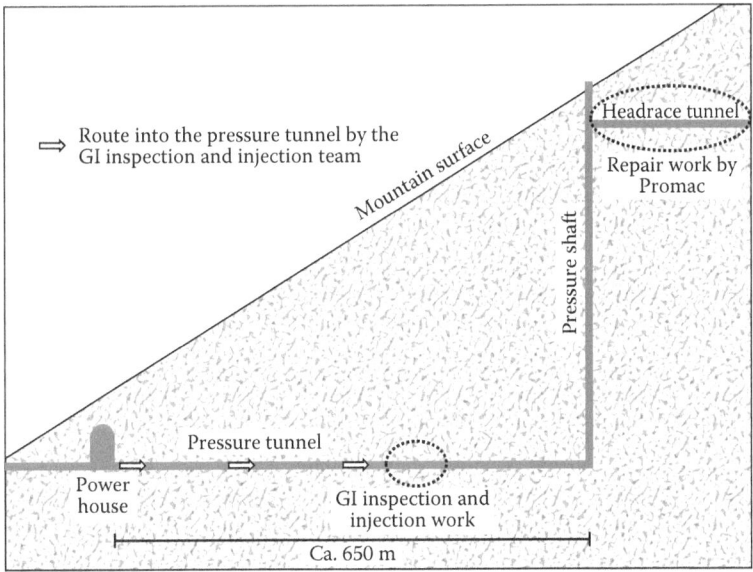

Figure 13.9 Simplified cross-section of the hydropower plant showing power-house and tunnel system.

At the same time, five workers and a supervisor from GI were working overtime in the pressure tunnel in order to finish injection work, which was scheduled for completion at 18:00.

At about 20:10, GI personnel noticed an increased water flow in the tunnel. The GI supervisor in the tunnel informed the watchman at the tunnel's access manhole. He made an operations field operator aware of the increased water level, who immediately ordered the personnel of GI to evacuate the pressure tunnel.

The five workers and the supervisor experienced an increase in the water level in the tunnel of almost 2 m and had to float along with the streaming water towards the powerhouse. They got out of the tunnel through the manhole around 20:35 helped by the GI watchman and a field operator through the only available exit (a manhole of diameter 1.2 m). All six GI personnel were sent to the regional hospital for medical examination.

13.6.2 *Establishing the sequence of events*

The simplest way to document the sequence of events is through a narrative description. This is done in sequential order and describes the different events and actions that resulted in the loss of control in a situation and eventually harm. In a Level 1 investigation, there are no particular requirements for such a description other than that it should make sense and give the reader a reasonable understanding of what happened.

Various analytic tools have been developed for the purpose of displaying the sequence of events in a graphical format that are helpful for improving the quality of the investigation into the sequence of events in a Level 2 investigation. They have different aims:

- To define the starting and end points of the description of the sequence of events.
- To help in checking that the relevant data about the sequence of events have been documented and that the data are logically coherent. In particular, it is important to document the different actors involved and their respective actions.
- To document deviations from the normal and faultless production process.
- To act as a communication tool when groups discuss the sequence of events and causal factors.
- To provide coded data as input to accident databases.

Figure 13.10 shows a method for the documentation of the sequence of events and causal factors in an Events and Causal Factors Chart. It is used in different methods for accident analysis, such as in man, technology, and organisation (MTO) analysis.

Table 13.4 shows a similar way of displaying the sequence of events through a timeline. It is favoured here because it allows for the documentation of each event in detail.

A sequentially timed event plotting (STEP) diagram displays the sequence of events in a graphical format by applying the Multilinear Events Chartering Method (Hendrick and Benner 1987). An example is shown in Figure 13.11. The procedure to establish a STEP diagram in connection with group work involves 14 activities:

1. Use a blank piece of paper (A3 or larger).
2. Identify all involved actors (persons, machines, materials, and so on) and list them in a column on the left of the sheet.
3. Mark the start and the end of the sequence of events.
4. Put each event on a Post-it® slip. The events include deviations as well as expected events in accordance with plans and so on. Identify which actor is 'responsible' for the event.
5. Check that all relevant events have been documented.
6. Locate the event slips at the correct positions on the sheet in relation to the actors and the time sequence.
7. Describe the end state for the actors.
8. Extend with more actors and events if necessary.

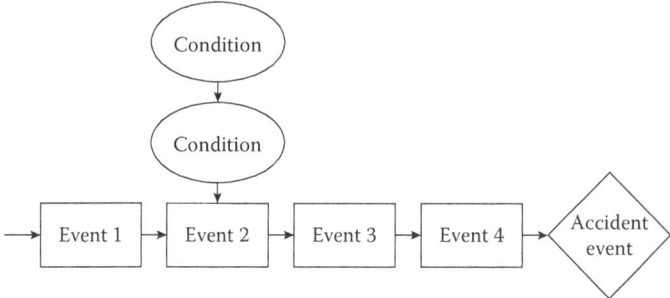

Figure 13.10 Events and Causal Factors Chart. (Adapted from DOE, *DOE Workbook: Conducting Accident Investigations,* U.S. Department of Energy, Washington, DC, 2000.)

Table 13.4 Timeline of the incident in Section 13.6.1

Day, time	Event
Jan 15	Promac mobilises at site and starts repair work on the headrace tunnel. They receive a permit-to-work (PTW) for the repair work with a duration of three months.
Jan 28	Plant operations, in a meeting with the project team, instructs the project to ensure that the excess water in the headrace tunnel shall be removed through an access tunnel and not through the pressure shaft to avoid getting debris into the plant.
Feb 3	GI is instructed in a meeting with the project manager to start repairing cracks in the lining of the pressure tunnel.
Feb 4	Promac is instructed in an e-mail from the project manager to complete removing water from the headrace tunnel before February 8.
Feb 5, 08:00	Plant operations receives an application for PTW from GI for February 6–7 for inspection of the pressure tunnel and injection of cement to repair identified cracks.
17:00	GI sends a quote by e-mail for the inspection and injection work.
20:15	The project manager approves the injection work according to GI's e-mail.
Feb 6, 07:00	In the morning meeting, Promac decides to postpone dewatering of the headrace tunnel due to their lack of a tractor with a front-loader.
07:00	GI personnel mobilise at the site and start the inspection and injection work in the pressure tunnel.
Feb 7, 07:00	GI personnel start a new shift of inspection and injection work in the pressure tunnel. The work is planned to be completed at 18:00.
18:00	The GI supervisor decides to continue with the same shift working overtime until 22:00 to complete the injection work.

(Continued)

Table 13.4 (Continued) Timeline of the incident in Section 13.6.1

Day, Time	Event
19:00	Promac starts removing water from the headrace tunnel using a front-loader.
20:10	GI personnel in the pressure tunnel experience a water flow increase. The GI supervisor notifies the GI watchman at the manhole outside the pressure tunnel by radio.
20:15	A plant field operator performing an inspection at the main inlet valve (MIV) floor is notified by the GI watchman about the increased water flow in the tunnel.
20:17	The field operator notices the increased water level at the manhole, which is 10 cm below overflow level. He immediately instructs the GI watchman to start evacuation of the pressure tunnel.
20:20	The field operator notifies the plant control room operator about the increased water level. The control room operator immediately instructs Promac to stop the dewatering work.
20:20	The first GI workers reach the manhole and get help from the GI watchman and the field operator to evacuate.
20:40	The GI workers and supervisor have completed their evacuation.

9. A further breakdown of the events may prove necessary.
10. Identify missing information and fill in with new events. Consult witnesses if necessary.
11. Identify events that represent deviations.
12. Determine which effect each event has had on actors other than those responsible for the event.
13. Show relations by using arrows.
14. Review the sequence of events and identify conditions that have affected the events.

In other settings, PC software such as Excel can replace the use of Post-its.

We recommend using graphical techniques in the display of accident sequences on an exceptional basis only. A cost–benefit consideration must be made, where the time and attention needed to establish and analyse the diagram are considered as well. Experience shows that the establishment of a STEP diagram is merited in the investigation of accidents involving interaction and communication among different actors.

There are two critical factors to be addressed when establishing a timeline or a STEP diagram:

- The start and the end point of the sequence of events
- The level of detail of the description of events and actors

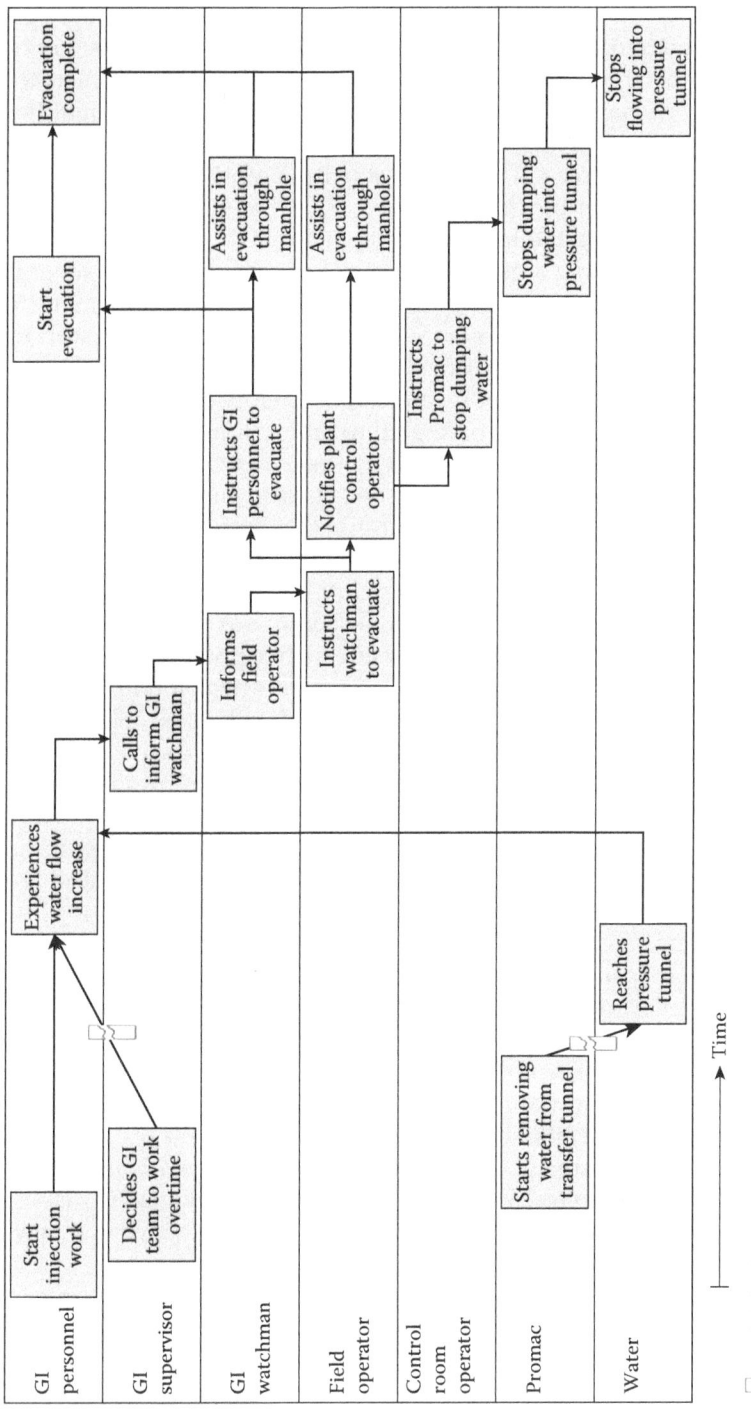

Figure 13.11 STEP diagram of the incident in Section 13.6.1.

Concerning the end point, the sequence of events description ought to include emergency handling when this may have a critical effect on the outcome of an accident. The most natural end point is when the injured persons have been transferred to a hospital that is qualified to treat them. The starting point is sometimes more difficult to determine (see also Section 5.1). 'Standard' starting points are at the start of an activity where the incident occurred or at the start of the shift of the crew experiencing the incident. In the case here, we have included critical decisions during a period of three weeks prior to the incident in the timeline. In the example with the robot accident in Section 5.1, the investigation included the sequence of events during the day shift on the previous day, when an interlocking guard was put out of operation during test production and not restored again before the start of normal production. For accidents involving contractors, it may be relevant to include critical decisions leading up to contract award as well.

In the so-called Swiss cheese model, Reason makes a distinction between active and latent holes or failures in the defence-in-depth to prevent accidents (Reason 1997). The latent failures may have existed for a long time and may sometimes be traced back to design. Latent failures with a long history should be regarded as conditions rather than events and should not be included in the events description.

In deciding on the level of detail in the events description, there is a need to balance between giving the necessary details to be able to make a viable analysis without losing the overview. Graphic presentations usually require some simplification, whereas the timeline allows for a more detailed description of each event. This is especially valuable when an accident sequence has evolved gradually through an extended initial phase of increasing lack of control before the sudden loss of control and development of loss occur.

Tripod beta is based on the Tripod model of Chapter 4 (Reason 1997). Figure 13.12 displays accidents as consisting of two separate event sequences similar to the Finnish accident analysis model (Tuominen and Saari 1982). One of these considers the build-up of hazardous energy and the second the movements of the victim in relation to this energy source. The accident occurs when the person and the released hazard come in contact. Associated personal computing (PC) software makes it possible for the analyst to document the findings of the investigation and to produce a graphic incident tree with the immediate circumstances of the accident (hazards and barrier failures) and the underlying active and latent failures and preconditions.

Data collection in connection with the establishment of the sequence of events follows the principle of triangulation. The three most important sources of evidence are interviews with personnel involved in the incident, physical evidence at the site of the incident, and documented records.

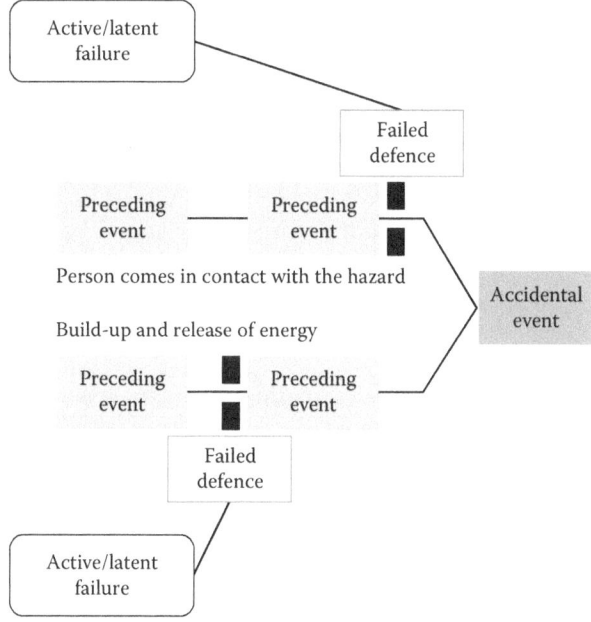

Figure 13.12 The Tripod beta analysis model (simplified), separating the sequence of events related to the hazardous energy from that of the person. (Adapted from Reason, J., *Managing the Risks of Organizational Accidents*, Ashgate, Aldershot, UK, 1997.)

Interviews start with an open question where the witness is asked to tell what he/she has experienced, followed by more detailed questions to fill in holes of missing information.

13.6.3 Identification and assessment of deviations

Deviation analysis (i.e. the identification and assessment of deviations from formal requirements or from the normal and planned sequence of work) is an important part of the Level 2 investigation. The norm that the deviation violates has to be identified in the analysis. This can be a regulatory or company requirement or a deviation from what is the normal and accepted practice (informal norm). Refer to Appendix B for the SMORT checklist on deviations (Tier 1).

A deviation analysis of the case in Section 13.6.1 is shown in Table 13.5. The deviations illustrate the different types of norms.

13.6.4 Barrier analysis

Here, we will apply the definition of a barrier from Chapter 9 as a set of system elements that, as a whole, provide a function with the ability to intervene in an accidental energy flow to change the intensity or direction of it and thereby eliminate or reduce loss. Figure 9.7 showed how barriers may fail, an important aspect in incident investigations. The events and causal factors charting and the STEP diagrams are suited for the visualisation of barrier performance. The symbols used to indicate barrier performance are shown in Figure 13.13.

Table 13.6 shows a more detailed barrier analysis of the incident in Section 13.6.1. Some of the results representing failed barriers coincide with the previous deviation analysis.

Table 13.5 Deviation analysis of the incident in Section 13.6.1

No.	Deviation	Reference	SMORT checklist (Appendix B)
1	GI personnel working unscheduled overtime after 18:00. PTW did not reflect this extension of working hours.	Plant PTW procedure	1.1.6
2	Dumping of water into pressure shaft, resulting in flow of water into pressure tunnel. Oral instructions by plant operation not to dump water into the pressure shaft had not been forwarded by the project to Promac.	Oral instructions by plant operation	1.1.1
3	The method for water removal from the headrace tunnel had not been defined in any methods statement/procedure.	Plant HSE management plan; draft ISO 45001, §6.1.2	1.1.5
4	The job safety analysis (JSA) for GI's work was generic for injection work and did not identify hazards related to the tunnel, including a zone for drowning hazards.	Plant HSE management plan; draft ISO 45001, §6.1.2	1.1.5
5	There was no contract between the repair project and GI with defined plant-specific health and safety requirements for the work. The start of work was agreed upon in a meeting with the project manager and was confirmed by e-mail.	Plant procurement procedure; draft ISO 45001, §8.4	1.1.5

(Continued)

Table 13.5 (Continued) Deviation analysis of the incident in Section 13.6.1

No.	Deviation	Reference	SMORT checklist (Appendix B)
6	The PTWs for the GI work and for the repair work by Promac did not identify risks related to simultaneous activities, including possible water removal and needs for barriers.	Plant PTW procedure	1.1.5
7	Plant operations did not coordinate the PTWs for GI and Promac. Operations did not receive any daily work list from Promac, and there was no check for possible hazards in simultaneous activities.	Plant PTW procedure	1.2.1

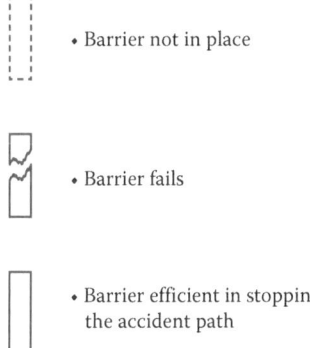

• Barrier not in place

• Barrier fails

• Barrier efficient in stopping the accident path

Figure 13.13 Symbols to indicate barrier performance.

13.6.5 Analysis of contributing factors in the man–machine system and at the place of work

The next step in the Level 2 investigation involves the identification of contributing factors. These are normal and planned conditions (as opposed to deviations) of a human, technical, or organisational nature at the place of work that are judged to affect the risk of accident in an increasing direction. The SMORT Tier 2 checklist of contributing factors at the department and in the work system is presented in Appendix B.

The *fishbone* or *Ishikawa diagram* in Section 4.5 is also suitable for the analysis of contributing factors to identify deviations and barrier failures. Figure 13.14 shows an outline of a fishbone diagram. It may be used in combination with the SMORT Tier 2 checklist.

Table 13.6 Form used for barrier analysis with the results of an analysis
of the incident in Section 13.6.1 shown for illustration

No.	Type of barrier	Situation at the time of the accident	Status of the barrier
1	Preventing build-up of energy	Not applicable.	
2	Preventing uncontrolled release of energy	The instructions to the project that excess water in the headrace tunnel should be evacuated through an access tunnel and not through the pressure shaft were never passed on to Promac.	Not implemented
3	Separate people from the hazard in time or space (avoid danger zone)	The PTWs for GI and Promac were not coordinated in order to avoid hazards due to simultaneous work.	Not implemented
4	Physical barriers between the hazard and people	A rock trap at the exit of the headrace tunnel to the pressure shaft had been removed to allow dumping of water into the pressure shaft.	Removed
5	Use of personal protective equipment	Not applicable.	

Table 13.7 shows the results of an analysis of contributing factors for the case in Section 13.6.1.

13.6.5.1   *Group problem-solving*

Various investigation methods have been developed that use team-organisation and group-problem-solving techniques (Kjellén 1983; Carter and Menckel 1985; Hale et al. 1997). The aim is to overcome some of the problems in accident investigations discussed earlier by ensuring:

- Access to the necessary knowledge and skills during the investigation by involving persons representing different areas of knowledge and competence.
- Organisational learning whereby the team members' detailed knowledge about conditions at the workplace of relevance to the

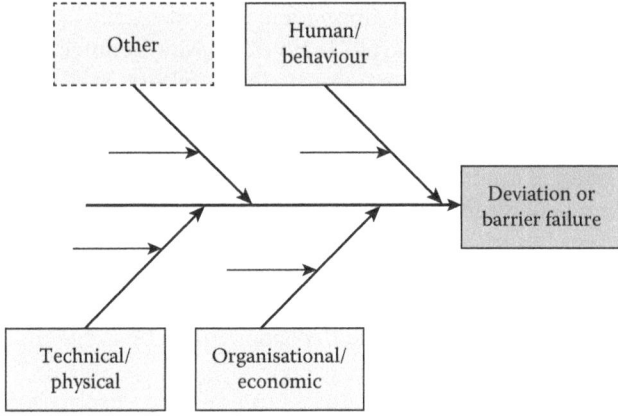

Figure 13.14 Fishbone diagram for use in combination with an MTO analysis or the SMORT Tier 2 checklist.

Table 13.7 Analysis of contributing factors for the incident in Section 13.6.1

SMORT category	Contributing factor	SMORT checklist (Appendix B)
Management of operation	Rely on oral communication for critical instructions.	2.3.5
Management of operation	Inadequate coordination between operations and project.	2.3.7
Management of contractors	There was no contract between the repair project and GI with defined plant-specific health and safety requirements for the work. Start of work was agreed in a meeting with the project manager and was confirmed by e-mail.	2.4.2
Human resources	Inadequate training of plant personnel in the management of the PTW system.	2.5.2
Fixed assets	Poor quality of lining in the pressure tunnel.	2.6.1
	Swelling clay causing collapse of segments of the headrace tunnel.	2.6.1
Maintenance of fixed assets	The PTW form is not providing adequate guidance in identifying the necessary barriers (e.g. in entry into confined space).	2.7.5

investigation are shared among team members and are made available in the decision-making. Team discussions contribute to a transfer of tacit knowledge among team members in a process called socialisation and to the development of a shared understanding of causal mechanisms. These processes also result in learning and attitude change among team members.

- Exchange of opinions on alternative interpretations among team members in order to avoid bias and arrive at judgements of a satisfactory quality.
- A basis for influencing the decisions on the part of the employees.
- Ownership of and loyalty to decisions made by the team.

Because an investigation involving group problem-solving is resource-demanding, there need to be criteria for selection of incidents for team investigation. Examples of criteria:

- Accidents resulting in recordable injury
- Incidents with potential for severe consequences or incidents of a frequently recurring type
- Incidents in new production systems

A typical team consists of three to five members representing the workers and supervisors from the workplace and HSE experts (e.g. a safety engineer or an occupational health nurse). In cases where an accident involves more than one department, each department has to be represented on the team.

The recommendations here imply that the team uses a well-structured problem-solving technique in its work. Ideally, the accident investigation process is sequential. In practice, there is some overlapping and readjustment. As soon as the investigators have accumulated enough information to have an adequate overview of the accident sequence and the conditions at the workplace, they start to analyse the causes. This analysis may lead back to a search for more facts to fill in gaps. They continue by developing remedial actions. These should be of a nature that provides long-term effects and represents second to fourth order of feedback according to Van Court Hare's hierarchy. The team's report should be kept simple and document the basic recommendations, the sequence of events, and the analysis of basic causes.

13.6.6   *Investigating the human factor*

Example: Fatal accident due to fall from a transmission line. At 09:00 on the day of the accident, Manny Hart and David Emden, employees of a contractor working for a power company, were assigned to climb a transmission line tower to fix a pulley block to start pulling an earth wire between this tower and a neighbouring tower. After fixing the pulley block at the desired location at 10:50, Mr. Emden descended from Leg A of the tower using the vertical lifeline (ensuring one of the two lanyards was hooked to the vertical lifeline all the time while descending) and safely landed on the ground, as shown in Figure 13.15. Subsequently, Mr. Hart moved on a horizontal beam to Leg B. He detached both the lanyards and started to descend. Leg B did not have the footsteps with a vertical lifeline. Instantly, Mr. Emden slipped, lost his grip, fell approximately 10 m from the tower, and landed on the ground. He sustained a fatal injury.

The accident was investigated by a team from the contractor led by the HSE manager. The team made the following conclusion: 'On the basis of the site investigation and interviews of witnesses and supervisor, it appears that the victim did not follow the standard practice of securing the harness while descending from the tower. As per the rule, one of the

Figure 13.15 Transmission line illustrating location of lifeline and where the fall took place.

lanyards of the harness shall always be hooked to the vertical lifeline/ tower structure all the times while climbing up or down the tower'. As per the terminology used by the contractor, the team identified the following 'immediate and contributing causes' and remedial actions:

- *Immediate cause:* Slip from the bracing of the tower and subsequent impact with tower component and ground
- *Contributing causes:*
  - Not hooking the full body harness on vertical lifeline while descending.
  - Taking shortcut (using Leg B for descending).
  - Slippery surface because of dew.
  - Loose chin strap, which resulted in helmet being displaced or dislodged from the head.
  - Complacency due to repetitive work of similar kind.
- *Remedial actions:*
  - All personnel shall be re-trained on the use of full body harnesses.
  - Site supervisor must ensure that towers are free from dew/slippery agent before allowing any one to climb the tower.
  - The contractor's site management will be notified immediately about any incident that occurs onsite.

This example illustrates the so-called blame culture where, in this case, the diseased employee was considered as having caused the accident by not taking the necessary actions to prevent it. In a company with a blame culture, people act defensively, and it will be difficult to find out what has happened (Kletz 2001). This example illustrates the 'attractiveness' of single-loop learning, where the employee at the sharp end is focused and failures higher up in the organisation or in design are overlooked for stakeholder or cost considerations (Perrow 1984). A 'no-blame' culture in accident investigations means that it is more important to analyse and understand in order to come up with efficient remedial actions, than to find a culprit of the accident.

Here, we will take the following approach in investigations into human errors:

- Try to understand why the victim behaved in the way he/she did (i.e. put one's self in the victim's situation).
- Were there any specific circumstances at the time of the accident that could explain his/her behaviour?
- Some important 'formal' questions:
  - Were the technical and safety arrangements adequate?
  - Was the safety standard adequately defined (i.e. in JSA, work instructions, method statements)?

- Was the standard adequately communicated through safety induction, formal training, and toolbox meetings?
- Was the standard adequately enforced through direct supervision and follow-up, and what were the consequences in case of non-compliance?
- Did the incentives (e.g. pay system) promote adherence to the standard?
- A key question is whether rule violation was an exception or occurred frequently and was accepted by the supervisors in practice

Example (continued): Further investigations showed that rule violations were the 'rule' rather than the exception in the lining of the transmission towers. The teams involved in this work were paid based on a piece rate. The particular team experiencing the accident showed high productivity. Because it was critical that the transmission towers were completed in time for the start-up of a power plant, the supervisors looked the other way when the team climbed the towers. After the accident, work at the transmission towers was suspended for a period to establish a new work and safety standard, and start-up of work was tightly supervised to ensure compliance. The time gained through looking the other way was thus lost many times over due to the fatal accident.

13.6.7   *Developing accident prevention measures*

There is a need for evaluation of the proposed remedial actions. Different schemes are offered for such evaluations. The following criteria for the evaluation of remedial actions at the workplace level are recommended.

1. These criteria follow from the energy model:
   a. Has the hazard been eliminated or reduced? Here, we are referring to such measures as exchanging toxic chemicals with less toxic substitutes or eliminating height differences to prevent falling.
   b. What has been done to prevent the triggering of the hazard? These types of measures usually are concerned with the prevention of incidents and with deviations such as human errors or mechanical failures.
   c. What has been done to interrupt the uncontrolled energy flow? This is a question of passive barriers (guards) or active safety systems such as fire detection and sprinkler systems.
   d. Is it necessary to reduce harm by means of personal protection?
   e. Is there a need for improved first-aid routines?

2. Additional considerations in deciding priorities:
   a. Is it possible to implement the measure within a reasonable period of time? If the identified hazard needs immediate attention, short-term measures should be chosen pending a more permanent solution.
   b. Are many people affected? A measure to reduce a hazard that currently puts many people at risk should be given high priority.
   c. Does the measure have lasting effects?
   d. Are any new hazards or other unwanted secondary effects created?
   e. Is the measure feasible from an economic point of view?
   f. Will those who are affected by it (workers and supervisors) accept the measure?

13.6.8   Quality assurance of the Level 2 investigation report

The example in the previous section illustrates the importance of quality assurance in the Level 2 investigation reports. Following is an example of questions that address this issue. They are based on the criteria developed for incident investigations in Chapter 8.

Questions for quality assurance of Level 2 investigation reports:

1. Is the description of the sequence of events adequate and is it complete? Is it possible to understand? Does it include both the phase before the injury occurred and emergency handling (if relevant)?
2. Does the activity where the accident occurred require a documented risk assessment? Has the investigation checked: Whether the risk assessment has been done? The quality of the assessment? That the identified remedial actions have been implemented?
3. Has the investigation checked: The status of barriers when executing the activity? What barriers are required? That the barriers have been implemented and work as planned?
4. Has the investigation done a deviation analysis (beyond checking barriers)? Have the deviations been identified with proper reference to the requirements (regulatory/internal)?
5. Have human, technical, and organisational contributing factors been identified? Do they make sense in relation to identified deviations and barrier failures? Is something critical missing? Is the investigation balanced and free from 'blame'?
6. Have recommendations been identified? Do they address the deviations, barrier failures, and contributing factors? Will they be efficient in preventing similar accidents from occurring?
7. Has the investigation been executed by a multidisciplinary team covering the different needs of expertise? Has it been signed off by a manager at the right level?

13.6.9 Costs of accidents

Table 13.8 shows a simple scheme for the registration of accident costs. It is based on the market-pricing model and is intended for use by the investigators as a complement to the ordinary investigation form. In most practical circumstances, the lost working hours due to sickness leave will dominate the monetary losses. A simple alternative is to include the first line of the table in the form for routine incident reporting (Level 1).

13.6.10 Computer-supported investigation, reporting and follow-up

Figure 13.8 showed an example of a traditional fixed form for the investigation and documentation of accidents and near accidents. In designing such a form, a trade-off has to be made between the need for comprehensive and detailed information on the one hand and the need for a form that

Table 13.8 Form for registration of accident costs

1.	Lost working hours for the victim:				
	40	×	€ 50		= € 2000
	Number of hours		Hourly salary		
2.	Lost working hours, other employees:				
	3	×	€ 50		= € 150
	Number of hours		Hourly salary		
3.	Capital costs due to production stop:				
	1	×	€ 200		= € 200
	Number of hours		Costs per hour		
4.	Material costs:				
	0	+	0		= 0
	Scrap costs		Repair costs		
5.	Costs of medical treatment:				
	€ 100	+	0	+ 0	= € 100
	Medical doctor		Nurse	Physiotherapist	
6.	Transportation:				
	15	×	€ 1	+ 0	= € 15
	Number of km		Costs per km	Public transportation	
7.	Other costs:				
	0				= 0
	Total costs for the company				**= € 2465**

Source: Sklet, S. and Mostue, B.A., *Kostnader ved arbeidsulykker i prosess- og verkstedsindustrien* [Costs Associated with Occupational Accidents in the Process Industry], SINTEF Report STF75 A92032, Trondheim, 1993. In Norwegian.

is simple and easy to take on and use on the other. Developments in information technology offer possibilities for the use of intranet to collect and store incident data that represent a way out of this dilemma. Investigation results are successively fed into a computer, and an interactive program will choose the new questions depending on previous answers (Van der Schaaf et al. 1991; Vinnem 2014). Such programs are built up as a logical tree structure (see Section 4.5).

Example: A transportation firm has developed an interactive program for the registration and analysis of accidents involving material damages. They are especially concerned with motor vehicle and material-handling accidents. The interactive program first asks questions about the type of accident. If the answer is 'a motor vehicle accident', the program will proceed by asking questions about:

- Date and time of occurrence
- Type of transportation task
- Place of occurrence and physical environment
- Driver's age and experience
- Types of vehicles involved and condition of vehicles
- Extent of personal injury and material damages

Next, the program identifies typical traffic conflict situations, and the investigator has to choose among these. Depending on the answer, the program generates detailed questions of relevance to the chosen situation.

An advantage of these types of applications is that the results will immediately become available in a database for statistical analyses and for follow-up. Computer support is best suited for companies where the accidents mainly fall within a few well-defined categories. The program must not be too closed in the sense that it makes the investigator overlook vital information and uncommon causal factors.

The widespread use of information technology allows for new means of distribution and follow-up of investigation reports. It is now possible to distribute reports conveniently and promptly to all concerned parties via electronic mail. It is also possible to store the reports in a shared database, where different persons can access the documents and comment on and edit them. These types of information technology solutions provide several advantages:

- Enhanced quality of accident and near-accident reports and remedial actions by ensuring input and quality checks from people at different levels of the organisation and geographical locations.
- Transparency by making it possible for the reporter and other concerned parties (e.g. management and safety representatives) to

monitor the further handling of the report and the actions that have been identified as well as their status.
- Administrative simplification of distribution routines, approval of actions, and close-out.

Although the developments in computer applications have many advantages from a HSE management perspective, there are also some concerns. It is necessary to ensure that the reports have passed through an adequate quality control before they are made accessible to all concerned parties through e-mail or a shared database. Commenting and editing has to be carefully organised in order to avoid contamination of the database with irrelevant or distorted information. Security routines to avoid misuse of information and corruption of the data are also concerns.

Accident reports may easily be distributed to a large number of people as e-mails. This distribution will add to the general information load within the organisation and may create problems of overload and inexpedient reactions. The problem may occur in large companies and in companies with efficient reporting of near accidents. It is important to establish routines for the distribution of accident and near-accident reports inside companies, based on the criticality of events and receivers' actual needs.

13.6.11   Procedure for Level 2 investigations

The description of Level 2 investigations in Section 13.6 is summarised in Table 13.9.

The accident analysis framework presented in Chapter 5 represents a theoretical foundation for the described approach to Level 2 accident investigations. The steps of both Level 2 and Level 3 accident investigation follow the accident analysis framework by proceeding backward from the output and the incident and continuing until covering the contributing factors (Level 2 investigation) or root causes (Level 3 investigation) as shown in Figure 13.16.

13.7   Independent (Level 3) investigation

The commission or team in a Level 3 investigation applies many of the same methods and tools that we recommend here for use in a Level 2 investigation. The main differences between a Level 2 and a Level 3 investigation are that the latter is carried out by an independent team, and the investigation is extended to include root causes in project management (as relevant) and in general management. Thus, the steps of a Level 2 investigation (data collection, describing the sequence of events,

Table 13.9 Procedure for Level 2 investigations

Main steps	Detailed steps and tasks
1. Establish investigation team	1.1 Decide whether to investigate the incident and client 1.1.1 Normally, all recordable injuries and HIPO incidents shall be investigated at Level 2 1.2 Nominate team and team leader

2. Reconstruct the accident sequence, including emergency response, and establish factual circumstances	2.1 Before site visit
	    2.1.1 Secure information before the investigation (supervisors, first report, organisation, procedures, and so on)
	    2.1.2 Plan the investigation (inspections, interviews)
	2.2 Investigation
	    2.2.1 Inspections
	    2.2.2 Interviews
	    2.2.3 Secure documentation

3. Analysis	3.1 Analyse the investigation materials
	    3.1.1 Display the accident sequence
	    3.1.2 Analyse deviations
	    3.1.3 Analyse barrier performance
	    3.1.4 Analyse contribution factors

4. Reporting	4.1 Establish facts and interpretations and document the results in a report
	    4.1.1 Report writing
	    4.1.2 Quality assurance and de-briefing
	    4.1.3 Distribution of report

5. Follow-up and close-out	5.1 Follow-up is undertaken by the line organisation

Following are detailed instructions for some of the steps not described elsewhere in the chapter:

2.2.1 Inspections at the site of the accident
  a. Guided by a person knowledgeable about the site and the accident
  b. Ask questions about the activity at the site, technical and organisational conditions, and about accident facts to understand the sequence of events, possible deviations, and barrier failures
  c. Document by using camera

2.2.2 Interviews
  a. Explain the aim of the interview (facts, not allocating blame)
  b. Start with open questions (explain what happened...., who, what, where, when)
  c. Do not interrupt
  d. Note follow-up questions

(Continued)

Table 13.9 (Continued) Procedure for Level 2 investigations

Main steps	Detailed steps and tasks

2.2.3 Secure documentation; examples of documentation that may be relevant:
  a. Job descriptions, procedures
  b. Risk assessments, JSAs
  c. Plans (action plans, lifting plan, emergency response plan)
  d. PTW
  e. Documented competence (certificates, training records)
  f. Inspection reports, third-party verifications (e.g. lifting equipment)
4.1 Contents of a Level 2 investigation report
  a. Executive summary
    i. Short description of the sequence of events
    ii. Decision to investigate, investigation team
    iii. Conclusions and recommendations
  b. Facts about the occurrence
    i. The situation at the start of the events
    ii. Description of the sequence of events, STEP analysis
    iii. Identified deviations and barrier failures with reference to internal and external requirements
  c. Findings and recommendations
    i. List of findings from the MTO analysis
    ii. Recommendations
  d. Attachments
    i. Pictures
    ii. People interviewed

deviation analysis, barrier analysis, and analysis of contributing factors) are included in a Level 3 investigation. Here, we will describe a technique for this investigation based on the guidelines in ISO 19011 for auditing management systems (ISO 2011). Section 12.3 gives a more detailed description of important elements of this standard.

An immediate and necessary task for the team is to establish the facts about the event sequence and an explanation of why it happened. The team will then be able to recommend actions to prevent a recurrence of this particular event in the future. Severe accidents, however, usually are the result of an unusual combination of circumstances, and it is unlikely that this particular combination will recur. Therefore, an important task for the team is to make a systematic and independent examination of the organisational, technical, and individual circumstances around the accident. By decontextualising these, the team will be able to develop a basis for more generally applicable improvements.

The team has the duty of looking into the root causes of the investigated event related to weaknesses in the general and HSE management systems of the company. In this way, the team's recommendations not only will affect the probability of a recurrence of the particular event in

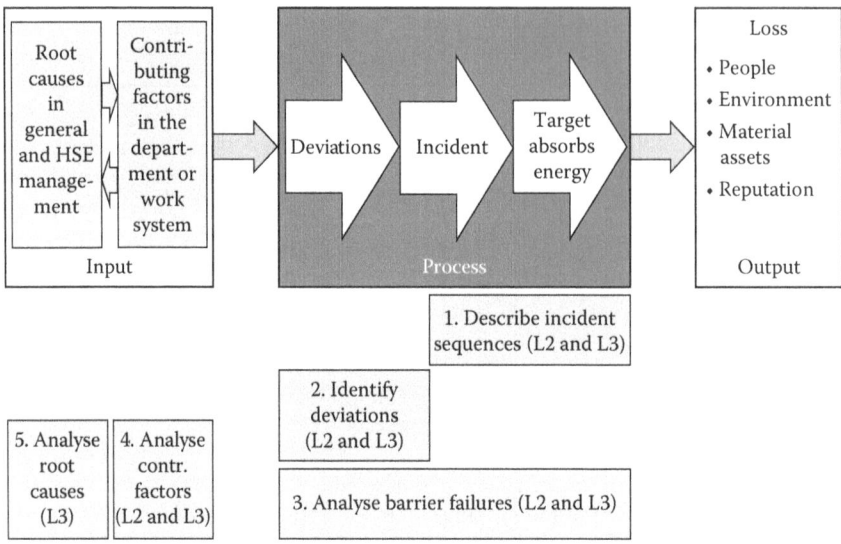

Figure 13.16 The steps of Level 2 and Level 3 investigation and the accident analysis framework.

question but the general safety level of the company. The team will receive its mandate from high-level management, and this will allow the team to balance interest groups inside the company that are benefited by a limited investigation with the situational factors discussed in Section 7.7. An independent team will also be able to explore such aspects as management's commitment to HSE and the prevailing accident perceptions within the organisation.

The approach selected here for in-depth accident and near-accident investigation should not be confused with police investigations of such events. A Level 3 investigation differs from police investigations in that the latter aim to discover violations of regulatory and company requirements – who violated them and who is liable. The focus in a Level 3 investigation is on learning and improvements.

13.7.1 The steps in an in-depth investigation

Here, we will outline the typical steps in a Level 3 investigation into a severe accident or near accident that have been adapted from the basic steps in an audit according to ISO 19011 (see Figure 12.5). They include:

1. Securing the scene
2. Appointing an investigation team
3. Introductory meeting and planning the team's work

4. Collecting information
5. Evaluation and organising information
6. Preparing the team's report
7. Follow-up meeting
8. Follow-up and close-out

Whereas line management is responsible for Steps 1, 2, and 8, the team has principal responsibility for the other steps.

An important methodological issue in carrying out in-depth investigations is whether the investigation should be undertaken on the scene or after the event. When the investigation is carried out *on the scene*, the investigation team arrives at the scene shortly after the occurrence, when physical objects are unchanged and witnesses are available. The team starts its work after the emergency response and health-care needs have been satisfied. There are obvious advantages with this immediate mobilisation of the team:

- Time-dependent physical data such as weather and lighting conditions are directly observable.
- Physical evidence needed for reconstruction of the sequence of events is likely to be unchanged.
- Witness descriptions are largely unaffected by 'rationalisation'. Experience shows that the people involved wish to tell about their experiences and that the statements given by witnesses and involved persons immediately after an accident are different from those given at a later stage.

When it is not feasible to mobilise an investigation team on short notice, the investigation is carried out *after the event* (i.e. in practice, one to a few days after the occurrence – or sometimes more). For the in-depth investigation to be viable, parts of or the whole team must visit the accident site and carry out investigations on the spot. A practical solution that utilises the advantages of an 'on the scene' investigation is to make one member of the team available at short notice to carry out the initial investigation immediately after the occurrence. The whole team will then assemble at the site of the incident to review the findings and continue the investigation.

13.7.1.1   Securing the scene

Irrespective of whether the investigation is carried out on the scene or after the event, the scene of the accident has to be made safe and preserved for the forthcoming investigation. This is the responsibility of the immediate supervisor. The first task is to take prompt corrective actions to secure the site in order to avoid secondary accidents. It is then necessary to prevent physical items from being operated or moved.

It is advisable at this stage to take photos of any physical evidence. When an investigation team has been appointed, the chairperson of the team will take over responsibility for the site. It will not be released for clean-up and returned to normal operation until all physical evidence has been secured.

13.7.1.2 Appointing an investigation team

The appointment of the investigation team is the responsibility of line management, usually at the business area level. The decision needs to be endorsed by the company's joint management–labour health and safety committee (if one exists). The team receives a mandate defining its members and describing its authority, the aim and scope of the investigation, resources, reporting schedule, and distribution of the report. The team has to have access to any relevant documentation and personnel for interviews.

The mandate needs to address:

- The scope of the investigation, defining the incident and circumstances to be investigated
- The client of the investigation
- Investigation team leader and other members
- The objective of the investigation
- The authority of the team, access to information, and personnel
- The timeframe for conducting the investigation and reporting
- Report confidentiality and distribution

    Example: Authority for a Level 3 investigation team, 'The Investigation Team has been given all necessary authority, accreditations, and cooperation required to perform the investigation with efficiency by the 'Client' and in accordance with Company procedures. It shall execute the investigations that the Team finds necessary in order to establish the factual circumstances around the accident and contributing factors and the root causes. The Investigation Team shall have full access to any information or human resources that it may find necessary or useful to conduct the investigation. It is further expected that the Construction Project, Engineer, Contractor, and Subcontractors and any other party involved in the accidents and the circumstances around it are informed about the Team's mandate and fully cooperate with it'.

The following conditions need to be considered in designing the team:

- *Credibility and competence:* The team members need to enjoy an adequate credibility in management's and employees' eyes. The members of the team must represent competence in the technical and management aspects of the event as well as in accident-investigation techniques. Accident investigators should be formally trained and certified in a manner similar to that of accredited quality auditors (cf. ISO 19011 in ISO 2011). The chairperson and other members must come from a different organisational unit than that affected by the event. The total number of members should not exceed five. Three members is a recommended size.

Example: A Level 3 team for the investigation of a fatal rock slide accident during scaling of a road section under construction consisted of the following persons:

- A senior HSE manager with extensive experience in investigations and audits of construction sites led the team.
- A civil engineer represented road construction and project management competence on the team.
- An engineering geologist represented the necessary competence to assess the risk of rock slides and to design slope stabilisation measures.
- A supervisor for the rock-scaling department of a contractor other than the one involved in the project being investigated represented job planning and execution competence and competence regarding qualifications of personnel and requirements as to equipment.

All team members came from organisations outside of the project.

- *Time away from competing duties:* A serious incident will bring lots of attention and will also produce anxiety among the involved personnel in the period immediately following the event. To be able to act on these circumstances, the team should be freed from other duties to focus on the investigation and to come up with their report and recommendations promptly. This priority will also have a beneficial impact on the relationships with the affected workplace. The team will, through their presence, demonstrate that they take the problems of the local management and employees seriously.
- *Power:* The independent team will not have the power to implement their recommendations. Authority and power are, however, important in another respect. The team should have full access to all relevant sources of information, including the right to interview personnel and to consult written and electronic files at their discretion.

13.7.1.3 *Introductory meeting and planning the team's work*

The team starts its work by taking control of the scene of the event and by ensuring that it has been preserved. After that, the chairperson calls a meeting and invites persons knowledgeable about the event and the affected organisation to give a briefing to the team on the status of the affected personnel, the status of the scene of the accident, the results of preliminary investigations, and background information about the activity where the incident occurred. The team then decides about:

- Investigating the scene of the event
- Conducting interviews with participants, witnesses, line management, and supporting staff
- Obtaining documentation such as event logs, arrangement drawings and/or maps of the site, and work and safety instructions
- Practical arrangements such as a meeting room, scheduling activities, contact person at the site, and so on.

The investigation team needs to contribute to an atmosphere of openness and trust. The team members' conduct during the introductory meeting, interviews, and follow-up meeting is decisive. During the introductory and follow-up meetings, it is important to emphasise the learning perspective. Interviews have to be carried out in an atmosphere of mutual respect and recognition of the interviewees' legitimate self-interest. As in audits, the follow-up meeting represents an opportunity for the creation of ownership of the results among those affected by them.

13.7.1.4   Collection of information

Information collection starts with the establishment of facts related to the sequence of events and circumstances around the incident. Information-gathering involves inspecting the scene and other locations if relevant, interviewing participants and witnesses, expert examination of physical objects, document reviews, verification of calculations, and so on. It is important to make accurate notes of all gathered information concerning source, date, time, place, person responsible for data collection, and type of information (factual or interpretive).

In inspecting the scene, care must be taken to note all relevant details such as debris, scratches, tools left at the scene, switch and handle positions, and so forth. The team should establish sketches of the scene, showing the location of different objects. This should be accompanied by general and detailed photographs.

The interviews should start as soon as possible after the event and initially should focus on the immediate sequence of events associated with the accident or near accident and the emergency response. Initially, the interviewee is asked to give a narrative of the event from his or her perspective without intervention by the interviewer. The interview results are then used to examine detailed facts and to confirm information received earlier. First, participants and witnesses are interviewed in order to establish an accurate description of the sequence of events. Next, the line organisation and staff officers are interviewed to identify contributing factors and root causes.

13.7.1.5   Evaluation and organising of information

Evaluation and organising of the information starts after the initial inspection of the site and the interviews with participants and witnesses. Here, the aim is to establish a description of the sequence of events and to compare it with existing information to determine whether there are any inconsistencies. The methods for displaying the sequence of events presented in Section 13.6.2 will support this work. Here, it is important to identify and evaluate deviations from normal practices and failures of existing barriers and controls. Often, the team will find it necessary to return to the scene or to conduct additional interviews to check and complement previously gathered information.

In the subsequent analysis of contributing factors, the team will look into specific aspects related to the workplace (see Section 13.6.5). The team will evaluate such factors as:

- *Hardware:* It is important to evaluate whether equipment has functioned as intended and whether the intended function was adequate. The team will also determine if the design, inspection, testing, and maintenance routines were adequate.
- *Human performance:* The team first reviews the tasks performed by the operators to identify human errors and to evaluate the adequacy of recovery actions and emergency response. Influences from design, procedures, and other circumstances (including emotions) should be evaluated (see Chapter 10). The team also has to address whether the involved personnel had adequate qualifications and the necessary instructions.
- *Procedures and accepted practice:* The team will evaluate the adequacy and coverage of procedures for the work affected by the event. It is also important to review accepted practice and whether or not this was followed. Identified discrepancies between procedures and accepted practice have to be addressed and evaluated.

Any changes from normal routines may contribute to an accident and have to be evaluated. Such changes may over-stress the organisation and make it more accident-prone. The team may apply a formal *change analysis* involving (Johnson 1980):

1. Description of the accident (or near accident) and a comparable accident-free situation.

2. Comparison of the two situations and identification of any changes in the situation of the accident as compared to the accident-free situation.
3. Evaluation of the changes to determine their effects on the accident. Here, it is important to consider obscure and indirect relationships.

In the subsequent analysis, the team will focus on contribution factors at the workplace and root causes in the general and HSE management systems. Here different techniques are available to support the investigation (see Section 5.6). In the next section, we will go into detail about the use of SMORT in root cause analysis. The team will apply the process of arriving at conclusions according to ISO 19011 when addressing management issues in the analysis (see Section 12.3.1 and Figure 12.7).

The team members must be aware of the risk of getting into conflicts with management when pursuing underlying management and organisational factors. Company managers usually are overburdened with daily duties and routines and want to avoid adding more tasks to their agenda, especially those perceived as difficult to manage. They expect the team to come up with immediate causes of the incident and tangible and effective countermeasures. Management also wants to avoid critical comments regarding their area of responsibility and management style.

The investigation team, on the other hand, is responsible for a thorough analysis of the underlying root causes. The team may also feel compelled to put forward many recommendations without considering the costs and benefits adequately. A gap between the expectations of management and the team is a breeding ground for frustration, anxiety, and self-censorship on the part of the team. At the end of the day, the team may find that management overlooks their recommendations. To circumvent such problems, the team has to act in such a way that its integrity is maintained. The team must at the same time strive for an open dialogue based on mutual support and respect.

13.7.1.6 *Preparing the team's report*

The results of the investigation are summarised in a report. The main results and conclusions in the report are presented at the follow-up meeting for quality check (see example that follows). The report is then forwarded to the client (the manager responsible for appointing the team) for approval, implementation, and follow-up. It should meet the general requirements of documentation of quality audits; see ISO 19011 (ISO 2011). A typical table of contents for a Level 3 investigation report is shown below:

1. *Executive summary,* presenting the essential facts and findings, probable causes and contributing factors, recommendations for managerial controls and safety measures that are necessary to prevent recurrence, and possibly, the need for further investigation.

2. *The Investigation Team composition and mandate.*
3. *Facts* (i.e. a description of the workplace, the conditions at the time of the accident, and the sequence of events). A timeline or STEP diagram may be used to present the results of the mapping of the sequence of events. This section also includes a list of identified deviations in the execution of the work where the incident occurred and an analysis of barrier failures.
4. *Conclusions* (i.e. the team's judgements regarding contributing factors and root causes of the accident). Here, it is possible to use the format for documentation of audit observations according to Section 12.3.
5. *Recommendations* regarding remedial actions, further investigations, and so on.
6. *Attachments*
   - Investigation program and a list of interviewed persons
   - A list of received documents used in the investigation
   - Photographic evidence

13.7.1.7 Follow-up meeting
The team presents its conclusions and recommendations to the affected organisation in a follow-up meeting. The aim is to give the involved personnel a chance to clarify misunderstandings and to give further explanations and comments. The follow-up meeting also fills the purpose of creating an atmosphere of trust and willingness to learn from the event.

13.7.1.8 Follow-up and close-out
The line manager who has appointed the team will be responsible for distribution of the report and for the initiation of the required actions.

13.7.2 Applying SMORT in Level 3 investigations

13.7.2.1 Outline of a SMORT analysis
The stepwise procedure of a SMORT analysis follows the steps described in Section 13.7.1 and is similar to that of a quality audit (cf. ISO 19011 in ISO 2011). Here, we will highlight those aspects of special relevance to SMORT:

1. *Appointing an investigation team:* One of the team members must be acquainted with the SMORT method.
2. *Introductory meeting and planning:* SMORT may be used to present the scope of the investigation in the introductory meeting. SMORT questions are selected and adapted to the specific circumstances of the event.
3. *Collection of information:* Results are documented in the SMORT questionnaire in order to facilitate analysis and reporting.

4. *Evaluation and organising information:* Evaluations of results from interviews, inspections, document reviews, and so on, are summarised by the use of the SMORT checklist and the following colour codes:
   Black: Critical, serious weakness needing immediate attention
   Red: Nonconformity, does not meet minimum standard
   Yellow: Observation, meeting minimum standard but should be improved
   Green: Checked and found acceptable
   Blue: Information is missing
5. *Preparing the report and follow-up meeting:* SMORT may be applied to display the results in a summary way. Reference should be made to SMORT in the report to show that a systematic method has been applied.

SMORT has primarily been designed for use in the investigation of severe accidents and near accidents. It complements existing routines for accident and near-accident investigations. SMORT provides a systematic and stepwise means of unfolding all relevant causal factors by starting with the specific and proceeding through the various managerial levels of the organisation.

An accident investigation is a search process and can only partly be planned in advance. It is thus recommended that the SMORT analysis of an accident be divided into different phases and that the aforementioned procedure is repeated for each phase. A full SMORT analysis may involve three phases: Tiers 1 plus 2, Tier 3, and finally, Tier 4. The report from one phase is used as input to the SMORT analyses in subsequent phases.

The SMORT checklist and questionnaires are intended for use in planning the investigation and in interviews and document checks. The questions may need to be rephrased during the interview in order to cover the actual situation. Checklists and questionnaires help to ensure comprehensiveness and objectivity in the investigation. The investigation aims to reveal causal factors. Here, it is important to distinguish between facts (i.e. causal relations that may be objectively documented) and judgements about contributing causes. The accident investigation will also go far beyond specified requirements.

13.7.2.2 Case: Incident involving flow of water into tunnel during repair work

Here, we will build on the analysis from the Level 2 investigation into the case presented in Section 13.6.1. An investigation team was appointed by the head of the project division of the hydropower company. The incident was selected for a Level 3 investigation because of its high potential and because it involved important learning aspects regarding project work in a plant in operation. The investigation team consisted of three members, a lead investigator with senior HSE management and investigation/auditing experience, a senior manager for operations from another company, and a senior project manager. The team had the mandate to examine the sequence of events and the underlying contributing factors and root causes in the management system. The aim was to determine learning and recommend improvements.

Initially, the team met with plant and project management and safety personnel to review the event logs and the results of the preliminary investigation by the plant personnel. The timeline and STEP diagram in Section 13.6 originally were established by this Level 3 investigation team. After this initial Tier 1 SMORT analysis, the investigation team decided to focus on planning hazardous work as well as coordination among simultaneous activities, including the use of a PTW system and management of contractors. The identified contributing factors are shown in Section 13.6.

Figure 13.17 shows an arrangement of the team's findings at SMORT, Tiers 1–4 in a diagnostic diagram, starting with the top event and tracing the underlying events and conditions back to the root causes in the hydropower company. The arrows do not represent causal relationships in a strict sense. Rather, they represent the investigators' interpretations of the available facts and their arguments regarding the need for changes. In carrying out the investigation, evidence is collected in a stepwise fashion, starting with the top event and proceeding downward as shown in the diagram.

The investigation report was handed over to the plant and project management, who initiated a number of actions. One immediate action was to arrange weekly meetings between project and operations to review and coordinate activities. The PTW routines were also reviewed. A PTW training program was established for the supervisors, operators, and contractors. A 'standard' contract for repair work was established that included plant rules regarding health and safety and emergency response, and requirements to permit to work.

13.7.3   Legal aspects of the team's report

In a criminal investigation of a severe accident, the police will look for breaches of statutory duties and other evidence of criminal negligence or wilful criminal acts. When addressing the question of culpability, the investigation will produce defensive reactions on behalf of the individuals and the organisation concerned. These reactions will counteract openness and trust and will obstruct the learning process. The Labour Inspectorate may in parallel perform a supervisory investigation into breaches of the health and safety legislation and regulations.

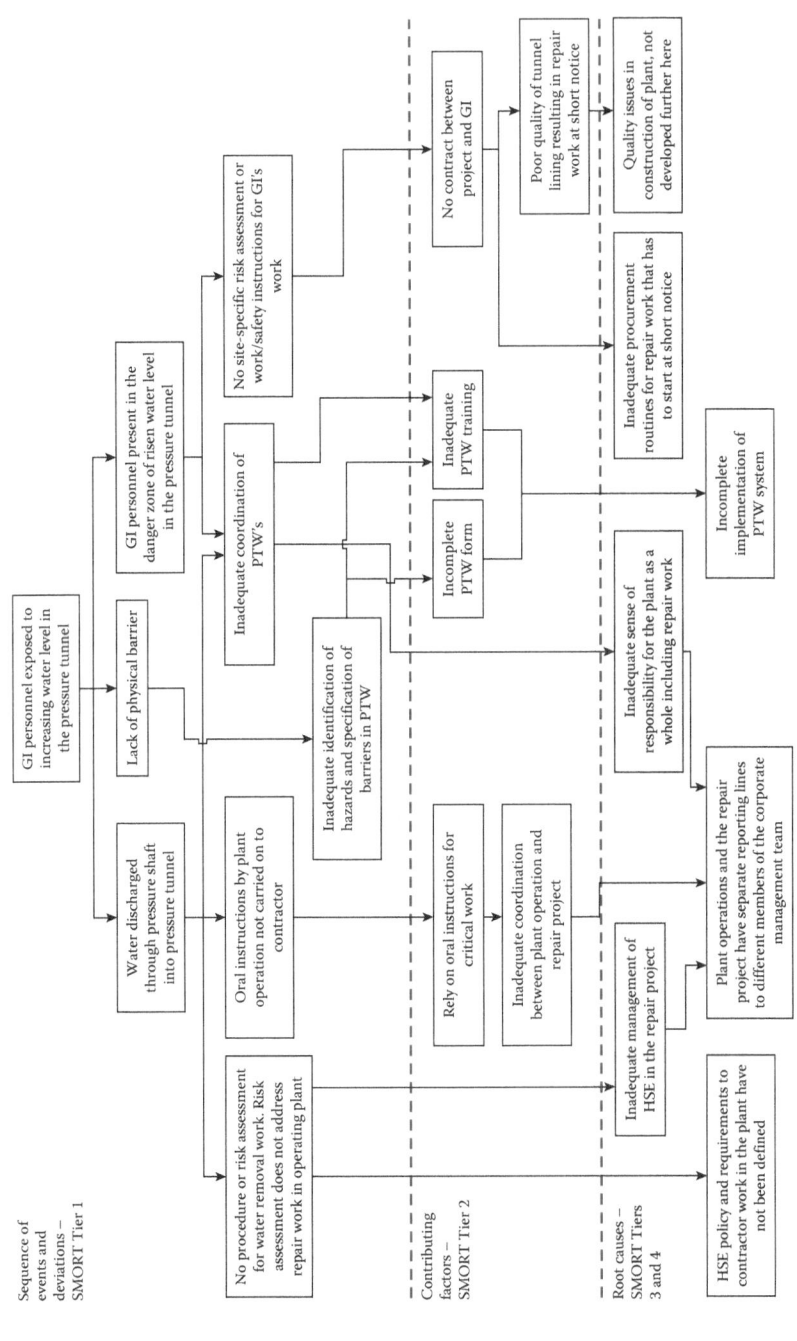

Figure 13.17 The results of the SMORT investigation displayed in a diagnostic diagram.

Earlier, it was common to pin the blame on the persons closest to the scene of the accident and with direct influence on the accidental event (i.e. workers and first-line supervisors). Legislation on requirements for companies' HSE management systems has made it possible to impose infringement charges on the company or penalties on its higher management such as fines or even imprisonment. The legislation on liability for compensatory payment in the area of product safety works in the same direction.

Police investigations and subsequent trials and negative publicity have deterring effects. Under certain circumstances, they will contribute to an improved safety level by focusing management's attention on its responsibilities and may counteract complacency. It must be acknowledged, however, that organisational learning from accidents and deterrence are separate goals that often are in conflict. The industrial organisation has to address this conflict in its internal procedures for accident investigations.

At the same time, the authorities must be aware of the negative side effects of police investigations. It has been proposed that the authorities' law enforcement role be played down in cases where the company demonstrates a clear willingness to learn from the accident. This principle should also apply when there is significant new learning involved as compared to a case when the same type of accident happens again.

The team's in-depth accident report often reveals circumstances that are sensitive from a personal, legal, or public relations point of view. Examples of such circumstances are (Hale et al. 1997):

- Violations of statutory requirements
- Evidence of hazards that have been known to management but have remained uncorrected
- Failure to ensure compliance with company rules and procedures
- Deliberate concealment of facts

It usually is the client that decides on the status and distribution of the report. The client may decide to restrict the distribution of parts of the report or of the full report. It is, however, important that the main purpose of the report is to clarify the facts and learn from mistakes not to perform 'criminal investigation'. The report has to reflect this purpose.

It is common practice for the company's legal department to review the report to ensure that such purpose is duly reflected and that individuals are not unnecessarily exposed. The intention of the review is not to advise the investigation team to remove facts that are important for the full picture of an event. However, the legal department cannot prevent the company or individuals from being exposed to police investigation and possibly penalties.

When the accident is under police investigation, the company management and employees are protected against self-incrimination and have the right to remain silent. This right does not apply in internal investigations by a company-appointed team because of the employer's management prerogative. Because a prosecutor has the right to require company documentation, including internal investigation reports, self-incriminating information may in this way be made available to the court. Knowledge of this fact will affect the team's work and may hamper openness and trust. A careful balance must be kept between the need for openness, to take full advantage of the accident experience, and the involved persons' legitimate right of self-protection when they face the risk of legal action. It may also be relevant to establish the right for personnel involved in an investigation to be assisted by, for example, an employee representative.

chapter fourteen

Accumulated incident experience

In this chapter, we will look into the use of databases of incidents. These are experience databases with compilations of data about accidents and near accidents and sometimes also about unwanted occurrences including unsafe acts and conditions. After incidents happen, the reporting system for the individual incident represents an immediate feedback loop. There is also a feedback loop based on aggregated data. By application of aggregated data, learning from multiple incidents is ensured. Such an approach is helpful for uncovering trends – frequent incidents and patterns among multiple incidents.

The stored incident data typically consist of facts about the incident, including time and place, activity, the sequences of events, type of incident (energy involved), and consequences (Table 14.1). The database may also include remedial actions and status of implementation. Here, we are focusing on the use of incident databases at the company level. Organisations and authorities also maintain incident databases used to support decisions in a similar manner to what we describe here.

We will here highlight seven purposes for an incident database. First, it provides *input to decision-making* by answering questions such as:

- What are the characteristics of our incidents? What can we learn from them?
- Has this type of incident occurred before?
- Are there any particular incidents with this machine that we need to consider when buying a new machine?
- Where shall we allocate resources to prevent incidents?
- What are the causes of our incidents?
- What has been done to prevent the incidents? What is the status of the remedial actions that have been decided?

Second, it is an important tool for the *establishment of periodic summary reports* on incident statistics. Such summaries are distributed periodically (e.g. monthly, quarterly, or yearly) depending on the size of the company and the level of risk. Their aim is to provide feedback on safety performance and to provide information about incident cases and incident distributions.

Table 14.1 Proposed 'smallest efficient set' of data elements in accident and near-accident records for computer storage

Category	Data elements and data types	
	Information from the reporter of the incident at the site	**Information added by and activities for the safety expert**
Administrative conditions	• Date, time • Department/site (free-text description or fixed alternatives) • Place (fixed alternatives)	Quality assurance of the information from the site
Injured persons	• Information about injured persons (role, occupation)	Quality assurance of the information from the site
Work situation	• Activity (fixed alternatives) System/equipment involved (free-text description, eventually tag number)	Quality assurance of the information from the site
Sequence of events	• Free-text description of sequence of events	• Incident-type classification (fixed alternatives) • Free-text description of deviations Quality assurance of the information from the site
Consequences/ losses	• Description of injury/damage • Type of injury/damage (free-text or fixed alternatives) • Type of event (fixed alternatives, i.e. accident, near accident, material damage, fire, including high potential [HIPO])	• Actual and potential loss (fixed alternatives) Quality assurance of the information from the site
HIPO incident	• Yes/no	Quality assurance of the information from the site
Remedial actions	• Free-text description • Responsible (name/position) • Due date	Follow-up

Third, the incident database supports *follow-up actions*. It provides simple feedback to decision-makers on responsibilities, target dates, and status of implementation in order to ensure timely implementation. A more advanced application is to study the long-term effects of implemented measures (e.g. by monitoring developments in the frequency of a specific type of incident).

Fourth, the database provides *input for other safety management tools and methods* as an experience database. Incident data are, for example, important for hazard identification in risk assessments. An incident database can also provide important input for incident investigation to check whether similar incidents have occurred previously.

Fifth, it supports *intra-organisational experience transfer* as a mean of sharing experience with other projects and plants.

Sixth, the accumulated incident experience is dependent on employee reporting. Input to the database is thus fostering *employee participation* and can also promote employees' feelings of ownership in the company and confirmation that the company cares about their concerns.

Finally, an incident database provides a means of *administrative support* through storing and keeping track of incident records.

14.1   Applying incident data in a safety information system

In this section, we will elaborate on the model of a safety information system presented in Section 7.5. After an incident report has been submitted for processing, the first step is to assess its severity and potential consequences. Quality assurance of the report follows, which may include adding more information, if necessary, before the incident is entered into the database.

As described previously in this chapter, the incident database serves several purposes. These are accomplished through retrieving and analysing data in an interactive manner. The users of a database thus have to: (1) update the database with facts on new incidents and (2) query the database by retrieving relevant data and generating reports that present the results of queries. Safety professionals have a key role in the safety information system because it is they who assure the quality of the recorded data and who retrieve and analyse incident data for different purposes.

In Chapter 7, we described experience feedback as a fundamental principle of systematic safety management. An incident database plays an important role in experience feedback as a compilation of incident experience from different sources. Safety management based on the reporting of incidents follows the knowledge creation spiral described by Nonaka and Takeuchi (1995) and depicted in Figure 7.11. The incident database is based on tacit knowledge acquired by the operators at the sharp end. This knowledge is made explicit through documentation – information from which is eventually fed into the database.

A traditional means of implementing incident experiences is to update work instructions. These documents will successively accumulate incident experience and become experience carriers for use in the training of new employees and in the follow-up of safe behaviour. In this way, the explicit knowledge will be internalised by the individual worker and, in combination with work experiences, will change his/her tacit knowledge.

We have to interpret the experiences from individual incidents and from analyses of incident concentrations and translate them into specific requirements for use experience carriers such as work instructions and technical specifications. If the incident database is put into use in this way, it will promote involvement of management and employees as well as the development of a shared understanding and ownership of the reporting system and its use in the prevention of accidents. Experience shows that this requires a dedicated community of practice inside the company that 'owns' the system and acts as a learning agency towards other stakeholders; see Section 7.6 (Drupsteen and Guldenmund 2014).

14.2 Incident database

A database is a collection of related facts or data that have been designed, built, and populated with information for a specific purpose (Elmasri and Navathe 2007). In order to define an incident database, we have to specify the types of facts or *data elements* to be included in each incident record. For each element, a *data type* must be defined. The data can be coded according to a nominal, ordinal, interval, or ratio scale or stored in free-text (Section 5.2). Both these specifications, the data elements to be added and the data type for each element, are included in the so-called *database definition*. Each data element and its associated data-type specification are linked to a *field* in the database definition.

In establishing a database definition, we have to decide what accident model to apply. This is a crucial decision. We have to make a careful balance between two concerns:

1. The amount and quality of the required information, as defined by the accident model
2. The time and attention needed for investigations and quality assurance of the results before storage in the database

In Chapter 5, we discussed problems that the investigator faces when applying a complex accident causal model in the collection of incident data.

A careful balance has to be maintained among the amount of information required, the quality of the information, and the time needed for the investigation. This balance is problematic because the benefits of detailed information of a high quality are usually advocated by a part of the organisation other than that responsible for the collection of the data and that have the first-hand experience.

Table 14.1 presents a proposed 'smallest, efficient data set' on incidents. It is based on parts of the framework for accident analysis presented in Chapter 5. (For classification of nominal data – that is, fixed alternatives – see Chapter 5.) Facts about the incident process and losses are described. However, contributing factors and root causes are not included in the database because we are restricting the data stored in the database to observable facts.

The data consist of both fixed data and free-text descriptions. The free-text description of the sequence of events should neither be too short nor too long and at the same time provide the minimum amount of information needed for the safety expert and others to understand. To fulfil this requirement, criteria for free-text descriptions should be given. The reason why a safety expert is needed to feed additional data is to secure the reliability of the data. Checking the quality of the data from the supervisors is an important task of the HSE staff.

14.2.1   Coding of incident data

In Section 5.2, we distinguished between two different types of data on incidents. Incident data can either be qualitative (observations documented in free-text descriptions) or quantitative (we use a scale to quantify the observations).

Here, we will discuss the principal difference between qualitative or free-text data on the one hand and coded data on the other hand. We will focus on the trade-off between using free-text descriptions and coding the data according to a coding schedule (a nominal scale).

Free-text data are by nature unstructured and thus less suitable for analysis of the type found in standard periodic reports on incident statistics. By coding the data, it is structured and standardised for easy retrieval and presentation in tables and diagrams showing the distribution of incidents by part of the body, type of event, type of cause, and so on. These advantages are acquired at the cost of loss of richness of the information. Classification of data leads to a loss of information about the detailed circumstances of the incident. These details are often essential to our understanding of why the actual incident happened, especially when we study complex incidents. We also run the risk of distorting the data in the coding process and concealing uncertainties.

14.3   *Accessing the database*

The user queries the database in two steps. These are repeated until a satisfactory result has been achieved. They include:

   Step 1: Searches in the database to identify the subset of incidents that meet certain search criteria
   Step 2: Presentations of the results in the form of event descriptions, statistical summaries, and so forth

   Example: An incident database for an aluminium plant includes about 500 records. The foundry manager is interested in the number and types of recordable injuries in his department during the last year. He performs a search for events that fulfil three conditions – for example, foundry, recordable injury, and last year. The result is 58 hits. He then displays the events in three histograms showing the distribution of incidents by nature of injury, part of the body affected, and incident type.

   The search criteria are determined by the database definition. Basically, there are three different principles for searches in the database:

   1. Searches for events that meet criteria defined on an ordinal, interval, or ratio scale of measurement.
      Example: Find all incidents for a particular year.
      Example: Find all incidents where the injured person is younger than 25 years of age.
      Example: Find all incidents where the injured person has been absent for more than 30 days.
   2. Searches for events that meet criteria on a nominal scale of measurement.
      Example: Find all incidents where women have been the victim.
      Example: Find all 'person falling to a lower level' incidents.
   3. Searches for events that include certain words in the free-text descriptions of, for example, the sequence of events.
      Example: Find all incidents where 'truck' is mentioned in the description of the sequence of events.
      Example: Find all incidents where the victim has been hit by falling scaffolding materials by selecting 'falling object' incidents (principle 2) and searching for *scaffold* in the sequence of events.

   There are two types of errors associated with these searches (see Figure 14.1). Type I errors concern the *degree of retrieval* (i.e. wanted data that are not found). Type II errors concern the *degree of precision* (i.e. retrieved data that are not wanted). Reasons for these errors can be

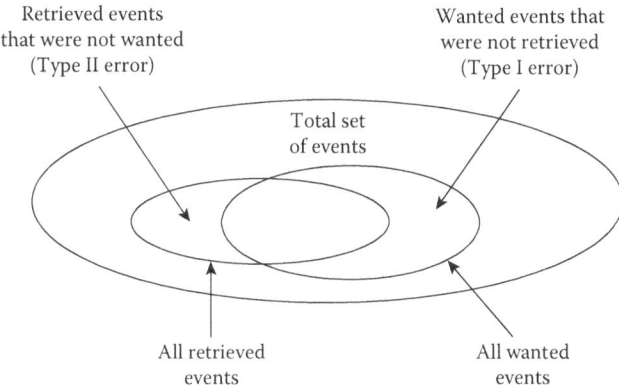

Figure 14.1 Errors related to the degree of retrieval and degree of precision.

misspellings in free-text descriptions or wrong categorisations (e.g. classifying an incident with falling objectives as an incident with person caught in or between objects). The way the database has been defined will affect the size of these two errors. Experience shows that a skilled user will achieve a higher degree of retrieval in free-text searches than in searches based on fixed alternatives (Kjellén 1987b). This high degree of retrieval may, however, be achieved at the cost of a low degree of precision.

There are certain qualities of an incident database that need to be in place if the database is to function as intended and to fulfil the different purposes described in the beginning of this chapter (Sepeda 2006):

- The database must be *accessible* and easy to access when needed.
- *User friendly* storage as well as retrieval
- Data entered must be *accurate,* that is, comprehensive, objective, fact-based, and valid.
- *Sufficient volume* (i.e. number of records). A larger database will strengthen the statistical significance but only if the data are accurate. A small database with accurate data will not provide a basis for statistical analysis but can serve as input for qualitative analysis.
- *Standardisation* on what data elements are registered in the database (cf. Table 14.1).
- *A comprehensive query system/search engine* that supports in answering the query.
- Ensuring confidentiality, integrity, and availability of the database (i.e. *information security* is necessary to prevent misuse, changes,

or manipulation of the information). Confidentiality of employees described in reports must be ensured. Reading and writing permissions must be established and maintained.

14.4   Analysis of incident data

The retrieved data are analysed and then presented as input for decision-makers or communicated to stakeholders. There are different approaches for the analysis of incident data. We will present some of them in the following sections.

14.4.1   Finding incident repeaters

An important question in connection with incident investigations is whether the same type of incident has occurred before. This question is easily answered by making a search in an incident database. The most efficient means of searching for similar occurrences is to utilise the free-text search facilities of the information retrieval program. It is important to note that different persons may use different words to describe the same work operation, equipment, and so on. There are also spelling errors and misconceptions that hinder efficient free-text search (i.e. a Type II error). The analyst must be well acquainted with the workplace in question and the different dialects in use. Following are some examples of applications of the incident database to find incident repeaters:

Example: There had been a truck accident at an airport. The HSE manager wanted to know whether there had been similar incidents with trucks before. She first looked into the coded field 'injury agent' for vehicles and found two reports, one of which was relevant. Next, she made a free-text search on 'truck' in the sequence-of-events description and found these two reports and another two reports. One of these was also relevant.

Example: A team on an offshore installation had performed a job safety analysis for entering a tank. The safety engineer wanted to check whether any important hazards had been omitted. In a free-text search, an accident with electrical equipment was identified that had not been identified as a hazard in the job safety analysis.

Example: The catering manager on an offshore platform was planning a first-aid course for kitchen personnel. He searched the database for accident cases illustrating the importance of first aid.

Example: One of the compressor trains on an offshore installation was due for replacement. The technical manager searched the database for previous incidents involving compressors in order to take information about experiences from these into account when purchasing the new compressor.

14.4.2   Uni- and bi-variate distribution analyses

Uni- and bi-variate distribution analyses use coded data. Results are displayed in tables, histograms, pie or radar charts, or other diagrams, showing the distribution of incident characteristics in absolute number or percentage. Uni-variate analyses show the distribution of one variable, and bi-variate analyses display cross-tables of two variables. This is a well-established data analysis method used in companies' periodical statistical summaries of accidents and near accidents as well as in companies' annual reports.

Example: The yearly accident and near-accident summary report of a medium-sized construction company displays results of various uni- and bi-variate distribution analyses, including:

- A bar chart of the distribution of reported events by type of event (i.e. a uni-variate distribution). This graph showed that there had been twenty total recordable injuries (TRIs), 274 first-aid injuries, 450 material-damage incidents, and 709 reports on unwanted occurrences.
- A bar chart showing the distribution of events by type of event and year for the last three years (i.e. a bi-variate distribution). It showed a decrease in the number of TRIs during the last year as compared to the previous two years (34 and 41, respectively). The frequency of reports of unwanted occurrences developed in a positive direction.
- A pie chart of the distribution of TRIs by activity. It showed that most incidents had been related to lifting operations and transportation.
- Separate tables of the distribution of lost-time, medical treatment injuries, restricted work cases, and first-aid injuries by type of injury and part of the body injured. Cuts in the finger were the most frequent combination.
- A bar chart showing the distribution of TRIs by incident type; see Figure 14.2.

Uni- and bi-variate distributions give an overview of the incidents and help prioritisation in relation to in-depth analyses and in selecting remedial actions. Graphs are easy to read but may lack precision. Figure 14.3 shows another illustrative example. Here, the analysis identifies the types of energies involved in the TRIs in different departments of

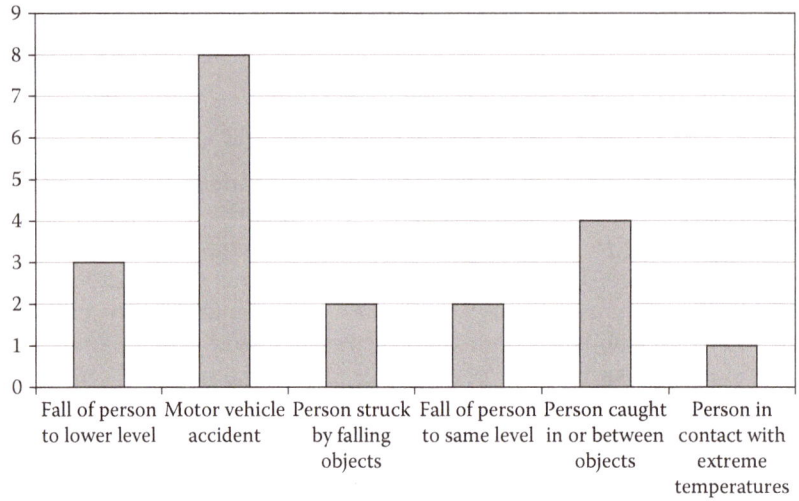

Figure 14.2 Uni-variate distribution of TRIs by incident type (N = 20).

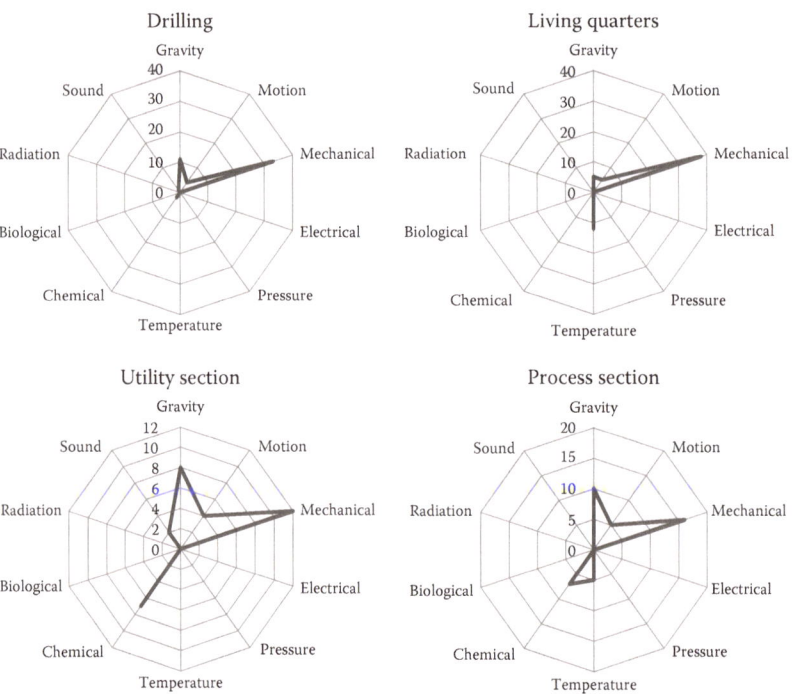

Figure 14.3 Distribution of type of energy for TRIs at two offshore installations by department.

an offshore installation. A further investigation shows that the incidents involving mechanical energy mainly are squeezing incidents in the drilling, process, and utility sections and cut injuries in the kitchen of the living quarters.

14.4.3   Incident concentration analysis

Incident concentration analysis makes use of several of the methods described in this subsection with the purpose of identifying clusters of incidents with common characteristics. The concentrations will indicate types of measures to prioritise. This method was first developed within road safety research and was denoted *black-spot analysis*. By analysing the geographical distribution of road traffic accidents, accident concentrations with respect to location were identified. Critical locations (such as crossings) with many accidents were assessed and modified to prevent future accidents.

The basic idea is to find concentrations or black spots that indicate where to prioritise safety measures. Black-spot analyses are used by many road authorities around the world but based on different models (Geuerts and Wets 2003). A common factor for these traffic safety models is that they limit the analysis to two dimensions.

In industrial safety, incident concentration analysis is carried out in several dimensions. Such an analysis will show that incidents concentrate around certain types of characteristics. The distribution will be a fingerprint for the analysed site/project, often displaying the energies involved in different activities. The identified concentration will provide decision-making support for remedial actions that will reduce the concentrations.

The steps of an incident concentration analysis are:

1. *Establish the context for the analysis:* What is the system to be studied? What is the time period being studied? What types of incidents should be studied? The types of incidents need to be relatively homogeneous. It is not meaningful to mix, for example, material damage incidents with incidents involving personal injury. The data set must not be too small and should include a minimum of 50 records.
2. *Establish uni- and bi-variate distributions for different dimensions of the total set of incidents for the system and time period to be analysed (do not include causal factors):* Location, activity, equipment, incident type (type of energy involved), type of injury, and part of the body affected. Identify incident concentrations along these different dimensions.

3. Select concentrations making up a significant portion of the total number of records (e.g. 5–10 out of 50 records) with similar characteristics and analyse these concentrations more in detail. Look for similar patterns in activities, sequence of events, and energy involved.
4. Make recommendations for measures to reduce incident concentrations.
5. Implement and follow up safety measures.

The analysis is facilitated if some of the data are coded, especially if large quantities of data are handled. An incident concentration normally involves the analysis of coded data and free-text descriptions that are 'coded' in the course of the analysis.

Example: This is a simple case to illustrate the use of incident concentration analysis. A construction project had experienced 53 incidents classified as 'high-potential' (i.e. with the potential to result in fatality or severe injury) over a period of three years. Of these incidents, 1 was a fatality, 24 were TRIs, 11 were material damage incidents, and 17 were near accidents. The first step of the analysis involved a uni-variate distribution analysis with respect to incident type (i.e. energy involved). Of the 53 incidents, 34 incidents (63%) belonged to two categories, 'movement of motor-vehicle' (18 incidents) and 'falling objects' (16 incidents). Both categories were subject to incident concentration analysis. For our purposes, we will focus on the first category.

A second step in the analysis was to establish new data elements in the database for the 18 incidents involving motor-vehicle movements with the following customised data:

- *Place:* Work area, site road, and public road
- *Activity:* Work, personnel transportation, and transportation of materials
- *Type of vehicle:* Pick-up, minibus, truck, and construction machine
- *Type of event:* Road departure, roll over, and uncontrolled movement (other)

The categories for each data element were based on the information in the reports. In the third step of the analysis, bi-variate distributions were analysed. Table 14.2 shows the distribution of the incidents with respect to activity and incident type.

Road or work area departure is the dominating event involving loss-of-control of energy, followed by roll over and other type of uncontrolled vehicle movement. It can involve significant amounts of energy when, for example, a truck continues off the road and into the valley in a mountainous area. There were no specific characteristics of the road departure incidents with respect to type of vehicle involved. Five of the road departure incidents involved trucks, four pick-ups, one minibus, and two construction machines (grader and roller).

Table 14.2 Distribution of HIPO incidents related to vehicle movement with respect to type of activity and incident type (N = 18)

| | Incident type | | |
Activity	Departure from road/work area	Roll over	Uncontrolled movement (other)
Personnel transportation	5	2	0
Material transportation	5	1	2
Construction activity	2	1	0

There were no further accident concentrations among the four roll-over incidents. Two of the incidents occurred when the driver drove off the road or out of the designated work area?

To sum up, road departure represents a significant accident concentration in the project. In a subsequent, detailed analysis of the road departure reports, they identified the following deviations and contributing factors:

- *Human factors* (six events): Inexperience, fatigue, and inattention
- *Environmental factors* (three events): Steep road inclination, bad weather conditions, and hole in the road
- *Technical factors* (two events): Ran out of fuel

Lack of roadside barriers were not mentioned but are an obvious measure to reduce the likelihood of this type of event.

14.4.4   Analysis of accident causes

In the example in the previous section on incident concentration analysis, we examined the 'accident causes' identified in the original reports. These were deviations such as fuel shortage and contributing factors such as steep road inclination and driver fatigue and inexperience. Of these, fuel shortage and road inclination represent observable facts, whereas the driver-related 'causal factors' are derived from interpretations of facts. The latter conclusions need to be treated with more care.

Results of uni- and bi-variate distribution analyses are often used as proof of causal relationships. Earlier, we quoted a study that shows that 88% of accidents are caused primarily by unsafe acts (Heinrich 1959). This statement is based on a uni-variate analysis of incidents by immediate cause. It is, for example, justifiable to speak about 'causes' of an incident if there is a logical relationship between the causes (specific events or conditions in the chain of events) and the loss. It is valid to state that, '25% of the injuries in the foundry were caused by contact with hot metal' or that 'a third of the injuries in the slaughterhouse were cut injuries due to contact with knives'.

'Unsafe acts' are examples of factors more distant from the injury event, where the logical relations are more uncertain. So-called ex post facto analyses are used to study whether a causal relationship may exist. In these analyses, we compare incident statistics with similar statistics for incident-free situations. The aim is to identify factors that are more common in the incident material than what is expected because of pure chance. In the next step, physical, physiological, and psychological theories are brought in to explain the actual causal relationships.

Example 1: The effect on the risk of accident from using a mobile telephone while driving can be studied in this way. Such a causal analysis requires access to statistics on the percentage of drivers involved in incidents that use mobile telephones and statistics on the percentage of time that drivers in general use such telephones. A significantly higher percentage in the first case indicates that the use of mobile telephones increases the risk of accidents. The analysis does not, however, say anything about the causal mechanisms leading to an increased risk of accidents when using a mobile telephone.

Example 2: Let us consider a hypothetical example of injuries at a hamburger chain. An analysis shows the distribution of injuries by the victim's job experience according to Table 14.3.

The conclusion was reached that inexperience was a primary 'cause' of accidents. A more detailed analysis revealed the picture according to Table 14.4, which rejects this hypothesis.

Both these examples illustrate the importance of considering exposure. Without considering exposure, an analyst will be biased due to insensitivity to the sample size (Tversky and Kahneman 1973); thus, there is a danger of jumping to the wrong conclusions.

Statistical hypothesis tests such as χ^2 are used to study the significance of statistical relationships. Basic textbooks on statistical hypothesis refer to such testing. Even if such tests show strong statistical correlation, the causal explanatory value of the results can be questioned.

Table 14.3 Distribution of burn injuries at a hamburger chain by job experience

Job experience	Number of injuries
<6 months	33
6–12 months	18
1–2 years	9
3–5 years	5
>5 years	2

Table 14.4 Distribution of burn injuries at a hamburger
chain by job experience

Job experience	Number of injuries	Number of employees	No of injuries per employee
<6 months	33	412	0.08
6–12 months	18	180	0.1
1–2 years	9	124	0.07
3–5 years	5	98	0.05
>5 years	2	37	0.06

A famous safety expert commented on Heinrich's results showing that 88% of accidents are caused by unsafe acts: 'To say that most of the accidents are caused by human error is like saying that falls are caused by gravity'.

14.4.5   Analysis of remedial actions

Statistical analyses of accident causes do not necessarily lead us closer to the goal of reducing the risk of accidents. The value lies in the possibility of predicting the effects of measures to reduce the risk of accidents. One of the data elements in an incident database is remedial action. The data on remedial actions can be used to indicate the level of learning in an organisation. Van Court Hare's (1967) order of feedback (see Chapter 7) is used for this purpose. By applying Van Court Hare's classification of order of feedback in an organisation, we can make an analysis of the degree of learning from incidents; see Table 14.5 for examples. For each incident, there can be several remedial actions – each representing an order of feedback.

The analysis can be used to produce meaningful statistics to indicate the level of learning. The analysis can also include links to severity and potential severity. Such an analysis will indicate whether the organisation learned from high-potential incidents.

14.4.6   Statistical methods for incident analysis as part of continuous improvement

We have already described the close link between safety management and quality management and have shown that many of the basic principles in safety management are rooted in quality management (e.g. feedback control, Deming's circle). Ishikawa (1976) has described seven basic quality tools for process improvement and problem solutions; most of these are relevant for the analysis of incident data (two of them, histograms and control

Table 14.5 Analysis of order of feedback of an injury with high potential

Sequence of event	Remedial actions	Order of feedback
The operator of a construction machine moved the machine to the left due to a slippery floor, trapping and squeezing the helper between the machine and the tunnel wall	Issue a new procedure for risk assessment of such operations	Level 2
	Develop instructions for control of personnel in the danger zone during mucking	Level 1
	Check the maintenance standard of the construction machine	Level 1
	Training for the operators of the machinery to ensure safe operation	Level 1

charts, are presented in other parts of this book). One of these tools is the Pareto diagram, which is a useful tool for identifying and evaluating the factors that contribute to incidents with the highest frequency (Janicak 2003). Figure 14.4 shows an example of a Pareto diagram for activities leading to personnel injuries in the Norwegian fishing fleet from 2000 to 2011, based on data from Aasjord et al. (2013). The aim of this diagram is to identify the activities that contribute to most of the incidents. It is important that all the studied factors are included in the diagram. The Pareto diagram is a helpful tool to identify the most frequent factors, but it does not consider severity.

Another quality tool is a scatter plot that looks for relationships between pairs of variables. Scatter plots present the correlation between a dependent and an independent variable. Severity-distribution analysis and extreme-value creation are examples of such diagrams.

Severity-distribution analysis can be used in estimations of the probability of severe incidents at a workplace and in comparing different workplaces with respect to the expected severity of incidents (Briscoe 1982). A severity-distribution analysis is based on incidents within a specified period in time and follows a stepwise procedure:

1. Arrange the incidents by consequence (e.g. the number of days of absence) in an ascending order.
2. Divide the highest registered consequence value into intervals such that each interval has approximately the same size on a logarithmic scale; see Table 14.6 for an example.

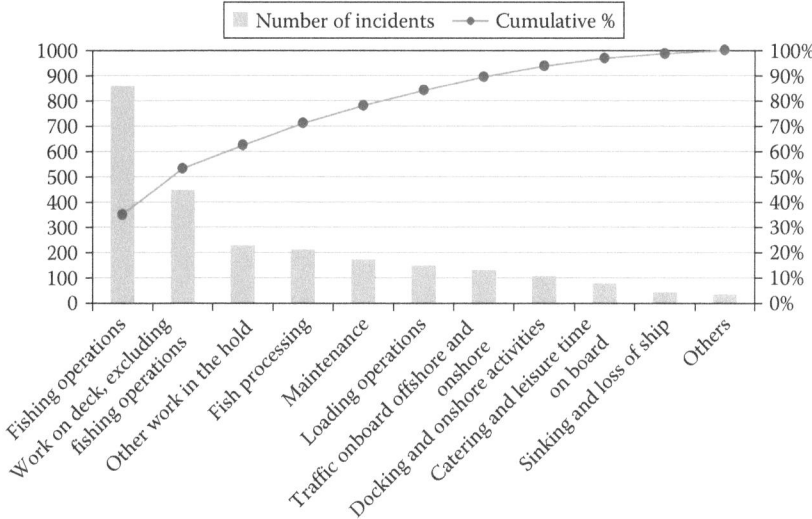

Figure 14.4 Pareto diagram of activities involved in injuries in the Norwegian fishing fleet from 2000 to 2011. (Based on data from Aasjord, H.L., et al., *Analyse av årsaksforhold ved dødsulykker og alvorlige personskader i norsk fiskeri* [Analysis of causes of Fatal Accident and Severe Occupational Accidents in Norwegian Fishery], SINTEF Report A23369, 2013. In Norwegian.)

Table 14.6 Severity-distribution analysis of accidents at a steel mill

Days of absence	Number of accidents	Cumulative number of accidents (N_1)	% of the accidents ($N_1/(N + 1))*100$)
<1	55	55	24.6
1–3	50	105	46.9
4–7	48	153	68.3
8–30	57	210	93.8
31–90	9	219	97.8
91–180	2	221	98.7
181–365	1	222	99.1
>365	1	223	99.6

Note: N is the total number of accidents.

3. Tally the number of incidents in each interval and the cumulative number.
4. Calculate the cumulative percentage of accidents for each interval and use a log-normal paper to plot the results (see Table 14.6 and Figure 14.5).

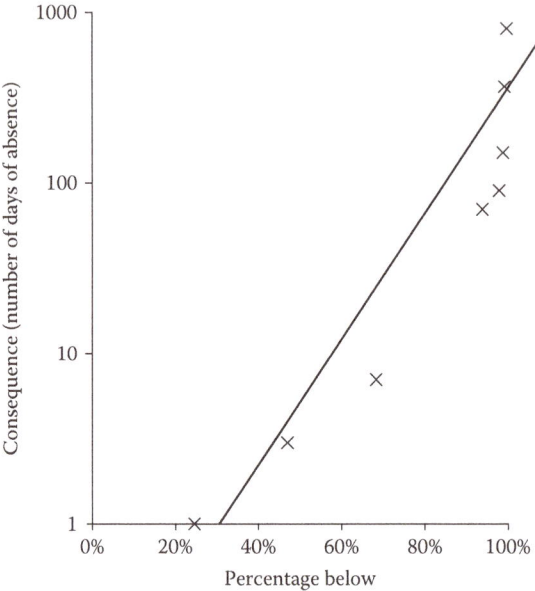

Figure 14.5 Severity-distribution analysis of the incidents at a steel mill.

Figure 14.5 shows an example of a severity-distribution analysis of the accidents at a steel mill. The slope of the straight regression line is an indicator of how probable it is that an accident will result in severe consequences. The higher the slope, the higher is this probability. In the case illustrated by the figure, there is a probability of about 0.02 that an accident will result in more than 100 days of absence.

The points with the highest severity lie above the straight line. This is an indication of the fact that accidents with severe consequences result from a different set of causes than those resulting in less severe accidents. It follows that these causes are not subject to the same strict management control. This is a warning signal that there is a need to look more closely into the severe accidents.

Janicak (2003) gives an overview of different statistical methods for the analysis of incident data. His overview includes the aforementioned methods, but he also adds methods involving probability calculations such as correlation analysis and regression analysis. These approaches will provide a statistical foundation for trending and forecasting safety performance. Janicak also shows different sample charts that can be used in control charts to monitor the safety performance.

section four

Monitoring of safety performance

In Section IV, we will present safety performance measures or indicators for use in the types of health, safety, and environment (HSE) management activities that involve feedback control (see also Section 7.3). A performance indicator is the measure of a key process in order to provide an objective basis for determining process performance (Daugherty 1999). Here, we will use the following definition of a safety performance indicator: a metric used to measure the organisation's safety performance in terms of its ability to control the risk of accidents in its activities. The risk of accidents is defined as a combined measure of the probability or frequency of accidents (per unit of exposure) and the consequence (extent of loss). We will also review safety performance indicators used to measure the results of safety processes and activities, tacitly assuming that they have a positive impact on the risk of accidents.

Section IV will focus on safety performance related to the control of accidents. Occupational diseases and environmental discharges as the result of planned activities will not be treated in any detail. Here, feedback control involves the establishment of safety performance goals and the follow-up of such goals through monitoring of actual performance and implementation of corrected actions if the actual performance deviates from the goal in the direction of too high a risk. The principles for establishing goals may differ:

- A fixed goal is established for a specific period (e.g. the following year). This type of goal usually is based on previously achieved results at a company or on results of other companies with which management wants to compare its company.
- The safety performance indicator must show continuous improvements from one period to the next.

Safety performance indicators play an important motivational role, especially at the higher management levels. Adequately designed performance indicators make safety visible in a summary way that is suited for communication, comparison, and benchmarking.

We will review different indicators of safety performance and their merits and shortcomings when applied in HSE management. We will base the review on the quality criteria for safety performance indicators developed in Chapter 8:

1. Observable and quantifiable
2. Valid indicator of the risk of loss
3. Sensitive to change
4. Compatible
5. Transparent and easily understood
6. Robust against manipulation

Chapter 15 gives an overview of safety performance indicators, based on the accident analysis framework in Chapter 5. The overview shows performance indicators for all phases of an accident. Chapter 15 also presents a review of different interpretations of the concepts of lagging and leading indicators. Chapter 16 presents different indicators using data about accidents and losses as the main input. Chapter 17 presents various indirect indicators based on data about the accident process as a precursor to loss. In Chapter 18, we will look into safety performance indicators, where the focus is on the underlying contributing factors at the workplace and root causes in general and HSE management systems.

Safety performance indicators are used in a system for safety performance measurement utilising all four sub-systems of a safety information system: data collection, processing, memory, and distribution. Chapter 19 addresses the selection of a combination of indicators for use in HSE management. It also discusses the potential (negative) side effects in safety performance monitoring that need to be considered in designing such systems.

chapter fifteen

Overview of safety performance indicators

Figure 15.1 presents an overview of the different safety performance indicators that we will discuss further in Section IV. This overview is based on the framework for accident analysis presented in Chapter 5. It limits the selection of safety performance indicators to those measuring the risk of accidents for use in feedback control (see Section 7.3). This separates performance measurement from auditing (Chapter 12), which is based on diagnosis. The distinction is not sharp, and some of the indicators at the input side in Figure 15.1 utilise data from audits.

Loss-based safety performance indicators will be our starting point. Among these, we find the most commonly used indicator, the total recordable injury frequency rate (TRI rate). We proceed by reviewing process-based indicators, which are used in feedback control according to principles similar to those of the fertiliser-plant case in Chapter 3. Finally, we will look into indicators relating to causal factors (i.e. indicators based on information about the contributing factors and root causes in the general and the health, safety, and environment [HSE] management systems of the organisation).

The validity of the safety performance indicators as measures of accident risk becomes more uncertain, the further to the left in Figure 15.1 that we move. At the far left (root causes), the performance indicators are dominated by measures of structural and cultural issues related to the management of HSE at the corporate level.

There are alternative ways of presenting overviews of safety performance indicators in the literature. The International Council on Mining and Metals (ICMM 2012) comes close to the model in Figure 15.1. In their overview of human-error focussed indicators, they distinguish between indicators representing outcome and proximate and distant causes. The outcome indicators are similar to the loss-based indicators in Figure 15.1. The proximate cause indicators include use of personal protective equipment (PPE), human error, and equipment and system failures. The indicators representing distant causes are divided into two categories, individual attitude and organisational culture.

The UK Health and Safety Executive (2001a) in their guide to measuring safety performance takes a risk-control approach in categorising

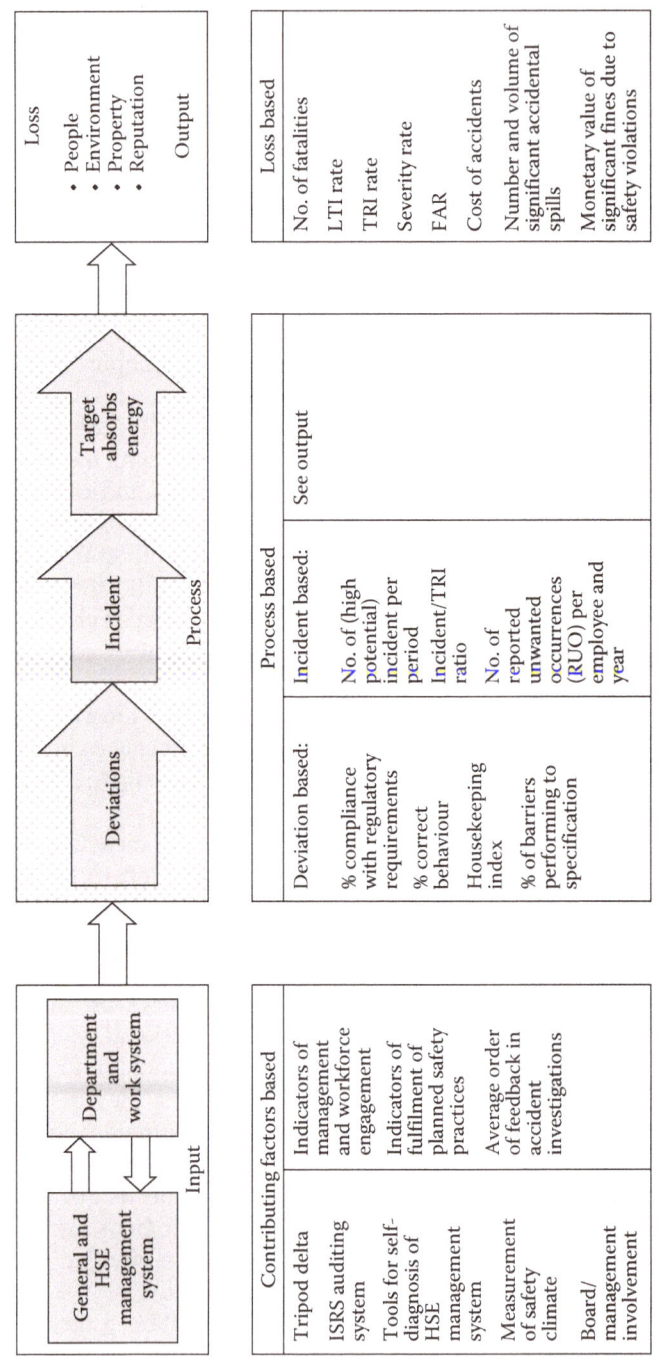

Figure 15.1 Overview of safety performance indicators.

performance indicators. They identify the following elements as subject to measurement:

- *Input:* 'Hazard burden', generally based on risk assessments.
- *Process:* Health and safety management systems and risk-control systems, workplace precautions, and health and safety culture. Here, the focus is on what works well rather than on failures.
- *Output:* Adverse outcome (injuries and so forth).

The UK Health and Safety Executive (2006) have published guidelines on process safety indicators. They relate the indicators to the different organisational levels of a company (corporate, site, facility). The lower level indicators are similar to those proposed by the International Association of Oil & Gas Producers (OGP 2011) in their publication on indicators in process safety (see Section 15.2).

15.1   Measures of risk

Generally, we use safety performance indicators to express an organisation's ability to control the risk of accidents in its activities. A subset of safety performance indicators, particularly those based on historic incident data, are intended to measure the risk of accidents in the future. The assumption here is that, unless you implement risk-reducing measures, the risk will remain about the same as in the past. The *risk of accidents* is defined as a combination of the probability or frequency of accidents involving losses in a specified activity and the extent of the losses (consequences). The measure of risk has three components:

1. A measure of the *exposure* to the activity involving accident risks.
   Example: In the TRI rate (see Chapter 16), the exposure is measured in terms of the number of man-hours for the activity in question.
2. A measure of the *frequency* of accidents (or injuries as a result of accidents) in the activity per unit of exposure.
   In the TRI rate, the number of injuries per million work-hours for the activity in question determines the frequency.
3. A measure of the extent of loss (consequence).
   The consequence in the TRI rate is measured on a dichotomous scale (yes/no). An injury is recordable if it fulfils any of four criteria: fatality, lost time beyond the day of the accident, medical treatment beyond first aid, or transfer to a less-demanding job (so-called restricted work).

There are two ways of reducing the risk of accidents in an activity. One way is to reduce the frequency of accidents per unit of exposure to the activity. The second way is to reduce the consequences. If we look at an organisation as a whole, there is also a third way of reducing the over-all risk of accidents. The risk of accidents is not the same for all activities in the organisation. Hence, we may reduce the overall risk of accidents by reducing the exposure to high-risk activities. Transferring personnel to less hazardous work does the job. It is important, however, that the risk of accidents does not increase for the remaining personnel performing the high-risk activity.

15.2   Leading and lagging safety performance indicators

The overview in Figure 15.1 does not distinguish between leading and lagging indicators, a distinction that has received increasing attention in the safety profession (Hopkins 2009). In the field of economics, the term *leading indicators* refers to predictors of economic development. These are indicators that change before the economy as a whole changes (O'Sullivan 2003). Lagging indicators, on the other hand, are those that usually change after the economy as a whole changes. For our purposes, we will define safety performance indicators in accordance with their use in economics:

- *Leading safety performance indicators* predict future developments in safety performance, that is, they change before the safety perfor-mance has changed (Janicak 2003; Kjellén 2009).
- *Lagging safety performance indicators* change after an activity's safety performance has changed.

The ability to respond ahead of time based on anticipation is a characteristic of resilience management systems (Hollnagel 2014; Podgórski 2015). A combination of leading and lagging indicators is required for such systems to sustain required operation through adjustment of their behaviour prior to, during, or following changes and disturbances.

The loss-based performance indicators to the right in Figure 15.1 are lagging in the sense that they, when published, estimate the risk of accidents based on statistics from events that occurred in a specific time period in the past. This time lag depends on the length of the period used for recording data and the time it takes to publish the results. The same applies to the process-based safety performance indicators, simi-larly derived from historic performance data from a past time period. The process-based indicators mainly measure frequency of incidents or degree of compliance with a standard, and it is tacitly assumed that these indicators correlate with the risk of accidents.

A high frequency of incidents and deviations is also a symptom of a bad 'control climate' in the organisation. According to Grimaldi (1970), this is correlated with an increased frequency of accidents, property damage, unscheduled production stops, accidental environmental discharges, and production outside of quality norms. Thus, it can be argued that a significant increase (or decrease) in the frequency of incidents and deviations results from a deteriorating (or improving) control climate. Consequently, they are leading indicators because they predict future changes in accident risk (see also Körvers et al. 2008).

In a similar way, it may be argued that process-based indicators determined from self-reported incidents and deviations of the employees, so-called unwanted occurrence reporting, represent leading indicators. They are a measure of the employees' willingness to share their experiences rather than 'true' measures of incident and deviation frequencies and thus are associated with a good learning culture in the organisation, which affects future accident risk in a positive way (see Chapter 17 for further details).

In more general terms, Podgórski (2015) argues that operational safety performance indicators that measure how well a given safety management system operates at the shop-floor level are early warning indicators that forecast an organisation's future behaviour and ability to adjust to changes and disturbances. These are characteristics of leading indicators.

There is no consistent use of the terms leading and lagging safety performance indicators in the safety literature, and there are interpretations that compete with the one we have presented here. The UK Health and Safety Executive (2006) uses the following definitions:

- *Leading indicators:* Relate to safety management activities and are either measures of those activities or the results of the activities.
  Examples: Measures of safety activities and measures of failures (deviations) discovered during routine check.
- *Lagging indicators:* Measures of unexpected failure occurring in normal operation.
  Examples: Accidents, near accidents, and deviations discovered following adverse events.

Janicak (2003) defines leading indicators as measures predicting future safety performance by 'assessing outcome actions taken before accidents occur and are measures of proactive efforts designed to minimise loss and prevent accidents'. He mentions measures of the quality of an audit program and behavioural sampling (percentage of safe behaviour) as examples of leading indicators; here, the latter is considered a lagging indicator.

In a guide for authorities and communities on safety performance indicators for chemical accidents, the Organisation for Economic Cooperation and Development (OECD 2008) makes the distinction between outcome (lagging) and activities (leading) indicators. Outcome indicators assess the effect of safety-related actions (policies, programs, procedures, and practices) on the likelihood of accidents and/or an accident's adverse impact on human health, the environment, and/or property. *Activities indicators* measure, on a regular and systematic basis, whether an organisation implements its priority actions as intended. Activities indicators help explain why a result (e.g. measured by an outcome indicator) has been achieved or not. Two examples from emergency response planning are call-out time after the occurrence of an accident (outcome indicator) and the extent to which an emergency response team has the necessary training and experience to deal with the various types of accidents that might occur (activities indicator). In Chapter 18, we will review outcome and activities indicators for safety systems and practices but will refrain from using the terms leading and lagging indicators in this context.

Based on experience from the UK oil and gas industry, the publication, *Leading Performance Indicators – Guidance for Effective Use*, defines a leading indicator as providing information about the current situation that may affect future performance (SCS 2003). The aim is to 'correct potential weaknesses without waiting for demonstrated failures'. This view coincides with that of Tarrants (1980), who pointed out the advantage of near-accident reporting as a proactive tool in the sense that it allows for actions to mitigate causal factors before they have resulted in accidents with loss. This is not consistent with the interpretation used here. Any indicator based on information about existing system weaknesses or incidents are here considered as lagging. This is irrespective of whether or not they have result in loss, which is a chance event.

The publication 'Step Change in Safety' presents a five-level Safety Culture Maturity Model for management commitment, responsibility, and consistency in the area of safety and proposes three levels of leading performance indicators to match the needs of companies at different levels of maturity (SCS 2003):

- *Level 1, Compliance:* Indicators of the organisation's degree of implementation of its management system and complying with regulatory requirements.
- *Level 2, Improvement:* Indicators of the effectiveness of the company's management system in indicating areas of weakness and need for improvements.
- *Level 3, Learning culture:* Indicators based around local (workforce) selected issues.

The ICMM (2012) provides a similar overview for the mining industry. There is no clear definition of a leading (or activities) indicator in this publication. Instead, it is described as 'measuring the direct and indirect precursors to harm, and giving advanced warning before an event occurs that might lead to an undesired outcome, providing the opportunity for a preventive action to be taken'. The ICMM adapts the maturity level approach for leading safety indicators of 'Step Change in Safety' and the publication gives examples of leading indicators at each level such as:

- *Level 1, Compliance:* Percent of occupational health and safety (OHS) legislation addressed by procedures; percent of complete statutory training requirements.
- *Level 2, Improvement:* Staff perception of management commitment to OHS; percent of planned behaviour-based task observations completed.
- *Level 3, Learning:* Percent of job tasks for which risk assessments have been performed; percent of staff having agreed responsibility and accountability (on OHS).

The OGP (2011) chooses a different approach in a publication on leading and lagging indicators in process safety – or more precisely in preventing accidents due to loss of primary containment (LOPC) of hazardous materials. They use the bow-tie model (discussed in Chapter 9) with LOPC as the primary event to define leading and lagging indicators in four tiers:

- *Tier 1:* Process safety events involving LOPC with greater consequences (e.g. major gas leakage)
- *Tier 2:* Process safety/LOPC events of lesser consequence
- *Tier 3:* Near-miss or unsafe act/condition related to LOPC below Tier 2 consequence thresholds or identified barrier or risk-control system weakness or failure.
  Examples: Number of process leaks identified during operation or down time; percent of management inspections delegated to subordinates
- *Tier 4:* Activities to maintain and improve barriers and risk-control systems.
  Examples: Percent of maintenance plan completed on time; percent of manager inspections of work locations completed

The OGP does not classify indicators as either lagging or leading; rather there is a sliding scale from Tier 1 to 4, where Tier 1 indicators are lagging and Tier 4 are predominantly leading. This interpretation is based on earlier publications by the UK Health and Safety Executive (2006) and the American Petroleum Institute (API 2010) and captures different dimensions:

- *Realised versus potential consequence:* Indicators are defined as lagging when they measure actual consequence, whereas indicators based on the potential (high) consequence of incidents are considered as leading to a greater extent.
- *Companywide versus local use:* Lagging indicators (especially Tier 1) are suited for companywide monitoring and benchmarking with other organisations, whereas leading indicators (especially Tier 4) are more suitable for local use.

The OGP points to the necessity of complementary use of leading and lagging indicators rather than basing the management of HSE on leading or lagging indicators only. They recommend 'dual assurance' from the UK Health and Safety Executive (2006), meaning the pairing of lagging indicators on barrier defects with leading indicators on tests of barrier strength.

It is obvious from this review that there is no consensus in the safety science community on the understanding of the difference between leading and lagging indicators. Here, we will apply the definition as used in the field of economics, where leading safety indicators are predictors of future developments in safety performance. This will limit the different types of indicators that fall into this category compared to the multitude of indicators claimed to be leading in the literature.

chapter sixteen

Loss-based safety performance indicators

16.1 The lost-time and total recordable injury frequency rates

The lost-time injury frequency rate (LTI rate) used to be the most common indicator of safety performance. It is defined as the number of lost-time injuries per one million hours of work. A lost-time injury is an injury due to an accident at work, where the injured person does not return to work on the next shift.

The LTI rate has been criticised on a number of grounds, among them that it is relatively easy to manipulate (e.g. by assigning an injured person to a lighter job to avoid lost days). Many companies use the so-called total recordable injury frequency rate (TRI rate) instead. One purpose is to increase the statistical basis by including types of injuries other than those resulting in lost time. A second purpose is to make the indicator more robust against the type of 'manipulation' where the injured employee is transferred to another job rather than being sent home on sick leave.

A recordable injury is an injury occurring during working hours resulting in fatality, lost time (as per the definition noted earlier), medical treatment other than first aid, or in restricted work (alternative job assignment or reduced work load). A medical treatment injury is an injury at work due to an accident or act of violence that requires treatment by a licensed doctor, or nurse in consultation with a doctor, before the injured person resumes normal work.

In applications of the LTI and TRI rates in health, safety, and environment (HSE) management, we calculate the rate for a period in time from statistics on accidents and working hours and compare the results with pre-established safety goals. We are also concerned with the development of the LTI or TRI rate over time.

In this chapter, we will focus on the use of the TRI rate. Where not explicitly stated, the same principles apply to the LTI rate. We will look more closely into the principles of analysing time series of the TRI rate before we discuss the merits and shortcomings of safety performance measures based on the LTI and TRI rates in general.

16.1.1   The control chart

In time-series analysis, we are interested in the development of the safety performance indicators over time. This is done in so-called *control charts*, where the safety performance indicator is calculated for consecutive periods and displayed in a diagram (Janicak 2003). These charts allow us to study changes in performance from one period to the next and trends over several periods. Here, we will look into the control chart for the TRI rate. Figure 16.1 shows an example.

The measured TRI rate for a period is an estimate of the underlying or 'true' TRI rate. From the statistical theory presented in Chapter 6, however, we know that this estimate is uncertain and that we must expect the rate to fluctuate from period to period due to pure chance. The upper and lower control limits define the range in between which the TRI rate will fall on average in 19 out of 20 periods (i.e. 95% confidence interval), provided that the underlying accident risk at a company is unchanged. These limits are calculated from the mean TRI rate during the studied periods because this is considered to be the best estimate of the 'true' TRI rate. We consider a change in the TRI rate for the latest period to be significant if its value falls outside the control limits for earlier periods.

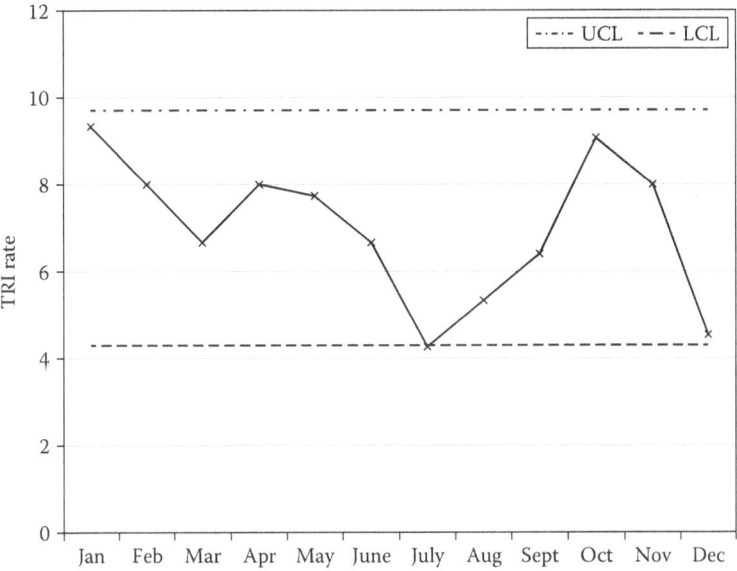

Figure 16.1 Control chart showing the development of the TRI rate for 12 consecutive periods (UCL = upper control limit; LCL = lower control limit).

Let us consider the example shown in Figure 16.1. The lower and upper control limits (LCL and UCL) have been calculated on the basis of the TRI rates for 12 consecutive periods. There is a 5% probability that the value for the thirteenth period (January of the following year) will fall outside these limits by pure chance (95% confidence interval). If the TRI rate for this period is 3, for example, we conclude that the safety performance has improved significantly.

The UCL and LCL are calculated by applying Formulas 16.1 through 16.3:

$$p = X_{tot} * \frac{10^6}{\left(N * e\right)} \tag{16.1}$$

$$UCL = p + 2 * \sqrt{\frac{p * 10^6}{e}} \tag{16.2}$$

$$LCL = p - 2 * \sqrt{\frac{p * 10^6}{e}} \tag{16.3}$$

where p is the mean value of the TRI rate for all periods that are included in the control chart, X_{tot} is the total number of recordable accidents, N is the number of periods, and e is the mean number of working hours per period. UCL and LCL are applied in evaluating the significance in changes in the TRI rate.

Example: An aluminium plant has 2000 employees, and each employee works 1800 hours a year. During the last 10 years, an average of 35 recordable accidents per year was registered. For a control chart with a periodicity of one year, we calculate the following:

$$\text{Average TRI rate} = 350 * \frac{1,000,000}{10 * 2000 * 1800} = 9.7$$

$$UCL = 9.7 + 2 * \sqrt{\left(9.7 * \frac{1,000,000}{2000 * 1800}\right)} = 9.7 + 3.3 = 13.0$$

$$LCL = 9.7 - 2 * \sqrt{\left(9.7 * \frac{1,000,000}{2000 * 1800}\right)} = 9.0 - 3.3 = 6.4$$

The procedure for the establishment and use of control charts is as follows:

1. Start establishing the control chart only when at least 100 accidents have been registered. Do not use data more than five years in age.
2. Decide about the period length; there must not be less than five accidents per period on average. An average of at least 10 accidents per period is preferred. Period lengths of more than one year are not meaningful.
3. Calculate the TRI rate for each period.
4. Calculate the mean TRI rate for all periods and the UCL and LCL according to Formulas 16.1 through 16.3 shown earlier.
5. Draw the control chart.
6. If one point falls outside the control limits, calculate new control limits without this point. If another one or more points fall outside the control limits, the safety performance is not stable. Causes of the unstable safety performance have to be identified.
7. Use the control chart to determine whether or not the TRI rate for the next period falls inside the control limits.
8. Use the control chart on a routine basis to evaluate the safety performance for new periods. Keep the control limits updated with accident and exposure data until the period before the last one. Avoid using data more than five years old or from earlier periods that for some specific reason are not representative of the current situation.

Example: Ship Yard, Inc., has 25,000 employees, and the number of working hours per employee and year equals 1800 on average. The following total recordable injuries (TRI) per month occurred during the last year:

January:	29
February:	36
March:	29
April:	26
May:	24
June:	30
July:	19
August:	17
September:	33
October:	25
November:	16
December:	31

To establish a control chart for the TRI rate, we calculate the control limits using Formulas 16.2 and 16.3. For the whole period, there are 315 TRIs, X_{tot} = 315. One period is a month; hence, mean number of working hours per month must be calculated, e = 25,000* 1800/12 = 3,750,000; p can then be calculated using Formula 16.1, which gives us p = 7.0. The control limits are calculated similarly to the earlier example: UCL = 9.7 and LCL = 4.3. The next step is to calculate the TRI rate for each period (i.e. month). We use Formula 16.4 to calculate the TRI rate for period i:

$$\text{TRI rate}_i = \frac{\text{TRI}_i * 10^6}{e_i} \qquad (16.4)$$

where TRI_i is the number of TRIs for period i and e_i is the number of worked hours for period i. For January, we have 29 TRIs and 3,750,000 hours worked, which gives us TRI rate$_{Jan}$ = 7.7. We calculate the TRI rate for each month and see that all months are within the control limits; we can thus conclude that the safety performance of the yard is stable.

In January of the following year, there are 10 TRIs, which gives us a TRI rate of 2.7. Based on the control chart from the year before, we can conclude that there has been a significant improvement in the yard's safety performance.

In safety practice, we rarely detect statistically significant changes in the TRI rate. Let us look at an example where management establishes a goal for the TRI rate for one year, which represents a 25% reduction as compared to the average TRI rate of the five previous years. When we apply the formula for the standard deviation of the Poisson distribution from Chapter 6, we can show that the mean number of recordable injuries per period must be at least 60 to be able to say that a 25% reduction in the TRI rate is significant. Only the largest companies are able to base their statistics on that number of accidents per year. It is also common practice to apply control charts of the TRI rate in this way when the goal does not represent a statistically significant change.

In quality control of production processes, control charts are applied to display results of statistical sampling tests of products. Basic textbooks propose the use of a confidence interval of ± 3 standard deviations in this application (see e.g. Tague 2005). For a stable distribution, 99.7% of the results of the tests will fall within this interval. The underlying normal distribution for the production process will determine the width of the area between the control limits. When the production series are large enough, the sampling size – and thus the degree of precision in the estimates of the mean value and the control limits – may be decided by the quality control engineer. In our application of the control chart, the underlying number of accidents per period determines the width of the area between the control limits. Often, we will struggle with the problem that the frequency of accidents is too low for us to be able to establish meaningful control charts. We have therefore relaxed the requirements to a confidence interval of ± 2 standard deviations.

16.1.2   Trends

We are not only interested in absolute changes in the TRI rate but also in changes over several consecutive periods. There is a positive or negative trend if the TRI rate increases or decreases from one period to the next during five or more consecutive periods. This figure is arrived at through simple probability calculus. The probability of an increase or decrease in two consecutive periods by pure chance is 0.5. Similarly, the probability for an increase or decrease during five consecutive periods is 0.06, which approximately coincides with our requirement as to significance. We very rarely observe trends based on consistent increase or decrease over five or more periods. Figure 16.2 shows an example where the fatal accident rate has decreased by 60% over a period of 10 years. Even in this case, there are no more than four periods with a consecutive decrease.

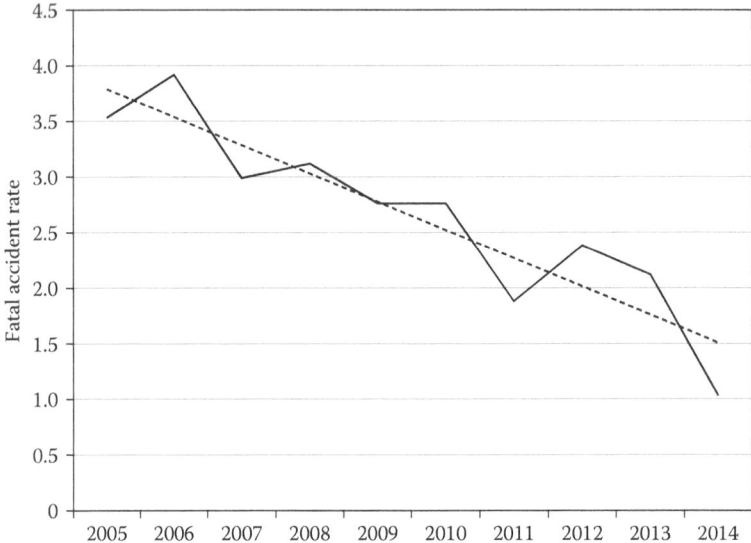

Figure 16.2 Chart displaying the fatal accident rate (number of fatal accidents per hundred million hours of work) for member companies of the International Association of Oil & Gas Producers. (Adapted from IOGP, *Safety Performance Indicators – 2014 Data*, International Association of Oil and Gas Producers, London, 2015). The figure also shows the linear trend line.

16.1.3   Comparison among plants

Figure 16.3 shows an example of the presentation of results of TRI-rate calculations for five plants. A presentation of this type not only allows for comparison of the total results but also allows us to check whether there are any significant differences in the reporting patterns among the plants. We would expect a distribution with a decreasing frequency of injuries as the consequence increases, but Figure 16.3 reveals deviations from the expected distribution.

A graph like this displays differences in reporting patterns – where Plant C did not report any restricted work injuries in the period in question. In cases where the TRI rate is totally dominated by lost-time injuries, we need to investigate further whether the reporting routines adequately capture medical treatment and restricted work injuries. If the answer is no, the plant is reporting a too low TRI rate.

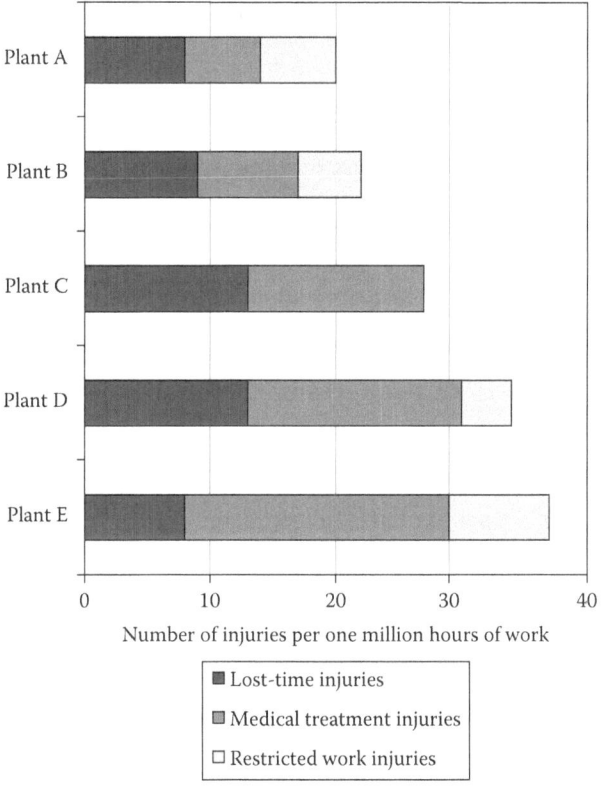

Figure 16.3 The TRI rates and their components for five plants of an aluminium company that produces car components.

16.1.4   The problems of safety performance measurement

The suitability of the LTI rate and later the TRI rate as safety performance measures has been extensively debated among researchers and practitioners in what is called the 'safety measurement problem' (see e.g. Rockwell 1959; Kjellén 2009). The 'problem' is visualised in an evaluation of the LTI and TRI rates by use of the requirements that safety performance indicators need to satisfy according to Table 8.1.

Let us take the TRI rate as the starting point. First, we ask whether it is quantifiable and measurable. This criterion is satisfied because the TRI rate is expressed on a ratio scale of measurement and draws from readily available statistics.

Next, we ask whether the TRI rate is a valid indicator of the risk of losses due to accidents. This requirement is more problematic because the TRI rate is insensitive to the severity of the injuries. An eye injury requiring treatment by a doctor to remove the foreign object and a severe fall injury with many months of sick leave count equally when calculating the TRI rate. The TRI rate thus scores low on validity. The same applies to the LTI rate, even if it does not include medical treatment and restricted work injuries. Other safety performance indicators that we will look into in the next section are better suited as to validity if they, for example, account for the degree of harm (e.g. number of days of absence).

Is the TRI rate robust against manipulation? This was a main reason why many companies decided to replace the LTI rate with the TRI rate. Let us first discuss this requirement for the LTI rate by looking at an example.

Example: Seven yards received contracts for the manufacturing of different modules for an offshore installation. The client (an oil company) monitored the LTI rates of the yards closely. Five of these ended up with an LTI rate below 20; the LTI rate was 55 and 70 for the two remaining yards. A closer analysis revealed that these latter yards had been more inclined to report minor accidents with only a few days of absence. Figure 16.4 shows the average percentage for five yards with an LTI rate below 20 and the actual values for the two remaining yards. 'Less severe' accident types include 'fall on same level', 'tripping', and 'fragment in eye'. Experience shows that these accident types often result in injuries with only a few days of absence. Further investigations showed that the five 'low LTI-rate yards' had established schemes for avoiding reporting minor injuries by, for example, offering the injured employees alternative work.

Experience shows that it is possible to manipulate the registration and classification of injuries and thus to affect the LTI rate as well as the TRI rate in a desired direction. This phenomenon may take place, for example, when top management unilaterally establishes ambitious safety goals for the workplaces. The effects occur when the safety

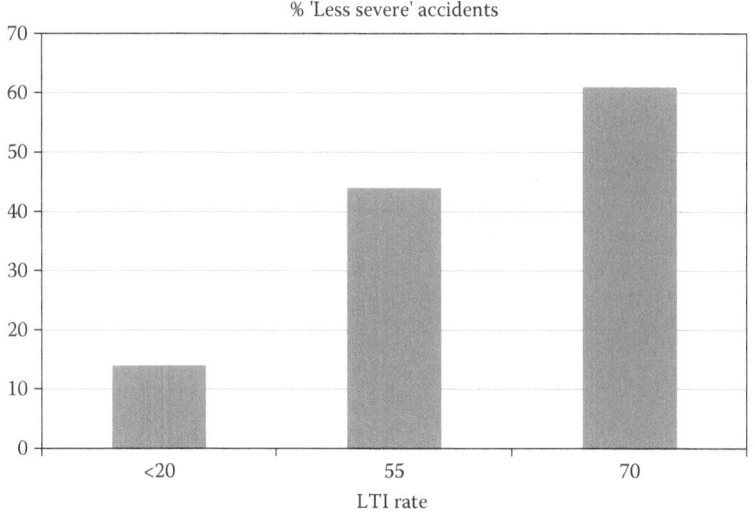

% 'Less severe' accidents

Figure 16.4 Percentage of the lost-time injuries at seven yards that were of a less severe type as a function of the LTI rate.

performance measure is closely monitored and negative results have consequences to the supervisors and the workforce. Examples of such consequences are top management attention and withdrawal of bonuses. A common means of reducing the number of reported lost-time injuries is to offer the injured person an alternative job that is less demanding. This will affect the LTI rate but not the TRI rate. We cannot expect to accomplish accident-risk reduction unless top management's ambitions to improve are shared by the whole organisation and affect the actual behaviour at the workplace level.

The issue of under-reporting is especially critical in contract work, where a high reported TRI rate may result in a bad reputation among potential customers and ultimately loss of further contracts (Kongsvik et al. 2012).

Is the TRI rate sensitive to change? To put the question in another way, will a change in the risk level at a workplace show up in the control chart as a significant jump in the TRI rate? The answer is most often 'no'. This has to do with the fact that recordable injuries are rare events. Five recordable injuries per period on average, for example, give statistical fluctuations in the TRI rate on the order of ±100% from period to period (see Chapter 6).

We will regard five recordable injuries per period as a minimum number in order to produce meaningful control charts. Figure 16.5 shows the minimum period length for a given number of employees that follows from this figure.

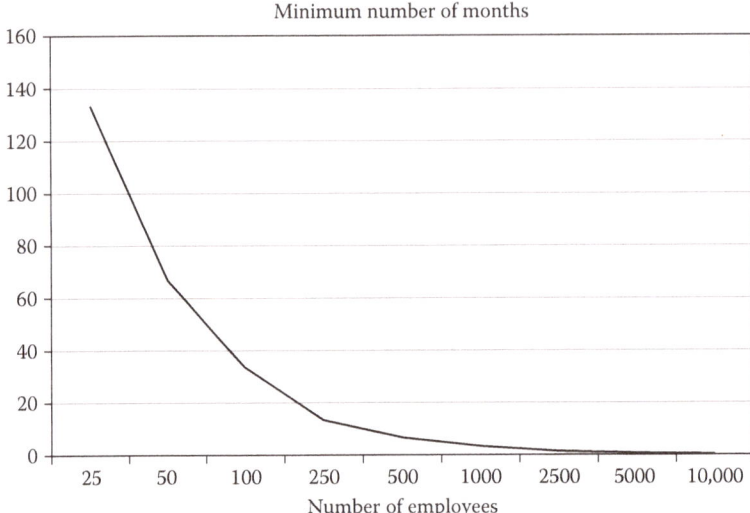

Figure 16.5 Relation between the minimum period length in months and the number of employees in order to produce meaningful control charts. A mean TRI rate of 10 is assumed.

For small companies, the period length will be too long in relation to our need for timely feedback. From Figure 16.5, we see that a company with 50 employees will have to use a period length of about six years in their control chart. Two methods are applied to increase the basis for the TRI-rate calculations for small companies:

1. In the *moving average* method, the mean TRI rate is calculated for a fixed number of periods ending with the last period. This method reduces not only the fluctuations but also the sensitivity to changes.
2. The *accumulated TRI rate* is calculated from one point in time and for the following periods. In project work, where the activities have a defined beginning and end, this method has its advantages.

Figure 16.6 displays these two methods together with a traditional display of the TRI rate.

First-aid injuries are more frequent events. It is not meaningful to include these events in the safety performance measures. First, they include many insignificant events, and the validity of the measure will decrease further. Second, the reporting of first-aid injuries is unreliable and affected by extraneous factors such as the availability of a medical clinic at the workplace.

Finally, we ask whether the TRI rate is easily understood and congruent with other performance indicators. Here, we see the two criteria that

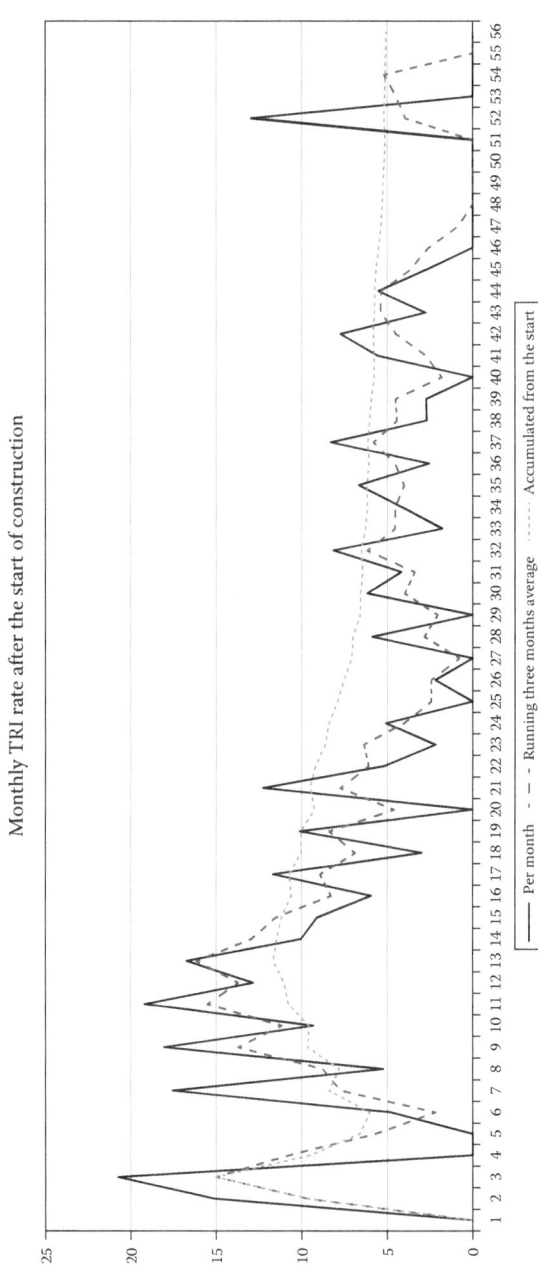

Figure 16.6 Chart showing the parallel developments of the TRI rate per month, the accumulated TRI rate from start of construction, and the moving three months' average at a construction site.

explain the success of the TRI rate in spite of its shortcomings. Experience shows that the TRI rate is easily communicated to the layman and that the merits of a low TRI rate are accepted by and large.

To summarise, we have discussed the following problems associated with the use of the TRI rate in control charts and in trend analyses:

1. Insensitivity to the severity of the accidents
2. Reporting inaccuracies
3. Statistical fluctuations

16.1.5 Zero-goal mindset

Many large companies, especially within major hazardous and construction industries, have adopted the so-called 'zero-goal mindset or philosophy' (Levitt and Samelson 1993). This philosophy is based on the belief that all occupational injuries and illnesses and process and environmental incidents are preventable and management's goal for them all is zero. It is an easily communicated goal with many implications that are not always fully understood.

The zero-goal mindset or philosophy is rooted in the 'continuous improvements' programs of the total-quality-management movement. The TRI rate is expected to improve by a certain percentage each year (e.g. 25%) as compared to the previous year. At the department level, where there may be only one or a few accidents each year, we often find goals of zero accidents. This type of goal is problematic because chance plays an important role at workplaces with a limited number of employees (Figure 16.5). In practice, the counting of accidents is reset at zero after an accident, and the updated goal is to have no accidents thereafter.

In striving for zero loss, management focuses its attention on minor incidents and injuries, motivated by the belief that a reduction in the TRI rate to virtually zero will be followed by a similar reduction in the risk of major accidents. Unfortunately, empirical evidence does not support this hypothesis. When the focus is on eliminating the causes of the few minor accidents that still occur, we cannot expect this to have any significant effect on the risk of major accidents. This is because there are different immediate causes of minor and major accidents.

16.2 Other loss-based safety performance indicators

Table 16.1 gives an overview of standard loss-based safety performance indicators. In calculating the frequency of accidents, the size of the activity for which performance is assessed has to be considered. We must expect more accidents to occur in a large company than in a small one, even if the activities are similar. It is necessary to standardise the indicators in

Table 16.1 Loss-based safety performance indicators

Type	Definition
Occupational accidents	
Total recordable injury frequency rate (TRI rate)	Total number of recordable injuries per 10^6 employee hours. The recordable injuries include fatalities, lost-time injuries, medical treatment injuries other than first aid, and injuries resulting in loss of consciousness, transfer to another job, or in restricted work.
Lost-time injury frequency rate (LTI rate)	Number of lost-time injuries per 10^6 employee hours for definition of lost-time injury.
Accumulated number of recordable injuries	Accumulated number of recordable injuries since the start of an activity (e.g. project).
Severity rate (S-rate)	Number of working days lost due to lost-time injuries per 10^6 employee-hours. Fatalities and 100% permanent disability account for 7500 days.
Average number of days lost	S-rate/LTI rate.
Days since last TRI/LTI	Number of calendar days since the last recordable or lost-time injury occurred.
Fatal accident rate (FAR)	Number of recordable fatalities per 10^8 working hours.
Accumulated number of fatalities	Accumulated number of fatalities since the start of an activity (e.g. project).
Environmental pollution	
Rate of emissions	Emissions due to accidents in kg or m^3 per ton production (e.g. emissions of fluorine to the air in kg per ton produced primary aluminium).
Material losses	
Loss rate	Number of accidents or loss in euros per produced unit (e.g. number of traffic accidents per 10^5 passenger-km).

relation to the *exposure* to the risk of accidents. For occupational accidents, the most common exposure measure is the number of employee hours. In traffic safety, for example, we use the number of vehicle kilometres or passenger kilometres as measures of exposure. By combining the frequency measures with the consequence measures, we can arrive at a measure of the risk.

Figure 16.7 shows a *Nelson–Aalen plot* of the accumulated number of accidents. The frequency of accidents appears as the inclination of curve. A related safety performance goal is to achieve a specific target at the end of the period. This shows up as a straight line in the plot.

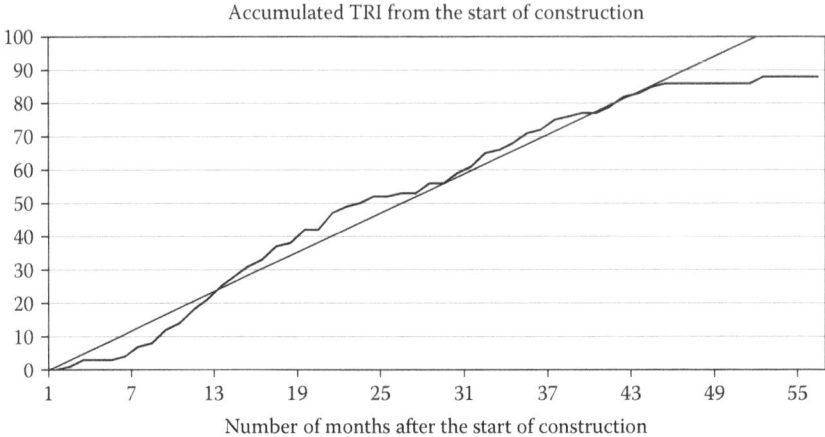

Figure 16.7 Nelson–Aalen plot of recordable accidents at a large construction project. The straight line shows the expected development, when the goal is a TRI rate less than 6, corresponding to a maximum of 103 accidents.

When the actual performance lies above this line, corrective actions need to be taken.

Two indicators the *number of days since last LTI* and the *accumulated number of lost-time injuries* both have the advantage of being readily understood. They are, however, insensitive to the size of the company and are thus not well suited for companies with significant changes in the number of employees.

The *S-rate* (severity rate) is less sensitive to reporting inaccuracies than the TRI and LTI rates. Single accidents resulting in a long period of sick leave may, however, dominate the statistics. It follows that the S-rate may vary considerably from period to period, especially in small companies. Another problem with the S-rate has to do with the fact that the sick leave for an injury may extend into the following periods. Thus, the true S-rate is not available until all injuries from a period have been closed.

Due to the low frequency of fatalities, the *fatal accident rate* (FAR) is rarely useful as a safety performance indicator other than for very large companies in hazardous branches of industry. Companies involved in activities where there is a significant risk of fatal accidents may simply choose to count the number of fatalities. Often, companies include more cases of fatal accidents than those occurring during working hours.

Example: A power company included fatal accidents among the following groups of people in their definition of recordable fatality: 'All people covered by the company's emergency response plan; all people under paid work employment; employees and consultants on business travel on behalf of the company; suppliers, visitors, trespassers, or other people detaining themselves on company premises (except for those intention-ally breaking and entering into the site). People that detain themselves in the surroundings of the company premises and are killed as a conse-quence of company activities (i.e. family awaiting workers, sales people, and so on) are also included'.

The types of performance indicators based on environmental and material losses vary from company to company depending on the type of production. Table 16.1 shows typical examples.

chapter seventeen

Process-based safety performance indicators

In the example in Chapter 3, we saw the advantages of feedback control through measurement of the process rather than the losses. This distinction between process- and loss-based measurements relates to the basic accident framework of Chapter 5. Here, we will further explore the potential of process-based safety performance indicators. An underlying assumption in using process-based safety performance indicators is that these are valid indicators of accident risk. We will discuss this issue further in connection with the individual measures.

17.1   Incident reporting

Table 17.1 shows different indicators based on the reporting of near accidents or incidents. Here, we will distinguish between performance indicators based on employees' self-reporting of incidents and incidents reported through line management or technical instrument and alarm systems. The latter is exemplified by process safety events in chemical, oil, and gas plants involving so-called loss of primary containment (LOPC) incidents, such as gas leakage (OGP 2011). In all incidents reporting, the potential consequence of the incident is an important consideration.

As opposed to the loss-based safety performance indicators, in the case of incidents reported by the employees, we are interested in attaining high rates or ratios because that is assumed to reflect the organisation's reporting culture and willingness to share and learn from the incidents. Exceptions are indicators based on the reporting of HIPO incidents. These are incidents that, under slightly different circumstances, could have resulted in severe loss, such as fatalities. They are usually highly visible and reported by the line management or through automatic monitoring or an alarm system.

The incident/TRI ratio gives an indication of the reliability in the reporting of incidents. Although the ratio between incidents and TRIs will vary for different types of incidents, a ratio of, say, less than one is interpreted as indicating a poor reporting culture.

Goal-setting and incentives are tools in use to acquire high reporting rates/ratios. Strong incentives may be counter-productive because a

Table 17.1 Examples of safety performance indicators based on the reporting of incidents

Type	Definition
Employee self-reporting	
Incident (near-miss) rate	Number of reported incidents (near misses) per period.
RUO rate	Average number of employee reported unwanted occurrences (RUO) including incidents, unsafe acts, and unsafe conditions per employee and year. This includes spontaneous reporting by the employees, excluding reports by health, safety, and environment (HSE) staff and from regular safety activities such as inspections.
Incident/TRI ratio	Ratio between the number of reported incidents in a period, and the number of total recordable injuries (TRIs) in the same period.
High-potential (HIPO) incidents	
HIPO incident rate	Number of high-potential incidents per period.
Process safety events (PSE)	
PSE rate	Total number of PSE, that is, events involving loss of primary containment of any material from primary containment above a defined threshold per million hours worked in the activity (OGP 2011).

superficially high reporting rate will be accomplished through the reporting of many incidents with a low learning value.

17.1.1 *Employees' self-reporting of incidents and unwanted occurrences*

Traditionally, initiatives to accomplish near-accident or incident reporting by the employees have represented a means of stimulating employee risk awareness and involvement in safety practice and improvement in experience feedback on accident risks in the workplace (Kjellén et al. 1986). In the so-called reporting of unwanted occurrences, the scope of employees' self-reporting is extended to include reporting of observed unsafe acts and conditions as well.

Figure 13.5 showed an example where a plant experienced a negative correlation between the developments in the LTI rate and the incident rate. Plant management made the interpretation that near-accident reporting had demonstrated its efficiency in promoting safety and preventing loss. The robustness against manipulation is not a concern here because it is the actual reporting behaviour that is our measure. 'Manipulation' is a means of acquiring a high reporting rate/ratio.

The *RUO rate* is measured as the number of reports of unwanted occurrences per employee and year. A RUO rate of 'one', for example, means that every employee on average submits one RUO per year, a very modest ambition. Companies with a good reporting culture achieve an order of magnitude in improvements compared to this level.

17.1.2   *Reporting of incidents related to process safety*

LOPC of significant quantities of flammable or toxic process media such as natural gas is a critical event in a process plant (e.g. chemical plant, refinery, oil and gas processing plant). The frequency of such events is an important indicator of the risk of major accidents. The reporting of LOPC events usually is not dependent on employee self-reporting but on the safety instrumentation and alarm system, although human operators may be more sensitive as detectors of, for example, small gas leakages. LOPC events are categorised by the release type, quantity, and rate, which are correlated with the potential for major accidents. As an example, American Petroleum Institute (API) RP 754 classifies a release of above 500 kg of natural gas into the environment as a Tier 1 (most severe) event (API 2010).

17.1.3   *Reporting of high-potential (HIPO) incidents*

A high or increasing *rate of incidents with high potential* (HIPO rate) is an alarm bell, indicating the possibility of an increased risk of large losses. HIPO incidents include those with the potential of resulting in fatalities (Grades 4–5, according to Table 5.5). Due to the large quantities of energy involved, they are usually difficult to hide, which means that we can expect an adequate reporting reliability either by line management or through employee self-reporting.

The HIPO incident rate may be used directly as a safety performance indicator as shown in Figure 17.1. The control chart approach is applicable here, where the number of incidents per period is plotted for consecutive periods. The upper and lower control limits are determined as $\pm 2 \times \sqrt{}$ (*mean number of incidents per period*), as discussed in Chapter 6. This approach is valid only when the activity level is relatively stable from period to period. It is highly relevant in high-risk industries, where specific types of accidents dominate the risk of fatalities and major material damage.

The use of goal-setting is applicable here but must be carefully evaluated because this approach may have a negative impact on the employees' willingness to report incidents (see Example 2 in Section 13.4.4).

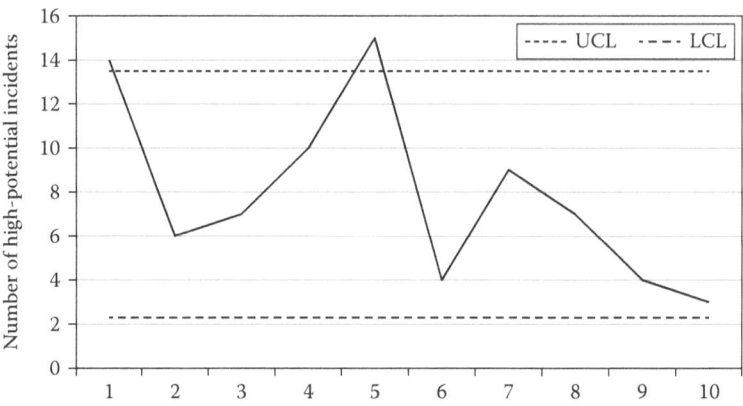

Figure 17.1 Control chart of the HIPO incidents in a refinery.

17.1.4   *Assessment relative to the quality criteria*

Generally, the incident-based performance indicators score the same or better than the loss-based indicators such as the TRI rate on many of the quality criteria for safety performance indicators. These indicators are observable and quantifiable and generally are more sensitive to change due to a higher frequency. They are also transparent and easily understood. Validity is a critical issue. The frequency of self-reported incidents by employees merely reflects their willingness to discuss experiences rather than the actual risk of incidents. It is tacitly assumed that this aspect is a positive characteristic of a company's learning culture and is correlated to safety. 'Manipulation' is regarded as an acceptable means of increasing the reporting frequency rather than a factor that reduces the value of the indicator.

The issue of validity is less complicated for HIPO incidents and PSEs. As indicated in Figure 5.3, HIPO incidents are dominated by only a few types of events. The same applies to the PSEs. It is possible, by the use of statistics or risk assessments, to calculate the likelihood that a HIPO incident or PSE will result in a severe accident. The bow-tie model in Chapter 9 is referenced.

17.2   *Reporting of deviations*

Table 17.2 shows examples of performance indicators based on reporting deviations. Most of the indicators measure % of compliance rather than % of deviation, but they correlate because the sum equals 100% of the observations on which the performance indicator is based. The safety literature discusses many different deviation-based performance indicators.

Table 17.2 Overview of types of deviations and associated performance indicators

Type of deviation and examples	Example of performance indicators	Reference
General	% of compliance with statutory requirements RUO rate (i.e. the average number of reported unwanted occurrences including incidents, unsafe acts, and unsafe conditions per employee and year)	ICMM 2012
Work situation		
1. *Human error* (e.g. wrong action, wrong sequence, omission)	Behavioural sampling (% of correct behaviour)	Krause et al. 1999
2. *Technical failure* (e.g. machine failure, missing equipment/ tools)	% of safety critical equipment that performs to specification when inspected/tested	Health and Safety Executive 2006
3. *Disturbance in material flow* (e.g. poor quality, delays)	Mean time between failures (MTBF)	Rausand and Høyland 2004
4. *Personnel deviation* (e.g. absence, inadequate qualifications, indisposed)	% of staff having required level of training/ qualifications when checked	
5. *Inadequate information* (e.g. inadequate or missing instructions/ procedures, permit-to-work [PTW])	% of PTW filled in correctly	Health and Safety Executive 2006
Environment		
1. *Intersecting or parallel activities* (e.g. disturbance from other work team)	% of simultaneous operations complying with simultaneous operations procedure	
2. *Bad housekeeping*	Housekeeping index (rating of housekeeping standard from 1 to 5 per plant area, see Section 17.2.3)	

(Continued)

Table 17.2 (Continued) Overview of types of deviations and associated performance indicators

Type of deviation and examples	Example of performance indicators	Reference
3. *Disturbances from the environment* (e.g. excessive noise, wind speed, precipitation, high or low temperature)		
Safety systems		
1. *Failure of active or passive safety systems*	% of safety systems/ barriers performing to specifications	Health and Safety Executive 2006
2. *Inadequate guarding*	Covered by failure of passive safety systems	
3. *Inadequate personal protective equipment or clothing*	Behavioural sampling	Krause et al. 1999
4. *Inadequate emergency response* (e.g. delayed notification and mobilisation, failure to evacuate)	Number of emergency response elements not fully functional when activated in exercise/real emergency	OGP 2011

Therefore, it is necessary to select the vital few that are of high relevance to the operation.

17.2.1   Statutory compliance

According to draft ISO 45001, organisations committed to comply with the standard shall keep an overview of applicable statutory requirements and be committed to comply. The degree of compliance is measured by establishing checklists of applicable regulatory requirements and checking the status through inspections and document reviews. A percentage of compliance is achieved by dividing the number of areas where the organisation meets the requirements by the total number of areas checked. This process may be repeated using adequate sampling techniques to monitor how the compliance rate develops in time.

17.2.2   Behavioural sampling

Behavioural sampling was developed in the 1950s to overcome some of the problems associated with the LTI rate as a safety performance measure.

It applies statistical sampling techniques in observing deviations from accepted safe work practices and conditions (Rockwell 1959; Tarrants 1980). The aim is to prevent losses by reducing the frequency of such deviations. The basic principle of accident prevention is the same as in the fertiliser plant case in Chapter 3. In behavioural sampling, we are concerned with the behaviour of people rather than with technical processes.

The measurement of safety performance through behavioural sampling includes the following steps:

1. Identification of safety critical behaviour by analysing accident reports, safety instructions, inspection reports, and so forth.
2. Selection of behaviour to be included in the performance indicator and establishment of a checklist with operational definitions of each item. The selected items have to be easily observable, and the distinction between safe and unsafe has to be clear.
3. Inspections of the workplaces at randomly selected intervals to observe the items and whether the performance is correct or not.
4. Plotting the safety performance index on a control chart. The safety performance index is defined as the percentage of the observed items that are judged as correct.

The next step in the development of behavioural sampling took place in the 1970s. Komaki et al. (1978) based their work on behaviour theory (see Chapter 10 and Section 13.4.3) and introduced feedback to the workers as a consequence (see Figure 17.2). The rationale behind this scheme was to increase the immediate and positive consequences of safe behaviour. After a 'secret' measurement of the baseline, the results were presented to the workforce in the department. (Preferably, this baseline should be around 50–60% to indicate potential for improvements without making the task to accomplish acceptably error-free performance seem insurmountable.) An agreement on improvements was arrived at, such as 90% safe performance. Next, the safety performance was measured and the results were displayed together with the mutually agreed-upon goal. Positive developments in safety performance as well as in the LTI rate have been reported following the introduction of behavioural sampling (Sulzer-Azaroff et al. 1994; Krause et al. 1999).

Figure 17.3 shows an example of a chart where the development of the so-called housekeeping index is displayed. The observation items in this application of behavioural sampling were observable conditions at the workplaces related to housekeeping (i.e. items that represent results of behaviours rather than the actual behaviours themselves).

Behavioural sampling in its original version has been criticised because it represents a view of workers as objects for control rather than as active participants in the work processes to improve safety. Thus, it is not

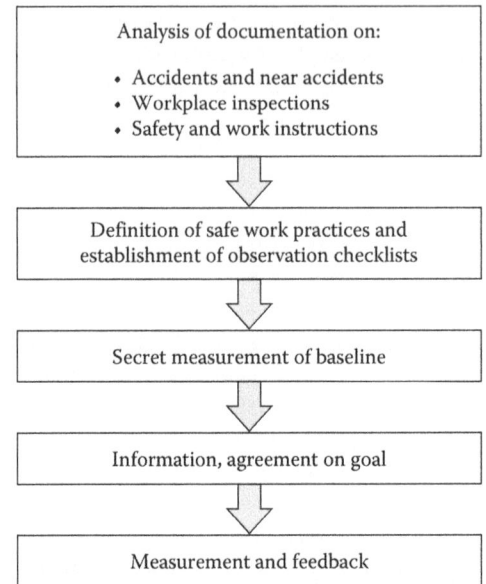

Figure 17.2 Application of behavioural sampling in controlling performance. (Adapted from Komaki, J., et al., *J. Appl. Psychol.*, 63, 434–445, 1978).

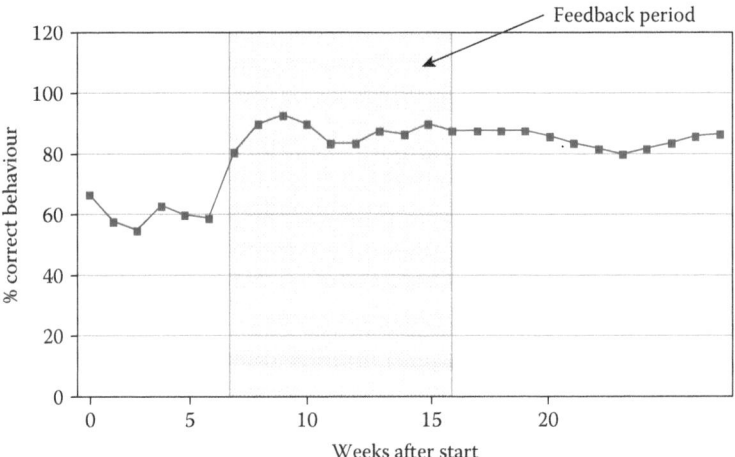

Figure 17.3 Results of behavioural sampling at a shipyard. Neither the control limits nor the performance target are shown. (Reprinted from *Int. J. Indus. Ergonomics*, 4, Saari, J. and Näsänen, M., The effect of positive feedback on industrial housekeeping and accidents; a long-term study at a shipyard, 201–211, Copyright (1989), with permission from Elsevier.)

well suited for application where a more participatory tradition in safety practice prevails. In Finland, a version of behavioural sampling called 'Tuttava' has been developed to make it more suited for these conditions. Tuttava addresses such problems as (Saari 1998):

1. The dichotomy between 'safe' and 'unsafe'. Originally, this has been defined through analysis of historical documents on accident and near-accident records, safety and work instructions, and so forth, and may be problematic. This is because the investigator may not be fully aware of the different consequences of 'safe' and 'unsafe' acts. In Tuttava, the definition is based on a thorough work analysis as far as the participation of the workers is concerned.

2. Outside investigators observing people. This may lead to blame and a feeling of 'big brother is watching you'. In Tuttava, the focus is on physical conditions such as housekeeping rather than on the behaviour creating certain conditions. The workers and supervisors make the observations themselves. By focusing on housekeeping, people at the workplaces can directly observe the positive results from better order.

3. Behavioural sampling based on Van Court Hare's principles of first-order feedback. No lasting effects other than those concerned with workers' behaviour and cognition are expected. In Tuttava, so-called implementation teams of workers and supervisors analyse the causes of deviations in order to come up with technical, organisational, and procedural measures to prevent recurrence and to facilitate correct behaviour. These teams represent the equivalent of quality improvement projects.

4. A narrow focus on safety. In Tuttava, performance items will affect not only safety but also productivity and quality. The aim is to make the program more interesting to management as well and to secure the necessary funding for improvements.

17.2.3   Housekeeping index

Behavioural sampling is a rather elaborate method. Following is an example of a simplified method that has been developed in industry.

Example: An aluminium plant has introduced a so-called housekeeping index. The plant is subdivided into housekeeping areas, and four times a year the safety engineer, the supervisor, and the safety representative from the area perform a safety inspection. A checklist with well-defined items is used. Examples of such items are:

- Are walkways free from obstacles and adequately marked?
- Are floors and stairs kept free from litter?

- Are waste containers available and adequately used?
- Are tools and equipment kept in the right place?
- Are goods and products kept in designated areas?
- Is the area around machinery kept tidy?
- Are signs in place?
- Are windows and lighting appliances cleaned and in order?
- Are toilets and coffee bars kept tidy?

Each area is evaluated by applying this checklist, and the results are given a score from 1 to 5, where 1 means that there are abundant deviations and 5 means that no deviations have been found. The average value is calculated for each area and displayed in a diagram.

17.2.4   Barrier performance

In Chapter 9, we discussed the vulnerability of barriers and how performance can be ensured through regular inspection, testing, and maintenance. The performance indicator '% of safety systems and barriers performing according to the specifications' helps in monitoring how well these systems fulfil their purpose of maintaining plant integrity. This is especially critical in chemical, oil and gas, and other major hazard industries (Health and Safety Executive 2006; OGP 2011). A problem with this type of indicator is that there are many different types of barriers and barrier elements in a plant, and barrier failures have very different effects on the total risk of major accidents depending on the types of barrier. Thus, it is very difficult to establish a common indicator of barrier performance for a plant based on measures of failure rates for the individual barriers (Koren 2006).

Example: Figure 9.10 illustrated four different barriers against exposure to hazardous chemicals. In a plant manufacturing aluminium rims, different types of solvents are used in the paint shop. These solvents represent a hazard to the operators, who may inhale them or be exposed through their skin. Following are two examples of indicators of barrier availability for this specific case:

- Proportion of working time when operators are exposed to a concentration of solvents in the working atmosphere of less than one-third of the threshold limit value
- Proportion of working time when operators wear required personal protective equipment

Both indicators may be measured through sampling techniques of the type described in Section 17.2.2.

Example: Similarly, we may define performance indicators for the availability of barriers against fires and explosions in a refinery such as:

- Frequency of significant gas leaks
- Availability in percentage of operating time of fire and gas detection and active fire protection systems
- Share of evacuation drills where the personnel are able to show up at the meeting point within a stipulated time period

17.2.5   Assessment relative to the quality criteria

How well do performance measures based on data on deviations meet the criteria in Section 8.1.1? The deviation-based performance indicators are observable and quantifiable to the extent that they have been operationally defined and that observations falling inside a norm (compliance) can be clearly separated from those falling outside the norm (deviation). This is especially critical when it comes to human behaviour; consequently, the documentation of conditions that represent a deviation (or unsafe practice) is an important part of – for example – behavioural sampling and the housekeeping index. This is not as critical in the RUO rate because it is based on the reporter's judgement and is not meant to represent any 'objective' frequency. Still, some companies have introduced the practice of scrutinising reports of unwanted occurrences to filter irrelevant reports.

There is no general answer to the validity of deviation-based performance indicators. Different sources are used in establishing the lists of deviations to be included in performance measures, including regulations, industry standards, accident experience, and risk assessments. It is tacitly assumed that these sources represent the best available knowledge in the selection process.

A major advantage with many of the deviation-based performance indicators is their sensitivity to change. This applies to the performance indicators based on input from regular activities such as observations and inspections. By applying an adequate sampling frequency, the width between the control limits may be reduced to a preferred level, and changes can be rapidly detected.

Deviation-based performance indicators dividing the number of observed deviations by the total number of observations are relatively easy to communicate as long as the types of deviations are relatively homogeneous. Problems arise when the performance indicators show, for example, a risk-weighted average, where it is not evident what they actually represent (Koren 2006; Podgórski 2015).

Performance indicators based on inspections, testing, and observations generally are more robust against manipulation than employee self-reporting. In the latter case, 'manipulation' is seen merely as a legitimate means of promoting a good reporting culture and creating safety awareness as well as the ability to recognise hazards and search for improvement. Typically, this will include:

- Information materials such as an RUO booklet for easy reporting and efficient routines for reporting and follow up (e.g. by use of information technology tools)
- Induction and regular (e.g. yearly) training of all employees, reminder in toolbox meetings
- Target setting and feedback at all levels on RUO rate results (employees, management), and quick feedback to the reporter
- Personal key performance indicator (KPI) for management, cash/gift incentives, and recognition for best RUO from the employees

The ideal overall condition is that a 'just culture' is established that is conductive to reporting, engagement, and safety improvements with employee involvement (Dekker and Breakey 2016). Controversial issues such as the processes for establishing rules and treatment of rule violations must be experienced as fair.

chapter eighteen

Causal factor–based safety performance indicators

This chapter takes us to safety performance indicators based on causal factors, according to our basic accident analysis framework in Chapter 5. Here, we will distinguish between performance indicators based on the measurement of (1) contributing factors at the workplace and (2) general and HSE management factors.

18.1 Performance indicators based on contributing factors at the workplace

Developments in the area of leading safety indicators have generated large sets of indicators for different levels of organisations. Table 18.1 shows a sample of indicators related to the workplace level. Here, we have chosen to classify them according to the man, technology, and organisation (MTO) model outlined in Section 4.3, which is similar to the classification of performance indicators of 'workplace precaution factors' according to the fishbone diagram (HSE 2001a). The sample in Table 18.1 is based on lists of performance indicators developed especially for process, oil and gas and mining industries but are of general applicability (SCS 2003; HSE 2006; OGP 2011; ICMM 2012).

18.1.1 Measuring the degree of learning from incident investigations

This type of indicator focuses on one aspect of HSE management (i.e. the extent to which the organisation uses the experiences from unwanted events to prevent recurrence). In Section 7.4.2, we introduced Van Court Hare's hierarchy of the order of feedback. This has been applied in the development of a measure based on an analysis of actions taken after accidents.

Figure 18.1 illustrates why this measure is interesting. It shows an example from an oil company where each action taken as a result of investigations has been analysed to determine the order of feedback that it represents. As indicated by the diagram, the company's organisation has behaved rationally by following up on severe accidents more seriously through the implementation of measures representing an order of

Table 18.1 Examples of safety performance indicators related to the different MTO elements

Type of contributing factor at the workplace	Example of performance indicators

M – Human/behavioural	Management and workforce engagement:
	  • % of inspections attended by management
	  • Staff attitude survey outcome
	  • % of workforce suggestions implemented
	% of staff assigned to hazardous tasks that meet task qualification requirements
	% of required safety induction of contractors conducted on schedule

T – Technical/physical	% of inspection and preventive maintenance tasks performed on schedule ('backlog')
O – Organisational/ economic	% of job procedures for hazardous work based on risk assessments % of regulatory requirements implemented in job procedures % of job descriptions defining safety responsibilities % of sampled permit-to-work (PTW) that identified the required controls % of plant design changes subject to risk assessment % of completed statutory training requirements % of planned safety training completed % of high-potential (HIPO) incidents not identified in risk assessment % of HIPO incidents subject to Level 3 investigation Degree of learning from incident investigations % of close-out of inspection findings (within one week) Number of stop-work events per employee and year % of emergency response elements that are fully functional when tested % of on-scene emergency staff that has participated in a drill per quarter/year

feedback above one (i.e. more permanent solutions). We want to use an indicator that shows the extent to which the organisation actually takes its severe accidents seriously in the way that Figure 18.1 illustrates.

The indicator on the average order of feedback used in accident investigations does this. It is based on an analysis of the actions documented in the accident reports. Each action is rated according to Van Court Hare's hierarchy from 0 to 4. If there is no action, a dummy action with zero order is introduced. 'Actions' concerning general information (e.g. 'to take care') are given the same rating. The average order of feedback is then calculated

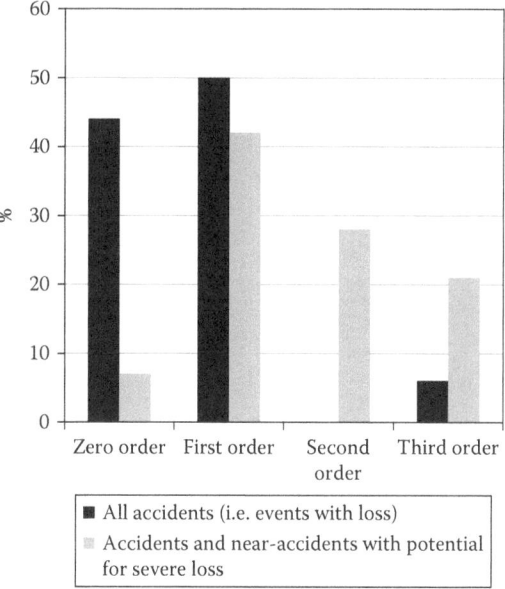

Figure 18.1 Distribution of actions taken after accidents by order of feedback. (Adapted from Blindheim, G. and Lindtvedt, J., *Erfaringsoverføring i sikkerhetsar-beidet ved hjelp av databasen Synergi* [Experience Feedback in Accident Prevention by the Incident Database Synergi], Project report, Norwegian University of Science and Technology, Trondheim, 1996. In Norwegian.)

for the total number of accidents within a defined severity category (e.g. lost-time accidents or accidents and near accidents with a potential for severe loss).

Figure 18.2 shows an example from three yards delivering to an off-shore project. The yard with the best performance in this respect, Yard C, also experienced the lowest lost-time injury rate (LTI rate).

The aim of introducing the average level of feedback as an indicator of safety performance is to improve an organisation's ability to use the experience from accidents to achieve continual improvements.

18.1.2   Assessment relative to quality criteria

The indicators in Table 18.1 have been selected from the literature and have been modified to some extent to contain a defined metric in order to be observable and quantifiable. There is, however, no systematic evidence on the validity of the different measures. This will be context dependent, and it will be necessary for the individual organisation to base its selection of indicators on its own accident statistics and risk assessments.

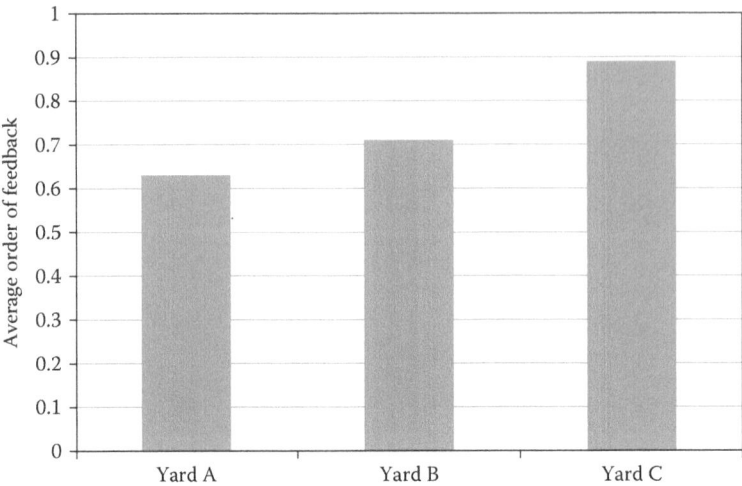

Figure 18.2 Weighted average order of feedback in accident reports from three yards delivering to an offshore project. (Adapted from Boe, K., *Evaluering av HMS-styring i byggeprosjekter* [Evaluation of HSE Management in Construction Projects], Master's Thesis, Norwegian University of Science and Technology, Trondheim, 1996. In Norwegian.)

Sensitivity to change is dependent on sampling frequency in performing the measurements. For most indicators, this is decided by the organisation – for example, when scheduling safety inspections, barrier performance tests, or attitude surveys. A few indicators depend on input that is determined by factors outside of HSE management – such as the frequency of design changes. Performance indicators on the quality of job procedures are hardly meaningful safety performance measurements, unless the procedures change frequently. Robustness against manipulation is a valid criterion in the sense that the performance measures aim at stimulating 'manipulation' on behalf of management to achieve good performance by introducing improvements that will directly affect the scores. Compared to the loss-based performance indicators, there is no intermediate statistic process between action and results.

18.2   Performance indicators based on factors in corporate management and in the HSE management system

In this area, we find performance measurement systems based on concepts or models of accident causation. In Table 18.2, we have subdivided

Table 18.2 Overview of performance indicators based on HSE management and safety culture factors

Corporate general and HSE management systems	Examples	Reference
General management		
	Tripod Delta, standardised questionnaire on 'general failure types' in the organisation	Van der Want 1997; HSE 2003
HSE management system		
Performance measurement systems	% of compliance with international standard (e.g. OHSAS 18001) ISRS Self-diagnostic tools	Alteren 1999; SCS 2003; Roy et al. 2008; DNV GL 2014
Measurement of specific elements (structure, activity)	HSE management elements established (policy) Resources devoted to HSE in % of budget % of HSE management activities completed as planned	
Safety culture		
Performance measurement systems	Employee safety culture perception survey Safety Climate Tool Safety Culture Maturity assessment	Energy Institute 2008c; Zohar 2010; HSE 2013
Measurement of specific elements	Staff perception of management commitment to safety Number of senior management site safety tours completed per individual and year Board and top-management visibility and involvement	SCS 2003; ICMM 2012

Note: ISRS = International Safety Rating System; OHSAS = Occupational Health and Safety Assessment Series.

these systems into three categories related to general management, HSE management, and safety culture aspects.

18.2.1 General management factors

Tripod Delta is a safety performance profile developed for Shell (Van der Want 1997; HSE 2003). Even if the aim is to improve safety, its approach goes far beyond the traditional HSE management system and thus falls under the heading of general management factors. It is based on the Tripod accident model presented in Chapter 4. Eleven so-called general failure types are identified in the Tripod accident model. Eight focus on conditions at the workplace, ordered with relation to their proximity to the victim, and thus are more related to the workplace factors of the previous section:

1. Defences (i.e. failure to provide the necessary measures designed to mitigate the effects of human or component failures)
2. Error-enforcing conditions (e.g. a shift schedule that produces adaptation problems, poor physical working conditions, or a remuneration system that promotes rule violations)
3. Training (i.e. deficiencies in the routines to provide the necessary knowledge, skills, and awareness)
4. Procedures (i.e. unclear, unavailable, or incorrect work instructions)
5. Design (i.e. poor plant layout or poor design of equipment and tools)
6. Hardware (i.e. material failure due to inadequate quality, equipment, and tools that are not available as well as failure due to ageing)
7. Maintenance management (i.e. failures in the system aimed at maintaining an adequate plant and equipment integrity)
8. Housekeeping (i.e. failures in the routines to keep the plant tidy and clean and to dispose of waste)

The remaining three general failure types concern the general and HSE management systems and include:

9. Communication (i.e. failure in the transmission of the necessary information throughout the different parts of the organisation for a safe and efficient operation)
10. Organisation (e.g. ill-defined safety responsibilities or other organisational deficiencies that allow warning signals to be overlooked)
11. Incompatible goals (i.e. failure to manage conflicting safety and production goals or incomparability between formal and informal rules)

The absence or presence of each of the 11 general failure types is determined through the utilisation of indicator questions, which are administered through a questionnaire given to the workforce. There are about 2200 such questions. Each time a questionnaire is distributed at a plant, a random selection of 10% of the questions covering all general failure types are selected and adapted to the conditions at the plant. The results are used to compute what is known as a failure-state profile. Figure 18.3 shows an example.

Tripod Delta is intended for use in plants where the accident frequency rate is too low to give adequate feedback for further improvements. It provides feedback on potential causes of incidents.

A test of Tripod Delta in the automotive sector in France has been reported (Cambon and Guarnieri 2008, as quoted by Podórski 2015). The implementation of the Tripod Delta questionnaire supported changes in areas such as improved employee empowerment and employee–management communication.

18.2.2 HSE management system

The international standards on HSE management systems such as ISO 14001 and draft ISO 45001 (ISO 2015b, 2016) have been accompanied by auditing tools to assess compliance. In case this results in a score such as % of compliance with the standard, it is treated here as a safety

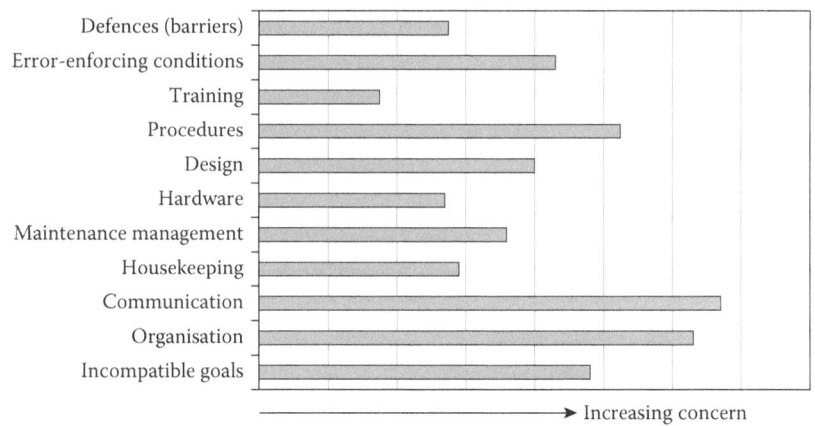

Figure 18.3 Example of a failure-state profile. (Reprinted from Van der Want, P.D.G., In *Safety Performance Measurement*, Institution of Chemical Engineers, Warwickshire, UK, Copyright (1997), with permission from European Process Safety Centre.)

performance indicator. The different rating systems in this area are built up around the following principles:

1. An 'ideal' HSE management model based on draft ISO 45001 or a similar standard defines the elements of the corporate HSE management system and the contents of each element.
2. A scale is established for measurement of each element with respect to degree of adequacy or degree of compliance with the standard.
3. A set of criteria is used in evaluating actual performance in relation to the ideal model.
4. Different data collection methods are used in the evaluations. They include:
   a. Self-evaluation by a company rating team.
   b. Rating by an independent assessor (third-party verification).
   c. Questionnaire to the work force.
5. The company establishes goals for comparison with actual performance.

18.2.2.1 International Safety Rating System

The International Safety Rating System (ISRS) uses standardised questions during data collection and fixed criteria in evaluating the results. Specially trained auditors, who usually come from an outside organisation, do the rating. The ISRS is based on the International Loss Control Institute (ILCI) model of accident causation (Chapter 4). The first edition (1978) of the ISRS manual was developed primarily as a safety-performance measurement system but is now applied as an audit tool. In the eighth edition of the manual (DNV GL 2014), 15 key processes are addressed:

- Leadership
- Planning and administration
- Identification and evaluation of risk
- Human resources
- Compliance assurance
- Project management
- Training and competence
- Communication and promotion
- Risk control
- Asset management
- Contractor management and purchasing
- Emergency preparedness
- Learning from events
- Risk monitoring
- Results and review

The 15 key processes are part of a continual improvement loop similar to Deming's circle (see Chapter 7). The processes include requirements from OHSAS 18001, ISO 14001, ISO 9001, and other relevant standards.

For each key process, a number of sub-processes, questions, and criteria for evaluation of the results have been developed. This makes it possible for the auditor to assess the company. Depending on the results, the company is given a score for each topic. An overall score can also be given. Verification processes are performed, and the company must be prepared to offer evidence for their answers.

The ISRS scores are subject to goal-setting similar to the way the TRI rate is utilised. Typically, top management sets goals for improvements in the ISRS scores for the following years. In order to meet the goals, middle management applies the ISRS manual as a comprehensive and detailed recipe for development of their internal HSE management system. This may be done at the cost of internal priorities and is an example of how a measurement system will affect the behaviour of the organisation being measured. Chaplin and Hale (1998) report that the introduction of ISRS auditing at a chemical processing plant allowed management to respond to clearly defined expectations for change in the HSE management system of the plant and thus to introduce significant changes in the system. They conclude that the ISRS auditing system has positive effects, especially in the early stage of development of a HSE management system. The accident frequency rate was already too low at this plant to allow for the identification of any significant changes as a result of the introduction of the auditing tool. Evaluations of ISRS in other studies suggest that it has a low effect on such safety performance indicators as the LTI rate (Eisner and Leger 1988; Guastello 1993).

ISRS has stimulated companies and branches of industry to develop their own rating systems adapted to their needs.

Example: An oil company has developed a contractor's HSE qualification guideline based on these principles. Seven elements in the contractor's HSE management system are considered. They include: Leadership and commitment; policy and objectives; organisation, resources, and documentation; evaluation and risk management; planning and procedures; implementation and performance monitoring; and auditing and review. For each element, a number of questions are asked. Table 18.3 shows an example of a question and the scale applied in evaluating the results. Results are displayed in a profile, showing the mean rating for each element.

Although adapting an off-the-shelf system such as ISRS to internal needs may result in indicators with an improved 'face validity' and acceptance by the users, it is uncertain whether validity actually has improved. One study has shown that production personnel, safety personnel, and managers tend to overestimate the effects of the quality of

Table 18.3 Scale applied in evaluation of commitment to safety through leadership

	A = Unacceptable	B = Poor	C = Acceptable	D = Excellent
How is the commitment to safety through leadership?	No commitment from senior management.	Safety compliance delegated to line management; no direct involvement by senior management.	Evidence of active senior management involvement in safety aspects.	Evidence of a positive safety culture in senior management and at all levels.

the safety management system elements on the risk of accidents. At the same time, they underestimate the effects of the quality of general and production-management elements (Tinmannsvik 1991). When these personnel participate in the development of safety rating systems, these biases will be transferred to the new system and will result in poor criterion validity.

18.2.2.2   *Self-rating as a means of improving HSE management*

The safety element method (SEM) is a rating system used internally in a company by a review team (Alteren 1999). The method uses a matrix with six HSE management elements and five stages or levels of performance, as shown in Table 18.4.

For each element, the review team defines at what stage the company is operating. The results are plotted in the matrix. The review team then decides about ambitions for improvements.

A number of interesting questions arise when we compare rating systems based on third-party verification such as ISRS with the safety element method. In each of the two cases, a different set of motivational factors at the management level applies:

- The introduction of ISRS requires a certain degree of commitment on the part of top management and the marshalling of adequate resources for the ISRS audits and for follow-up of results to meet management's goals. SEM does not involve commitment to the same extent in relation to resources and results.
- ISRS is prescriptive and gives detailed instructions on how to design a company's HSE management system. An advantage is that it is relatively easy to arrive at a solution by applying ISRS. SEM requires problem-solving and results in local solutions. We must expect that

Table 18.4 Safety element method

Element	Stage 1	Stage 2	Stage 3	Stage 4	Stage 5
Goals, ambitions	Missing	Ambitions to satisfy regulations.	Go beyond regulations.	Go beyond regulations and match the best companies.	Working to influence and improve regulations.
Management	Modest obligations to safety work.	Follow-up of accidents. Low accountability for weak HSE management. Mainly local treatment of risks.	Actively engaged in safety work. Reactions for breaches of safety instructions. Systematic safety work. Focus on technical and human failures.	Safety is equally prioritised and followed up as production and quality. Comprehensive views and systematic approaches. Strong focus on organisation and management factors.	Strong management commitment and obligations to improve safety culture. No self-satisfaction. Line management is a good model.
Feedback systems, learning	Causal transfer of experiences.	Simple statistics. Many short-term corrective actions.	Thorough statistics. Deviation control. Action plans and measures are worked out. Time schedules are kept.	Proactive seeking for improvement. Continuous preventive measures. Thorough processes when working out action plans.	Extensive and systematic exchange of experiences with other companies.

(Continued)

Table 18.4 (Continued) Safety element method

Element	Stage 1	Stage 2	Stage 3	Stage 4	Stage 5
Safety culture	Mastering risky challenges is the ideal.	Little extra done to work safely, more essential to finish fast.	Mainly seeking safe behaviour, sometimes chances are taken.	Safe behaviour is a matter of course. The employees are actively seeking each other's experiences.	Always safe working methods, never breach of routines. All employees work actively to obtain a working environment without losses.
Documentation	Small amount of formal routines.	Satisfies minimum requirements.	Comprehensive documentation. Improvements following audits.	Plain and practical documentation. Procedures are accepted and followed by most employees.	Documentation is well known, always up to date, and followed.
Result indicators	No result indicators on safety.	Absenteeism and accident statistics are the only indicators.	Extensive use of safety results as indicators.	Coordinated and integrated goals. Relations among injury, damage, and other losses are visualised.	Has received international awards for HSE and quality management.

Source: Alteren, B., *Saf. Sci.*, 31, 231–264, 1999.

management will feel a greater ownership of the solutions that follow SEM than those that follow from ISRS.

- ISRS is based on independent evaluations and on detailed instructions as to how to make the evaluation. It will be possible for management to compare the results of their own company with that of other companies applying ISRS. It is not possible to compare the results from SEM between companies in the same way. The application of ISRS will allow for benchmarking and will promote 'competition' among companies in a way that SEM does not.

Roy et al. (2008) have developed a more elaborate self-diagnostic tool on OHS management system performance. It consists of 67 indicators within nine OHS management system elements, including OHS organisation, management commitment, employee responsibilities, norms and behaviour, safety activities, communication, and workplace compliance. Respondents rate each indicator on a scale (1–10). Test results in three Canadian companies show that the tool needs further development to achieve adequate validity.

18.2.3 Measurement of safety climate

In Section 4.8, we introduced the concept of safety culture, which is 'the product of individual and group values, attitudes, perceptions, competencies, and patterns of behaviour that determine the commitment to, and the style and proficiency of, an organisation's health and safety management' (ACSNI 1993). Here, we will use the term 'safety climate' to denote such aspects of an organisation that are possible to measure by use of a questionnaire-based survey and where the results meet statistical criteria for aggregation to the organisational level (Cox and Flin 1998; Flin et al. 2000). Results of such attitude surveys are used as performance indicators at the organisational level.

There is no general agreement on what makes up the dimensions of the safety climate. Examples of dimensions that are mentioned in the research literature are (Zohar 1980; Brown and Holmes 1986; Dedobbeleer and Béland 1991; Niskanen 1994; Cox and Flin 1998; Flin 2000; Ghahramani and Khalkhali 2015):

- Management attitudes and commitment, demonstration of concern for the safety of the employees
- Involvement by the employees in safety work
- Communication on safety among different organisational levels and units
- Risk perceptions within the organisation and attribution of causes of accidents

- Perceived priorities of safety versus production goals
- Belief in effects of safety work
- Adherence to safety rules and acceptability of rule violations
- Active search for new risks

Research into safety climate has been ongoing for more than three decades. Positive correlations have been identified between safety climate scores and employee-related factors such as safety knowledge, motivation, compliance, and participation (Clarke 2006; Christian et al. 2009; Barbaranelli et al. 2015). Many studies show, however, that it has been difficult to establish a robust relationship between safety climate and safety performance, as measured by the frequency of injuries (Kongsvik et al. 2010).

Reason (1998) has analysed the effects of an inadequate safety culture on the risk of accidents from a barrier perspective. The three conditions that he focuses on are found in the earlier list of safety climate elements. First, a poor safety culture will increase the frequency of human errors and rule violations and thus also will increase active barrier failure. Second, it will result in complacency and in unwillingness to check and maintain passive barriers adequately. Third, the organisation that is characterised by an inadequate safety climate will be unwilling to report and follow up near accidents and to identify deficiencies in the barriers.

The Tripod Delta performance measurement system also includes a scale for measurement of top-management commitment, which is an important element in the measurement of the safety climate. Its scale ranges from pathological to generative proactive (see Section 4.8). This indicator has some aspects that are immediately attractive. The emotions and defence mechanisms that will follow from the introduction of such an indicator for feedback control purposes must, however, be carefully considered.

18.2.4   Assessment relative to the quality criteria

The performance indicators based on general and HSE management factors selected for this review are observable and quantifiable in order to allow for benchmarking and trend analysis. Validity is a more critical issue. This is discussed in the review of the research literature in connection with the presentation of each type of indicator, and there are no general conclusions. The performance measures in this category are sensitive to change but, due to the resources required for data collection, sampling frequency will be rather low (typically yearly). Compatibility with other corporate goals may be an issue, especially if implementation will require substantial resources and management attention. Many of the indicators are rather complex and of the aggregate type, meaning that they are based on the mean of many measurements (Podgórski 2015). This makes the indicators less transparent and easy to understand as compared to indicators based on a single measurement. All indicators in this category are sensitive to manipulation and, in most cases, this is regarded as a benefit. The use of the ISRS auditing tool is one example where a crescendo of activities to improve HSE management system elements just prior to an audit were reported (Chaplin and Hale 1998).

chapter nineteen

Selecting safety performance indicators

19.1 Issues to be addressed in the selection process

As we have seen in Chapters 16 through 18, there are many different safety performance indicators from which to choose. Each of them has its advantages and weaknesses; no single indicator is ideal and covers all the different needs in safety performance monitoring for an organisation. Instead, using a combination of such indicators is recommended.

The literature describes some critical issues that should be addressed when selecting safety performance indicators:

1. It is more advantageous to select a few key performance indicators rather than developing a complex measurement system based on a large number of indicators (Podgórski 2015). The latter would involve large investments in man-hours for training and for the collection and processing of data. The quality of decisions might deteriorate because of conflicting information and content overload.
2. Information needs at the different levels of an organisational hierarchy should be addressed (HSE 2006; OGP 2011). At the top management level, there is a need for aggregate data on safety performance for monitoring purposes, whereas facility management needs more detailed information on specific performance such as for safety activities and systems (as per Section 7.2).
3. Changes as to what safety performance data are needed as a company develops along the safety culture maturity dimension also should be addressed (SCS 2003; ICMM 2012). At the lowest level, a company focused on compliance would benefit from an auditing tool such as the International Safety Rating System (ISRS) in order to ensure that the internal HSE management system meets accepted standards. When this has been achieved, the application of the audit system may be terminated. Another example is behavioural sampling, which is applied for a period until adequate compliance has been achieved. At the highest level ('learning'), emphasis is put on indicators of workforce and management engagement such as employee self-reporting, site visits, and safety inspections performed by higher management.

4. Another important consideration is establishing a structured process for the identification/development and selection of key performance indicators similar to the structured decision-making process presented in Section 7.3 (HSE 2001a, 2006; OGP 2011; ICMM 2012). Quality criteria play an important role, as introduced in Chapter 8 and further discussed in connection with the different types of performance indicators in Chapters 16 through 18. The SMART (specific, measurable, achievable, relevant, and time-bound criteria), as discussed by Podgórski (2015), fulfil the same purpose. We find that all existing performance indicators have their limitations, and selecting key performance indicators will involve compromises.

19.2 Examples of combinations of performance indicators

19.2.1 Corporate level

Performance indicators at the corporate level must be of high aggregation. Typically, indicators are selected from the far right and far left areas of the overview of indicators presented in Figure 15.1. Following is a proposal for relevant indicators:

- *Total recordable injury (TRI) rate:* The limitations of the TRI rate are well known, but this indicator represents the 'industry standard' and is difficult to avoid. It is suited for benchmarking and for high-level performance monitoring. Company size and level of risk are critical factors in deciding sampling frequency.
- *High-potential (HIPO) incident frequency:* For companies in industries with a significant risk of accidents resulting in severe consequences, monitoring of the HIPO incident frequency is important. This will make it possible to identify negative developments in the control of major accident risks.
- *Fatal accident rate (FAR):* This is important in large companies involved in activities with a significant likelihood of fatalities involving employees and contractors.
- *Performance indicators on management and employee commitment and involvement:* These are factors that many studies have shown to be correlated with good safety performance. Each company needs to make a selection based on their own experience of what is practicable. Examples of widely used performance indicators in this area are the rate of unwanted occurrences (RUO) and senior management participation in visible safety activities such as site safety inspections.

19.2.2 Plant or facility level

The indicators at this level need to address specific risks and mitigation measures, and it is not possible to give general advice. Following is an example adapted from the International Association of Oil and Gas Producers (OGP 2011) on the steps for identifying performance indicators to monitor critical barriers:

1. Ensure management ownership and establish implementation team.
2. Establish HIPO incident monitoring to assess company performance (related to company level; see previous section).
3. Confirm that critical processes and barriers are in place to prevent major incidents.
4. Select indicators to monitor weaknesses in critical barriers.
5. Collect data, analyse performance to establish a baseline, and use this information to set performance targets and actions.
6. Regularly review critical barriers, actions, performance, and indicator effectiveness.

As discussed in Section 19.1, temporary performance indicators are needed to monitor the implementation of new safety processes or activities. (See the example given in Section 17.2.2 on behavioural sampling that was used to ensure workforce compliance with safety rules.) The introduction of a new permit-to-work or lockout-tag out system is another example where frequent monitoring of compliance may be required during the initial phase.

19.3 Considerations regarding potential unwanted effects from the application of safety performance monitoring

When making recommendations on the selection of safety performance indicators, we not only have to consider conditions outside the workplace, such as legislation and top-management priorities, but we also must consider how the application of such indicators may affect the behaviour of middle and lower management and the employees. Earlier, we discussed how unilateral control by top management will have negative effects on the behaviour of personnel at these levels (Argyris 1992). For example, top management may require compliance with unrealistic safety goals for which the lower levels do not feel any ownership. Different departments may be able to show positive safety performance results without really changing the basic conditions that control accident risk. We also discussed how top management, by exercising more unilateral control as a reaction to such signs of superficial adaptation, could further increase maladaptive behaviour at the lower levels.

Example: Following is a typical scenario, where the starting point is the application of the lost-time injuries (LTI) rate. Today, we find many companies with very ambitious safety goals (e.g. an LTI rate of zero). Each workplace is expected to meet this rate, and one single lost-time injury represents a deviation from the goal. Workplaces may respond by hiding accidents by transferring injured persons to other areas. Management may then introduce ambitious goals concerning the TRI rate, which also accounts for these types of accidents. Another approach is to introduce the ISRS. This implies a significant increase in the degree of workplace control by focusing on many different topics. According to Argyris (1992), such indicators, when introduced as unilateral management directives, will result in an organisation's superficial adaptation to the requirements without claiming responsibility or showing commitment. Management may then respond by introducing measurements directed at these conditions, such as safety climate measurement.

Our point is not to warn against the implementation of ambitious safety goals from the top down. Rather, the conditions for goal recognition and acceptance throughout an organisation have to be carefully considered. It is important that these goals are experienced as realistic and fair and that the associated measurement methods are regarded as meaningful. The ways the goals and indicators are introduced are equal in importance to their selection. The causes for deviations from these goals should be analysed at a sufficiently low level of the organisation. It is only here that adequate knowledge exists about the factual circumstances that explain the results.

Another issue is how the use of safety performance measures impacts the focus of an organisation's managerial capacity. The old proverb 'what gets measured gets managed' also applies to the area of safety performance monitoring. On an organisation's scorecard of performance, only a few measures will be allocated to safety. They will not be able to capture the complexity of an organisation's management of accident risk. This is exemplified by the Texas City refinery explosion in 2005 and by the Deepwater Horizon blow-out in the Gulf of Mexico in 2010, where management focused on personnel safety and the TRI rate, and plant integrity and the risk of major accidents were not given adequate attention. Each organisation needs to understand the limitations of their selected safety performance measures and ensure that vital safety factors are not overlooked.

section five

Risk assessment

This section reviews different methods for risk assessment. Risk assessment is a planned activity involving identification, analysis, and evaluation of accident risks. Here, we are mainly concerned with the assessment of occupational accident risks. We will start in Chapter 20 by establishing some common principles and steps of the different risk assessment methods. In Chapters 21 through 23, we will discuss three different methods that are used primarily in the analysis of the risk of occupational accidents: coarse analysis, job safety analysis, and comparison analysis. Chapter 24 is dedicated to the risk assessment of machinery. Later, in Section VI, risk assessment methods used in the planning phase of construction projects are elaborated in Chapters 25 through 27.

Several independent studies show that inadequate risk assessments and/or risk management are among the most frequent contributing factors to accidents in the construction industry (Haslam et al. 2005; Hale et al. 2012; Winge et al. 2015). We have reasons to assume that these results apply to many other industries as well. Although inadequate risk assessments are a frequent contributing factor to incidents, there are legal requirements for the performance of risk assessments in companies (see Section 2.2). A 2011 study performed by several safety authorities in Norway showed that one-quarter of the 800 investigated companies had not identified hazards and possible unwanted incidents. Small- and medium-sized companies, in particular, lacked risk assessments.

Risk assessments differ from other methods and tools presented in this book because they are proactive (i.e. they support prevention of accidents at the planning stage). This implies that, by using the results of risk assessments as input to decisions made before the execution of work activities, the risk of accidents can be reduced. This is illustrated in Chapters 25 through 27 where we demonstrate how decisions made prior to the operations phase result in less hazardous work activities.

The principal steps of risk assessments follow the diagnostic process in Section 7.3.2 and have many similarities to the methods presented in Section III. Risk assessment utilises the same sub-systems as a safety information system – data collection (risk identification), processing (risk analysis and risk evaluation), memory (e.g. use of historical accident data), distribution of results as input to decision-making, and decisions on risk handling.

chapter twenty

The risk assessment process

20.1   What is risk?

There are many different interpretations of the term *risk* among research-ers, practitioners, and laypeople. Rausand (2011) defines risk as the answer to three questions based on Kaplan and Garrick (1981): (1) What can go wrong? (2) What is the likelihood of that happening? (3) What are the consequences? ISO 31000 *Risk Management* gives a rather vague definition of risk as the effect of uncertainty on objectives (ISO 2009). This standard contains five notes on this definition. Among these are a note saying that risk is often characterised by a combination of the consequences of an inci-dent and the likelihood of occurrence, which is also the most applied defi-nition in occupational accident prevention in practice. In this book, we define risk as *a combination of the probability or frequency of occurrence of acci-dents and the extent of losses (consequence) of the accidents* (Section 15.1). The bow-tie diagram introduced in Figure 9.3 corresponds to these dimen-sions of risk. The centre of the bow-tie represents incidents. The left side represents causal chains leading to the incidents. The events in the causal chain and the incident itself are input to incident identification as well as to frequency analysis. The bow-tie also identifies barriers to prevent acci-dents. The quality of the barriers will influence the frequency of incidents as well as the consequences. The right side of the bow-tie represents con-sequence chains. These three main parts of the bow-tie thus correspond to the three questions raised by Kaplan and Garrick (1981) noted here.

In recent years, expressing risk using the term uncertainty, in terms of lack of knowledge, has received increased attention. Aven and Renn (2009) define risk as uncertainty about the occurrence of incidents and their consequences on an activity with respect to something that humans value. The argument behind this definition is intricate and mainly rel-evant to quantitative assessments of risk related to major accidents with low probability and severe consequences. Expressing the uncertainty of a risk can be valuable for decision-makers (Rausand 2011). To make good choices, decision-makers must understand the basis for their decisions, including the uncertainty of the results.

In practice and in daily language, the notion of risk is often mixed with the notion of hazard. Here, we define hazard as a source of energy that can lead to injury to personnel or damage to the environment or material assets. Risk is an expression of the likelihood and consequences of future losses due to a release of hazards.

20.2   The risk assessment process

Accidents and near accidents are unwanted events occurring at random points in time. They also represent opportunities to learn about hazards and causal factors at the workplace and within a company. We should take advantage of these opportunities when they occur and conduct adequate investigations. Workplace inspections and safety audits are pre-planned activities that also represent opportunities for learning. Their initial focus is on deviations, contributing factors, and root causes and not on hazards.

Risk assessment is complementary to these different activities. It is a planned activity in which we want to identify and remedy hazards before accidents and near accidents happen. Risk assessment is thus anticipatory in nature and supports feed-forward processes rather than the feedback processes of, for example, accident investigations. By applying risk assessment, we will speed up the learning process.

Risk assessment involves the following activities (ISO 2009; Rausand 2011):

1. *Establishing the context.* Defining the analysis object and determining its limits. The analysis object can be a geographical area of a plant, a machine, a job, and so on. Planning is another important part of this initial activity; become familiar with the object, establish risk acceptance criteria, establish the analysis team, make a time schedule, collect background information, and so on.
2. *Identification of hazards and unwanted occurrences* where people, the environment, or material assets may come into contact with a hazard.
3. *Risk analysis* (i.e. establishing a risk picture by determining the frequencies and the consequences of the unwanted incidents). Some standards and guidelines for risk assessment integrate hazard identification and risk analysis into the same activity.
4. *Risk evaluation* (i.e. comparing the results of the risk analysis with risk acceptance criteria to determine whether the risk is acceptable or not).
5. *Risk treatment* (i.e. decisions on and implementation of measures to eliminate or reduce risk where required).

Risk management means the total coordinated activities that take place to direct and control an organisation with regard to risk (ISO 2009). Thus, it includes risk assessment and risk treatment together with communication of information about risk as well as systematic monitoring

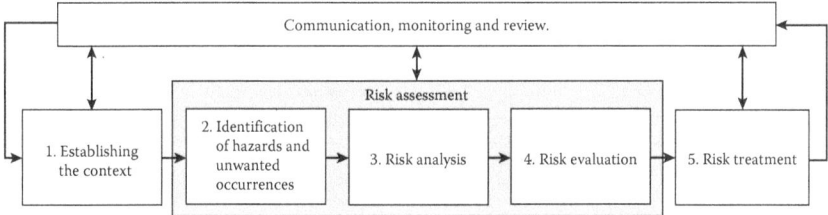

Figure 20.1 Risk management. (Adapted from ISO, *Risk Management. Principles and Guidelines*, ISO 31000:2009, International Organization for Standardization, Geneva, 2009.)

and review of processes. Figure 20.1 illustrates the relationship among the activities in risk management.

The outcome of a risk assessment process is by nature a feed-forward process. Instead of information on the actual or historic performance (i.e. feedback), information on expected results of a future activity is distributed to decision-makers as new input to modify and improve the activity.

Occupational accidents occur in what Rasmussen characterises as loosely coupled work systems with a relatively high accident frequency but low magnitude of loss (Rasmussen 1997). The risk assessment process itself utilises feedback mechanisms in applying historical data in combination with judgements of future conditions. At the other end of the scale, we find tightly coupled production systems with well-defined hazards such as chemical and nuclear power plants. The loss following an accidental nuclear reaction or release of toxic or flammable chemical compounds may be major, but the associated frequency of occurrence is low. Risk assessments in such systems must be based on an analytic strategy involving system modelling to predict future incidents.

Figure 20.2 shows examples of input to the risk assessment process based on experience feedback. A distinction is made between experience databases and experience carriers (Kjellén 2008). Experience databases are compilations of data from different sources of experience such as accident and incident reports, failure rates of different components, data on exposure such as number of hours of operation, and so on. Experience carriers represent explicit knowledge at a higher level of synthesis, where the raw experience data have been aggregated and decontextualised. Examples are drawings, procedures, standards or specifications, experience checklists on hazards, experience-based analytic models, and criteria for accepted risk. The latter are often experience-based and reflect the level of stakeholder tolerance.

Tacit knowledge plays an important role in all parts of risk assessment, and utilisation of this knowledge is accomplished by the risk assessment

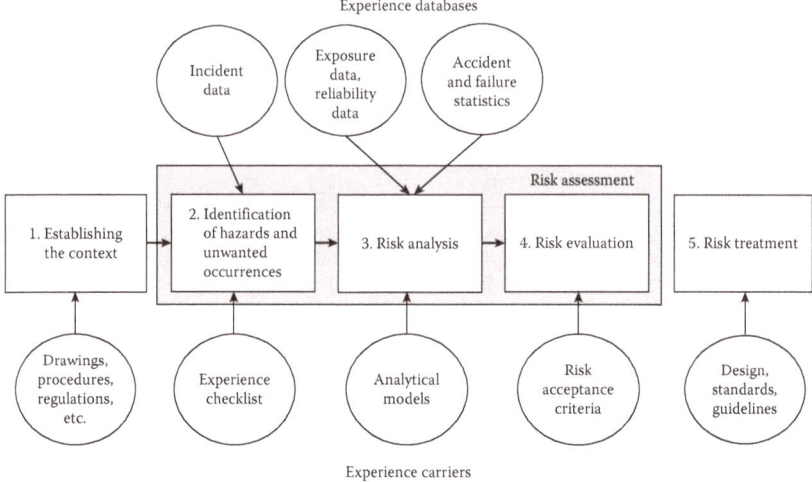

Figure 20.2 Examples of the use of explicit experience in risk assessments. (Adapted from Kjellén in Melnick and Everitt 2008.)

team serving as an arena of experience exchange. Individuals with operational experience from situations similar to those of the analysis object share tacit knowledge about hazards, causes, expected frequencies, and so on. This shared knowledge is converted into concepts suitable for the context of the risk assessment through group discussions and is documented in the risk assessment. Group processes allow for participation in risk assessment by those stakeholders that have an adequate variety of knowledge and values to promote validity, reliability, comprehensiveness, and accuracy of the results.

20.2.1   *Establishing a risk picture*

In this section, we will describe the steps necessary to establish the risk picture that is to be evaluated. This is the process of systematically using available information and knowledge to describe risk by identifying unwanted occurrences and causes, frequencies, and consequences of incidents. This does not include the evaluation of risk, which is the reason this process is not called risk assessment. Table 20.1 shows the main steps of this process (Rausand 2011).

There are various reasons for performing risk assessments. One basic reason is to provide support for decision-making. An important part of the planning steps is to understand the requirements of the decision that the assessment results will support. The team performing

Table 20.1 The process of establishing a risk picture

Main activity	Detailed steps
Plan and prepare	Define objectives and scope
	Clarify decision requirements
	Establish study team
	Establish project plan
	Describe and delimit study object
	Become familiar with the study object
	Provide background information
Hazard identification and development of incidents	Identify hazards and threats
	Develop and select incidents (accident scenarios)
Causal and frequency analysis	Determine causes of incidents
	Determine frequencies of incidents
Consequence analysis	Determine consequences of incidents
Establish risk picture	Establish risk picture
	Assess uncertainties of results
	Make risk analysis report

Source: Rausand, M., *Risk Assessment. Theory, Methods and Application,* Wiley, Hoboken, NJ, 2011.

the assessment should consist of people with different knowledge and from a variety of levels within the organisation. One person with knowledge about risk assessment should facilitate the process. To analyse something, you need to know what you are analysing. Becoming familiar with the study object is thus important. Experience carriers and databases are key sources of background information about the study object.

The second step is to identify relevant hazards of the study object and develop accident scenarios for further analysis. Analytical methods (e.g. hazard identification [HAZID], hazard and operability study [HAZOP], and failure modes, effects, and criticality analysis [FMECA]), checklists, and experience data are relevant experience carriers and databases for this process together with tacit knowledge held by the study team. The key question in this step is: What can go wrong?

The third and fourth steps are to analyse causes, frequencies, and consequences of the scenarios identified in the previous steps. Here, the key questions are: Why does the incident happen? How often do we think the incident is going to happen? What are the consequences of the incident?

In the fifth step, the estimated frequencies and consequences are used to establish a risk picture. For qualitative risk analysis, the risk picture often is established using a risk matrix. We will discuss the risk matrix in Chapter 21.

20.2.2   Risk evaluation

The established risk picture provides input for the risk evaluation (i.e. the risk picture is compared to risk acceptance criteria to determine whether the risks are acceptable or not). An *acceptance criterion* defines the highest accepted risk as documented, for example, in a company's health, safety, and environment (HSE) policy. The risk acceptance criteria are used as a reference in evaluating the risk analysis results and in determining the need for remedial actions. Therefore, these criteria have to be in place prior to the start of the analysis. Such an approach hinges on an equity-based criterion (HSE 2001b), that is, a fixed limit representing the maximum level of risk above which no individual can be exposed. One example is that no person performing paid work for Company A shall be exposed to a higher probability of experiencing a fatal accident than one in one thousand. In practice, equity-based criteria are combined with utility-based criteria and technology-based criteria (HSE 2001b). Utility-based criteria compare benefits of measures to prevent a risk and the cost of the measures, while technology-based criteria reflect the assumption that a satisfactory level of risk is attained when the best available technology is used to control the risk. The as low as reasonably practicable (ALARP) principle is a widespread approach for applying a utility-based criterion (similar concepts are so far as is reasonably practicable [SFAIRP] and as low as reasonably achievable [ALARA]).

Figure 20.3 illustrates the UK Health and Safety Executive's (2001b) framework for evaluation of the tolerability of risk, which applies to all of the aforementioned types of criteria. The figure also shows the relationship between the acceptance criterion and the HSE goal. There are three different regions in the framework that create input to decisions on risk reduction:

- *The unacceptable region (high risk):* These are risks that fall above the acceptance criterion. Risk-reducing measures must be implemented.
- *The tolerable or ALARP region (medium risk):* These are risks that fall between the acceptance criterion and the goal. Risk-reducing measures should be implemented as long as they are practicable and the associated costs are reasonable.
- *The broadly acceptable region (low risk):* These are risks that are accepted in a given context. Risk-reducing measures are not required.

One important part of the ALARP principle is that all identified risk reduction measures should be implemented, unless it can be demonstrated that there is a gross disproportion between costs and benefits.

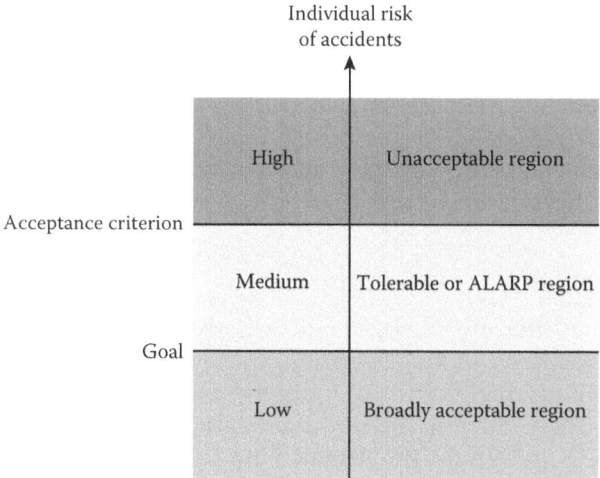

Figure 20.3 Regions for evaluation of the acceptance of risk.

However, the ALARP principle is more than just cost–benefit analysis (Vinnem et al. 2006). Inherently safe design can imply that the risk is ALARP. Furthermore, following known standards and best practice can demonstrate that the risk is ALARP. The latter are examples of technology-based criteria.

The risk acceptance criterion can be expressed both qualitatively and quantitatively. In a thorough literature review on risk acceptance criteria, Johansen (2010) classifies different expressions of risk acceptance criteria. Quantitative risk acceptance criteria are either individual or societal. Here, we will focus on criteria relevant to occupational accident prevention. The main expressions of individual risk are individual risk per annum (IRPA) and localised individual risk (LIRA). IRPA measures the probability that a specific individual is killed due to given hazards during one year's exposure (Rausand 2011). Examples of acceptance criteria are 10^{-3} for workers in the UK nuclear power sector and 10^{-3} for some Norwegian oil and gas offshore operators (Johansen 2010); 10^{-3} implies that individuals in an identified group of workers at highest risk shall not be exposed to a probability of experiencing a fatal work accident that exceeds 0.001 per year. There are many other types of risk acceptance criteria that use measures such as fatal accident rate (FAR) or potential loss of life (PLL). (See the Norwegian standard NORSOK Z-013, 2010, for an overview.)

Risk matrices (see Chapter 21 for examples) are a more qualitative way of expressing acceptance criteria. These matrices express risk in pairs of frequencies and consequences, and the risk acceptance levels usually are expressed by three zones following the categorisation shown in Figure 20.3.

Johansen (2010) underlines that acceptable levels of risk are neither absolute nor universal but are trade-offs and contextual premises. It is however important that the risk acceptance criteria are set prior to the performance of the risk assessment. Here, we will use acceptance criteria related to the risk of fatality or injury to personnel. These criteria must meet the same standards as those of the safety performance indicators discussed in Section III.

20.3   Methods of risk analysis

An important attribute of risk analysis is that it is systematic. Ideally, all possible hazards should be found and evaluated. The various risk analysis methods apply different tools and checklists to accomplish this goal. Often, they combine the use of analytic tools and checklists with brainstorming and group problem-solving techniques. There are many textbooks and standards describing various risk-analysis methods (see Rausand 2011; Harms-Ringdahl 2013). Table 20.2 summarises some of the methods.

The methods differ along five dimensions. In the beginning of this chapter, we defined risk as a combination of the probability or frequency of occurrence of accidents and the extent of losses (consequence) of the accidents. The definition was based on three questions as illustrated by the bow-tie model shown in Figure 20.4. Most of the methods categorised in Table 20.2 cover the whole range of the bow-tie model. However, some methods only focus on parts of the bow-tie.

Two of the methods, fault tree analysis and comparison analysis, are *deductive* in that they start with the accident. They proceed by analysing the underlying incidents and deviations (fault tree analysis) or contributing factors (comparison analysis). Some of the methods are mainly inductive in that they start with an incident and proceed by studying the effects of this incident. This applies to, for example, HAZOP and failure mode affect analysis (FMEA), although they also have a component of causal analysis. Other analyses (e.g. coarse analysis and job safety analysis) start with the hazard and use a combination of inductive and deductive analyses.

Table 20.2 The scope of different categories of risk-analysis methods relevant to occupational accident prevention

Method		Scope and aim	Analysis object	Basic characteristics	Measure of risk	Applicability
Coarse analysis	Coarse analysis is also called preliminary hazard analysis (PHA). A simplified PHA without expressing risk PHA is sometimes called HAZID.	Overview of hazards in an area by type, causes, consequences, and risk	A geographical area of an enterprise	Checklist of hazards	Risk matrix (low/medium/high)	Early design Design Modification
	Energy analysis	Overview of volume, sources, and stores of energy	A geographical area of an enterprise	Checklist of energies	Risk matrix (low/medium/high)	Design Modification
	Risk and vulnerability analysis	Overview of risk related to safety, environmental care, reputation, and finance of a unit	Operation plant, project	Checklist of risk categories including budget overrun, permit violations, HSE, integrity, and so on	Risk matrix (low/medium/high)	Early design

(Continued)

Table 20.2 (Continued) The scope of different categories of risk-analysis methods relevant to occupational accident prevention

Method	Scope and aim	Analysis object	Basic characteristics	Measure of risk	Applicability
Operative analysis					
Job safety analysis	List of hazards connected to each step of a job by type, causes, consequences, and risk	Individual jobs	Breakdown of job into steps; checklist of hazards	Risk matrix (low/ medium/ high)	Operation
Hazard identification					
HAZOP	Qualify process safety systems by identifying potential deviations, their causes, and consequences	A process plant documented in, for example, piping and instrument diagrams	Breakdown of plant, into process units (tanks, pipelines, reactors, and so on); checklist of process parameter deviations	Consequences of deviations	Early design Design Operation Modification
Structured what-if technique (SWIFT)	Overview of incidents and quality of safeguards given initial events	A geographical area of an enterprise	Checklist of what-if questions	Risk matrix (low/ medium/ high)	

(Continued)

Table 20.2 (Continued) The scope of different categories of risk-analysis methods relevant to occupational accident prevention

	Method	Scope and aim	Analysis object	Basic characteristics	Measure of risk	Applicability
Hazard identification (continued)	Failure mode and effect analysis FMEA/FMECA	Assessments of system's reliability	A technical system	Reliability model of the system; checklists of failure modes; failure data for components	System availability	Design Modification
	Hazard log	Dynamic overview of hazards	Operation plant, project	A living register of hazards of all kinds can be used in other risk analyses		Early design Design Operation Modification
Causal and frequency analysis	Fault tree analysis	Estimate the probability of severe accidents with consequences to people, materials, or the environment	A technical system	List of known top events; logical model (fault tree) of causes of failure data for components	Frequency of top event with known consequence	Design Operation Modification
	Cause and effect diagrams (Ishikawa diagrams)	Overview of causes of defined critical events	Operation plant, project Technical system	Fish-diagram and checklist of six categories of causes (6M)		Design Operation Modification

(Continued)

Table 20.2 (Continued) The scope of different categories of risk-analysis methods relevant to occupational accident prevention

	Method	Scope and aim	Analysis object	Basic characteristics	Measure of risk	Applicability
Development of accident scenarios	Event tree analysis	Establish frequency distributions of potential accidents	A technical system	Checklist of critical initial events; event frequencies and probability of distributions of different outcomes of each event	Combinations of frequencies and losses	Design Operation Modification
Human reliability analysis	Human reliability analysis (different methods available)	Overview of possible human errors and their consequences and causes	Tasks performed by an operator	Breakdown of a task into steps and identification of causes and consequences	Probability and consequences of human errors	Design Operation
Comparison analysis	Comparison analysis	Assess changes in accident frequencies for a new plant as compared to an existing plant	A new or modified plant	Overview of accident in existing plant by area, activity, and type of event; systematic assessments of effects of new design on exposure and probability	Accident frequency rate	Design Operation

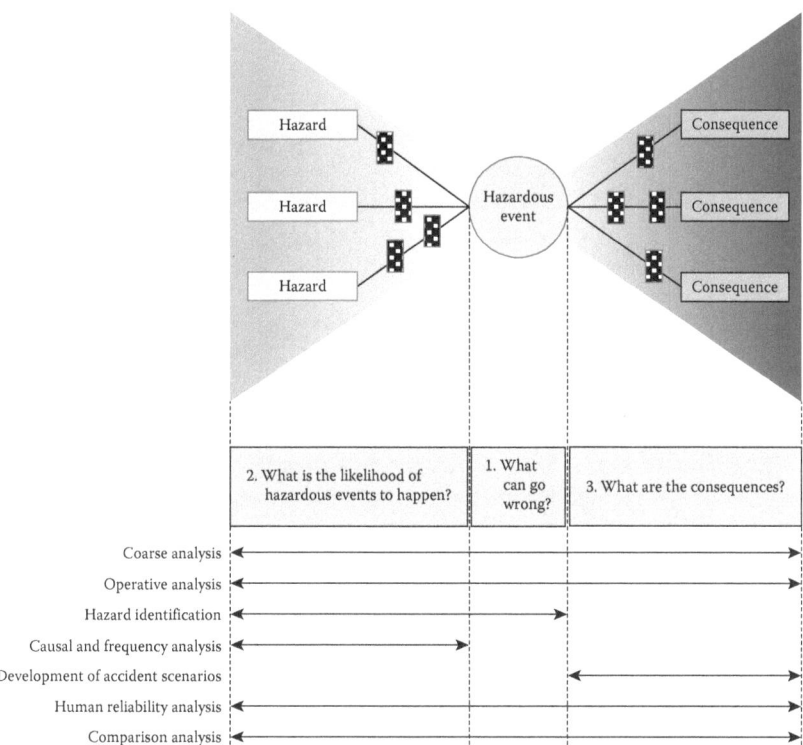

Figure 20.4 Categories of risk analysis methods and their scope.

chapter twenty-one

Coarse analysis

The aim of coarse analysis, also called preliminary hazard analysis (PHA), preliminary risk analysis, or energy analysis, is to identify and evaluate the hazards in a system. A simplified coarse analysis (without estimating the risk) is sometimes called hazard identification (HAZID; see Chapter 12). The coarse analysis is a simple method that generates a crude risk picture. The results of a coarse analysis are used for decisions on risk-reducing measures. The method is also often used as an initial analysis, which is followed by a more detailed risk analysis.

A coarse analysis typically follows a stepwise procedure:

1. Plan and prepare
   a. Establish the objective of the analysis
   b. Select and describe the analysis object
   c. Establish the analysis team
   d. Plan the analysis:
      * Time (working hours)
      * Training, access to facilitator
      * Access to information
      * Administrative support
      * Scheduling
   e. Collect background information and familiarise with the object
2. Perform the assessment
   a. Identification of hazards
   b. Identification of causes, frequencies, and consequences
   c. Establish risk picture, risk evaluation
3. Develop safety measures
4. Document the results of the analysis
5. Follow-up of results

This chapter gives a detailed description of each step of the coarse analysis, which is similar to the steps of risk assessment shown in Table 20.1.

21.1   Plan and prepare

21.1.1   Objectives and decision requirements

A coarse analysis can be performed for many purposes:

- To prepare an overview of existing risks and measures in the study object
- To prioritise remedial actions
- To identify needs for detailed risk assessments

It is important that the study team understands the objective of the coarse analysis prior to the analysis process. If the aim of the coarse analysis is not agreed upon, it may lead to answering other questions than those expected by the client.

21.1.2   Describe and delimit analysis object

The *analysis object* of a coarse analysis can be the activities in an area of a plant or a building, transportation on a road section, a construction site, or similar delimited objects. Subdividing the system to be analysed into reasonable parts is recommended by following (for example) the established subdivision of departments or activities.

Example: In a coarse analysis of an aluminium plant, the plant was broken down into the following 'analysis objects' – carbon section for manufacturing of cathodes, electrolysis, foundry, and transportation. The latter is spread geographically over the whole plant.

In the description of the analysis object, establishing a comprehensive list of all jobs within the area is recommended but not strictly necessary. The checklist in Table 21.1 is used to support this task. It is recommended to visit the actual workplace during this step.

Table 21.1 Checklist of typical jobs

- Operation and supervision of machinery
- Normal, manual operations
- Inspection, sampling
- Handling of production disturbances
- Maintenance, including maintenance handling
- Housekeeping
- Transportation of materials (manually/by crane, truck, and so on)
- Walking

21.1.3   Establish study team and analysis plan

A team is responsible for the analysis. Depending on the circumstances, the team will consist of the team leader, a secretary and experts with specialised backgrounds.

The *group or team leader* has to be well acquainted with the coarse analysis method and have management capabilities. The team leader is responsible for:

- Assisting with the selection of other team members
- Preparation of the analysis and its execution
- Representing the team in discussions with the client and the management responsible for the object being analysed
- Submitting the analysis report

The secretary of the team is responsible for documenting the team's analyses and for summarising these in the report. Often, the team leader takes on the responsibility of secretary as well.

The participants in the team have to ensure that all conditions of significance to safety are brought up in discussions and are assessed. A coarse analysis during the design phase aims at identifying and assessing hazards that may occur when the designed system has been put into operation. Participants will represent the engineering organisation (typically layout, mechanical, and safety disciplines) and personnel with experience from similar systems in operation (production, maintenance). In an analysis during the operations phase, the team will include persons having direct experience with the job(s) to be analysed. It is recommended that the team size does not exceed seven persons.

There may be a need for a *facilitator* in case the team is not fully acquainted with the analysis method. Responsibilities of the facilitator include:

- Guiding the team in the analysis method
- Assisting in collection of documentation
- Providing experience transfer from other teams

It is management's responsibility to ensure adequate allocation of resources. This includes time (working hours) for the analysis and training, access to a facilitator (if necessary), and access to documentation and personnel for interviews.

A typical coarse analysis of an area (e.g. the foundry in an aluminium plant) will require a total of 40–60 man-hours, including planning (four to eight hours), team meeting (six persons for six hours), and reporting (eight hours). Normally, one team meeting will be sufficient to execute a coarse analysis.

21.1.4   Background information and familiarisation

Preparation includes the collection and review of documents about the analysis object and familiarisation with it. A typical list of documents for a coarse analysis includes:

- System layout drawings
- Manning and organisation plans
- Training schedule for new employees
- General safety instructions for the department (if available)
- Accident and near-accident reports for the analysis object or similar systems
- Previous risk assessments on the same or similar systems
- Relevant regulations and standards

21.2   Hazard identification and analysis of causes

We reviewed the basic principles for hazard identification in Section 12.1. These are applicable in risk assessments, too. In coarse analysis, the team uses its experiences and imagination in the identification of hazards and possible incidents within an area. The checklist in Table 12.1 will support this task. Experience data on accidents and near accidents will provide valuable input for this process.

In the coarse analysis, we especially want to focus on hazards that may result in serious consequences. Focusing on hazards that may result in many (but less severe) accidents is also recommended. The accident statistics for the actual plant or similar plants for accident concentrations should be consulted. Results are documented on a record sheet according to Table 21.2. Section 21.3 will explain how the risk picture in the table is established.

Causes should be identified using the team's experience. The checklists of deviations and contributing factors at the workplace level presented in Chapter 5 can be used as an extra control. It may be necessary to ask someone with expertise to come to the team's meeting in order to delve deeply into special problems. Technical design is focused upon in analyses that are carried out during the design stage.

21.3   Establish the risk picture and evaluate the risk

Risk estimation serves as a basis for prioritisation in relation to safety measures and the need for detailed job safety analyses. The most serious consequence of the occurrence should be estimated from a realistic point of view of what may happen according to the scale shown in Table 21.3. The consequences are dependent on the amount and type of energy involved.

Table 21.2 Record sheet for documentation of the results of a coarse analysis; example is from a foundry at an aluminium plant, where Table 21.3 has been applied in the risk estimation

Analysis Object: Foundry						
Hazard	Causes	Consequences	C*	F*	R*	Recommendations
Explosion	Cold metal with moisture in contact with melted metal in furnaces	Fatality	D	C	High	Pre-heating of cold metal
Contact with hot surface	Splashing melted material during charging of furnaces	Serious burn injury	C	C	Medium	Remotely operated charging operation
Vehicles in movement collision	Insufficient line of sight in crossing	Crushing injury	D	B	Medium	Traffic regulation
Fall to lower level	Unsecured work at height on top of furnaces	Falling injury	C	C	Medium	Installation of work platform

* C = consequence, F = frequency, R = risk. Consequence and frequency range from A–E. Risk is either high, medium or low. See Table 21.3.

In establishing the risk estimation matrix according to Table 21.4, the following two anchorage points defining the acceptance limits have been applied:

- A lost-time injury (LTI) rate above 10 is considered to be too high and thus above the acceptance criterion. For a workplace with 100 employees working 2000 hours per year, an LTI rate of 10 corresponds to two lost-time injuries per year. It follows that 10 lost-time injuries per year is too high a number, but one is below the acceptance limit.
- A fatal accident rate (FAR) above 10 is unacceptable. With the same assumptions as in the previous item, this defines the acceptance limit as between one fatality per 10 years and one per 100 years.

Table 21.3 Matrix for risk estimation for a workplace with about 100 employees

Consequence		Frequency				
		1 per 1000 years (A)	1 per 100 year (B)	1 per 10 year (C)	1 per year (D)	10 per year (E)
First-aid injury	(A)	Low	Low	Low	Low	Medium
Lost-time injury	(B)	Low	Low	Medium	Medium	High
Permanent disability	(C)	Low	Low	Medium	High	High
Fatality, one person	(D)	Low	Medium	High	High	High
More than one fatality	(E)	Medium	High	High	High	High

Table 21.4 Matrix for estimation of the individual risk

Consequence		Frequency				
		1 per 1000 years (A)	1 per 100 years (B)	1 per 10 years (C)	1 per year (D)	10 per year (E)
First-aid injury	(A)	Low	Low	Low	Medium	High
Lost-time injury	(B)	Low	Low	Medium	High	High
Permanent disability	(C)	Low	Medium	High	High	High
Fatality, one person	(D)	Medium	High	High	High	High

If other acceptance criteria for the LTI rate and FAR are applied, the matrix has to be re-scaled to suit this application. Similarly, the matrix has to be modified when applied to considerably smaller or larger workplaces. Table 21.4 shows the matrix for the individual risk of an operator working full-time. It applies the following frequency limits for intolerable risk:

$>10^{-3}$ per year for fatality

$>10^{-2}$ per year for permanent disability

$>10^{-1}$ per year for lost-time injury

The expected *frequency* of incidents following from each identified hazard with the assumed consequence is estimated by using the team's experience and accident statistics (if available). The *risk estimation* follows from the team's assessments of consequences and frequencies.

The team is responsible for evaluating the need for remedial actions and further detailed job safety analyses. Tables 21.3 and 21.4 usually are interpreted in the following way: The risk regions (high, medium, and low) are described in Section 20.2, where a high risk implies that action must be taken to reduce the risk. Medium risk implies that risk-reducing measures should be implemented as long as they are practical and the associated costs are reasonable. For a low risk, no further risk reduction is required.

21.4   Risk treatment

21.4.1   Development of safety measures

The result of the risk assessment serves as a basis for the development of safety measures. Following Haddon's strategies for accident prevention (see Chapter 4), these preferences should be made in the selection of measures:

1. Will the risk be reduced or eliminated by design by removing the hazard and/or reducing the frequency?
2. Will safeguarding reduce the consequences?
3. Will the risk be reduced by access to personal protective equipment, safe work procedures, and/or information and training?

21.4.2   Documentation and follow-up of results

All findings from the analysis are documented on the coarse analysis record sheet according to Table 21.3. The following outline is proposed for the assessment report (Standards Norway 2008b):

- Aim and background.
- A description of the analysis object.
- Mandate and participants in the analysis group.
- Description of assumptions and simplifications.
- Description of the analytical method and of data used in the analysis (e.g. reviewed documents, observations, and so on).
- Results – hazard identification, risk analysis, and risk evaluation. Reference is given to the full analysis form, which is included as an attachment.
- Conclusions.

The report should be subject to quality control before it is approved and distributed. At a minimum, this should include a review by the team members. In project work, quality control is accomplished through interdisciplinary checks. For analyses of existing plants, the report should be sent to plant management for review. It is a management responsibility to follow up on results.

21.5   Establishing a database on potential accidents

A coarse analysis results in information on hazards and potential accidents similar to that obtained through accident investigations. Establishing a database with this information offers several advantages:

- It will be easy to compare results from different plants. This will facilitate checking the results in order to obtain common risk-estimation criteria.
- It will also be easier to learn from earlier findings and, as a result, to use accumulated experience on hazards and causal factors in new analyses.
- A joint analysis of the results from different departments with similar types of production will assist in the identification of potential accident concentrations.

Example: An aluminium company conducted a coarse analysis of its primary aluminium plants. An analysis team was established for each department that participated in the analysis. A health, safety, and environment (HSE) expert from the corporate staff led the teams. In addition, the teams included the department supervisors and operations and maintenance personnel.

The teams identified in all about 800 different potential accidents. A database was established that contained the following information about each incident:

- Plant and department
- Type of production and activity
- Possible sequence of events
- Type of incident
- Estimated frequency, consequence, and risk.

The corporate staff analysed the incidents. Figure 21.1 shows results for the electrolysis departments. There were fewer than 10 high-risk incidents requiring immediate attention and about 300 medium-risk incidents.

The identified accident risks were grouped, and the groups were ranked by accumulated risk level as a basis for prioritising further follow-up. For the electrolysis departments, the following types of incidents were considered to be of special concern:

1. Burn injury due to contact with melted aluminium
2. Vehicle hitting an operator working adjacent to an electrolysis cell

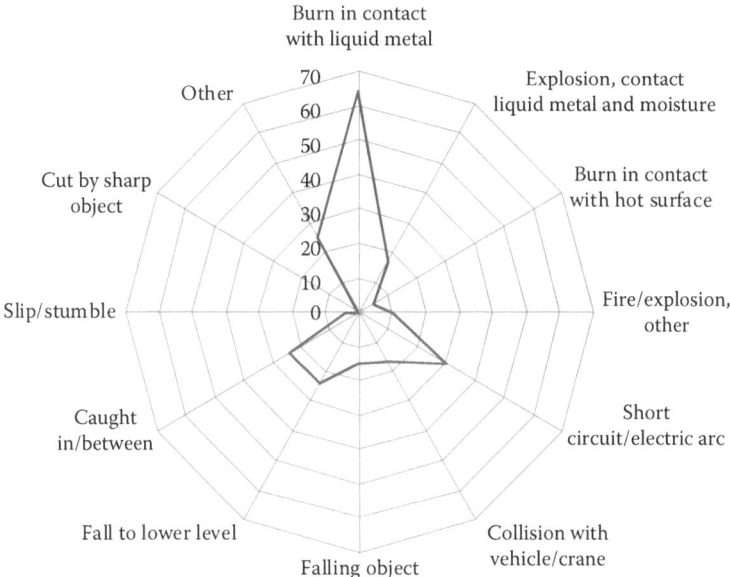

Figure 21.1 Distribution of all identified potential accidents in the electrolysis departments by type of hazard involved.

3. Burn injury due to electrical arc
4. Fall to a lower level during maintenance of electrolysis cells
5. Caught between heavy objects during crane operations

Each department was responsible for follow-up of its own results. The results of general significance were later implemented in a company technical standard for use in the design of new plants and modifications.

chapter twenty-two

Job safety analysis

Job safety analysis (JSA) is a systematic review and assessment of hazards to which employees are exposed when performing work activities. JSA is done prior to the execution of the work activity – for example, as part of the application of a permit-to-work (Chapter 9). The purpose of the analysis is to support decisions on selecting and implementing countermeasures to make the work activity safe.

There are different versions of JSA particularly related to the format of the JSA record sheet and whether the risk (frequency and consequences) is being assessed. In this section, we will present generic steps and principles for JSA and present an example of a JSA sheet. Other names for the same type of analysis are safe job analysis (SJA), job hazard analysis (JHA), and task hazard analysis (THA).

22.1 Analysis object and organisation of the analysis

A JSA is performed for work activities where hazards are present and are not adequately controlled by existing work procedures or by barriers. Situations where a JSA should be considered (based on the Norwegian Oil and Gas Association 2011) are:

- Activities with severe potential consequences
- Activities where the consequences for safety are uncertain
- Activities where experience data indicate frequent occurrence and/or severe consequences
- If the activity requires deviation from prescriptions of work in procedures or routines
- New work activities unknown to the workers performing them
- Activities where the teams of workers are not familiar with each other
- Activities that involve equipment with which the workers are unfamiliar
- Activities where the preconditions change (e.g. weather conditions, new time schedule, changed order of sub-activities, new interactions with parallel operations)

When performing a JSA of an activity before start up, the analysis normally is performed in a meeting with involved workers. If needed, experts can also participate. One person is responsible for the execution of the JSA, with the supervisor typically being responsible for the work activity. The person leading the group process should have knowledge of and experience with the JSA method. Involvement of the workers participating in the group is important because they have key knowledge about the practices regarding the job or similar jobs. Worker participation will also minimise the probability of missing hazards in the identification and will promote ownership of proposed countermeasures. The involved workers' risk awareness will also increase by participation in the JSA process.

22.2   The steps of a JSA

Figure 22.1 shows the basic steps of a JSA. The JSA is performed in a way similar to a coarse analysis but differs as to the level of detail by which the activities are broken down.

In the planning and preparation phase, the group of workers performing the JSA is decided and a meeting plan and milestones are established. Background information is collected (i.e. experience database

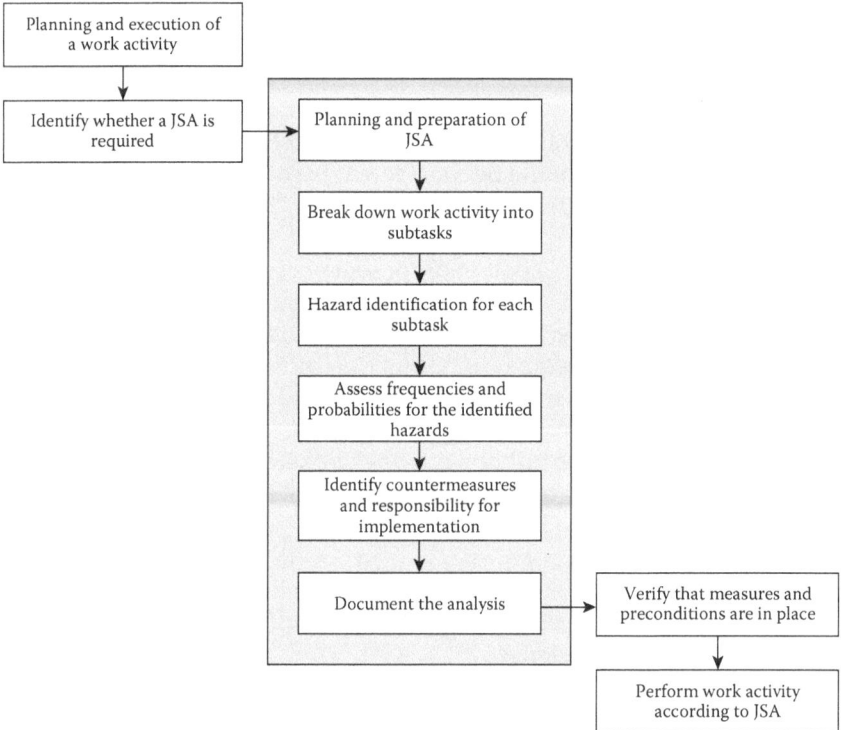

Figure 22.1 The basic steps of a job safety analysis (JSA).

and experience carriers) such as relevant information from the accident database, relevant accident investigation reports, and from similar JSAs performed previously. Work procedures, drawings, and plans are also relevant to fully understand the work activity. Additionally, the knowledge of the involved workers is important input for the JSA activities. A JSA sheet is being used to guide the JSA process and to document the results of the analysis. Table 22.1 shows an example of a sheet filled in with the results of a simple analysis of a lifting operation. The sheet is similar to that used for a coarse analysis; the only difference is the first column. The risk matrix in Table 21.4 is used to estimate the risk.

22.2.1 *Description of the steps of the job*

In this step, we find the largest differences between the coarse analysis and a JSA. The JSA team reviews and describes the activity's subtask – one after the other in a sequence. Thus, it is necessary that personnel with operational experience participate in the team. All subtasks need to be included as well as those outside normal practices. What disturbances can occur and what will then be done? The checklist in Table 21.1 is used as an extra control to ensure that no subtasks have been forgotten. Describe what is to be done in each subtask; use verbs such as 'lift, install, remove' in the descriptions.

The breakdown into activities should be performed at an adequate level of detail. Too coarse a breakdown makes the subsequent identification of hazards more arbitrary. If the breakdown is too detailed, the JSA will be time-consuming. Compare the activity description with the work instructions (if any). Do the instructions describe the actual performance of the work? Are 'dialects' present (i.e. does the way the work is done differ between different shifts). Preferably, the description of the job steps should be checked against the actual performance of the work.

22.2.2 *Subsequent steps*

The subsequent steps do not in principle differ from those of the coarse analysis (see Chapter 21). In JSA, there are requirements as to comprehensiveness, and no hazards are too insignificant to be documented. All hazards will sooner or later result in injuries or near accidents if they are not remedied.

In estimating the risk of accidents, working time has to be considered. Usually, the job is carried out by one or a few operators, and Table 21.4 is, in this case, most suitable. It represents the acceptance criteria for the total risk of an individual operator working about 2000 hours a year. The contribution from the analysed job has to be evaluated in relation to the total risk to which the operator is exposed.

Table 22.1 Record sheet for documentation of a JSA showing a simplified analysis from an offshore installation involving crane transportation of drill pipes from the pipe deck to a ramp. Table 21.4 has been used for risk estimation

Subtask	Hazard	Causes	Consequences	C*	F*	R*	Recommendations
1. Fix the slings to the drill pipes	Fall to lower level (fall from stack of pipes).	3 m height of pipe stack; icy conditions	Broken leg	B	C	Medium	Lower height of pipe stack.
2. Lift the pipes to the pipe ramp	Impact from or squeezed between moving objects (hands squeezed by sling).	Restricted view from crane	Hand injury	B	C	Medium	Replace slings by magnet lift.
	Falling object (hit by falling pipes).	Faulty slings	Crushing injury	C	B	Medium	Review routines for checking of slings.
3. Release the slings	Falling object (hit by pipes).	Restricted view	Crushing injury	B	B	Low	See previous.
	Fall to lower level (fall from pipe ramp).	Lack of fall protection	Broken leg	B	B	Low	Guard

*C = consequence, F = frequency, R = Risk.

Measures to remove or control the hazards must be identified. A person with the necessary authority must be given the responsibility of implementing the measures before start-up of the work activity.

22.2.3   Execution of work

Prior to the start-up of the work activity, the personnel involved are gathered for a review of the JSA results. Before starting the activity, the individual responsible for the work activity must verify that the identified measures are implemented. He/she must also check if there has been any change in the preconditions for the work compared to the preconditions when the JSA was performed. If there have been significant changes in the preconditions, a new JSA should be considered.

22.3   Merits and limitations

There are several benefits of the JSA as a method in systematic loss prevention (Glenn 2011; Crutchfield and Roughton 2013). By performing a JSA, both managers and employees obtain insight into task-specific hazards. If the results of the JSA are applied properly for risk reduction, the result will be better control of the hazards and more efficient work. Another advantage of the JSA is that the method is simple to use and is directly related to operative job tasks. This method promotes employee participation and thus also creates workers' ownership of measures and increases each participant's risk awareness. The JSA process gathers the team to perform the work activity. Thus, the process not only makes plans for safety, it can also work as a planning tool for efficiency and quality as the steps in the activity are analysed and discussed in the group.

A JSA requires time to execute and a budget for remedial actions. Thus, it may raise resistance and be met by arguments such as 'our employees know their job'. The result may be that it is used only occasionally (Crutchfield and Roughton 2013). For the construction industry in particular, Rozenfeld et al. (2010) point out that the JSA may not cover the dynamic nature of construction projects (e.g. change of work crews). Changes of preconditions should raise the possibility of performing a new JSA. Another problem with a JSA is that sometimes it is used too often, especially for well-known activities where the hazards already are controlled by barriers. In a balanced approach, the JSAs will focus on what is new and unique in the specific jobs to be analysed. Well-known job hazards are managed through the implementation of standard job and safety procedures. Glenn (2011) reasons that the JSA is embedded in safety practice, but little attention is devoted to the rationale, development, and use of the method. The JSA is a simple method that represents both strength and weakness. Its simplicity makes it user friendly.

However, when the systems to be analysed become complex, the method becomes too simplistic.

Job briefing (also known as toolbox talk) is a simple tool used at the beginning of the day or prior to the start-up of a work operation. Job briefings follow the same steps as the job safety analysis but in a more informal way. These briefings are used to identify and communicate existing and potential hazards before a job starts. The job briefing is performed by a supervisor and his work team at the beginning of the day and takes no more than 5–10 minutes.

A form for a job briefing is used as the basis for the process. It highlights issues for the work task such as:

- How can we perform today's work in a safe and efficient way?
- What can go wrong? How can this be avoided?
- Is there need for a detailed JSA?

A supervisor leads a discussion with his work team based on the questions in the form. It usually starts with covering today's work tasks step by step. The discussion will address what work is planned and how it should be done. How safe work activities are achieved is then discussed as well as which hazards are present and how they can be controlled.

Job briefing is a simple and user-friendly approach for identification of hazards and mitigation of risk. It promotes employee participation and strengthens workers' risk awareness.

22.4   Systematic mapping of hazards within an organisation

A JSA is applied in the analysis of individual work activities. It is also applicable in a systematic and detailed mapping of a department's hazards. The following is an example of how this may be accomplished:

1. All activities in the department are identified and listed. This is done, for example, by reviewing the responsibilities of the different positions in the department.
2. All activities are reviewed by performing coarse analysis. The results are used as a basis for decisions on which activities to focus upon in a detailed JSA.
3. The critical activities are analysed.
4. Measures are developed and implemented for each activity.
5. After the measures have been implemented, each activity is subject to an update of the analysis by observing the activity actually being performed. The aim is to check that the analysis describes the actual situation adequately and that the measures are adequate.

The analyses seen together will represent a 'total map' of the hazards in the different department activities. This map will be a powerful tool in the work to reduce the risk of accidents in the department. It is important that this map is updated with changes in the activities and with new accident experiences. If this is done systematically, the map will serve as an important carrier of the department's collective experiences on accident risks.

chapter twenty-three

Risk assessments of machinery

23.1   Requirements as to risk assessments

The European Union (EU) machinery directive defines goal-oriented safety requirements for machinery (European Council 2006). To document compliance with these requirements, risk assessments have to be carried out. The documentation on risk assessments has to be available before the manufacturer can issue a Declaration of Conformity and label the machinery with the Conformity Example (CE) marking. The harmonised standard ISO 12100 describes the basic principles for achieving machinery safety, including requirements as to risk assessments (ISO 2010).

This legislation accounts for the fact that machines interact and form production systems of varying complexity. In ISO 12100, machinery is defined as an assembly fitted with or intended to be fitted with a drive system consisting of linked parts of components, at least one of which moves, and that are joined together for a specific application. The directive is also valid for other products than those meeting the definition of machinery such as safety components fulfilling safety functions; lifting equipment and accessories; chains, ropes, and webbing for use in or with machinery; interchangeable equipment; as well as partly completed machinery and removable mechanical transmission devices. Assemblies of machinery also have to meet the requirements in the directive. These consist of machinery arranged and controlled so that it can function as an integral whole for a specific application. A sub-assembly is a mechanised assembly that cannot function independently and that is intended for incorporation into a machine.

In Section 9.5, we discussed the basic strategy for the selection of safety measures according to the European machinery safety legislation. It clearly defined the responsibilities of the designer for ensuring machinery safety by applying a prioritised list of measures based on Haddon's strategies (see Chapter 4). These start with the elimination of hazards, followed by the implementation of barriers and final information for the user on how to manage residual risks.

23.2   Method for risk assessment of machinery

ISO 12100 details the requirements for a manufacturer's risk assessment to ensure machinery safety. This risk assessment follows the same basic steps as those described in Chapter 20 and should be documented as part of the conformity process. The main steps in machinery risk assessment are (ISO 2010):

1. Establishing the context
2. Determination of limits of machinery
3. Hazard identification
4. Risk estimation
5. Risk evaluation

The results of the risk assessment are applied as decision-making support for risk reduction. Risk reduction should follow the prioritised steps described here; the complete risk assessment is illustrated in Figure 23.1. A group similar to the coarse analysis team in Chapter 21 is responsible for the risk assessment, and the results are documented in a record sheet according to Table 23.3.

All team members need to have a basic understanding of the machinery and its use. In addition, the individual team members should represent the following knowledge and competence:

- Design engineers with detailed technical knowledge of mechanical design and of the design of control and power supply systems
- Personnel with experience on how similar machines are operated and maintained
- Safety engineers with knowledge and experience about the risk analysis method to be used, safety regulations, and about accidents likely to occur in practice

In establishing the context for risk assessment, the following items must be documented prior to the start of any such assessment (Figure 23.1):

- Machinery descriptions (model, type, number, and so on), layout drawings, and other relevant technical documentation.
- Description of the machines' use for all phases of the machinery life.
- Lifespan, including mean time between changes of components.
- Ergonomic principles.

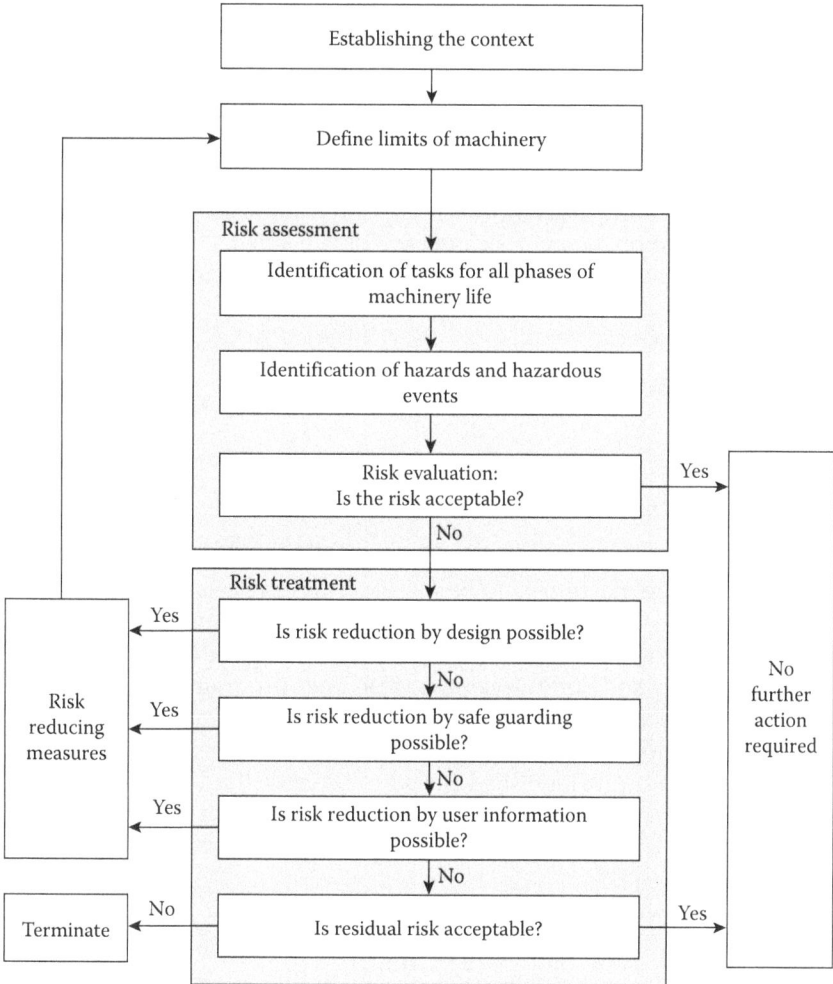

Figure 23.1 Risk assessment of machinery. (Reprinted from ISO, *Safety of Machinery. General Principles for Design. Risk Assessment and Risk Reduction*, International Standard ISO 12100:2010, International Organization for Standardization, Geneva, Copyright (2010), with permission from ISO.)

- Space requirements as defined by the footprint and need for access during operation, inspection, and maintenance.
- Regulations and standards.
- Ambient conditions at the place of installation (indoor/outdoor), lighting, and so on.
- Interfaces with the rest of the plant concerning physical arrangements, control systems, and energy supply. In some cases, the

machinery manufacturer will have to secure the necessary interface information from the main contractor.

- Accident and health-impairment statistics for similar machines, if available.
- For machines that have been documented by risk assessments before, these do not need to be redone. It is important, however, to consider all changes in design and assumptions about uses before deciding whether or not a new risk assessment is needed.
- Other experience information, such as experiences of users with similar machines and malfunction histories.

This information should be documented as part of the risk assessment result.

23.2.1   Determination of the limits of the machinery

The purpose of determining the limits of the machinery is to become familiar with the machinery to be assessed and to describe the system in a reasonable way for further assessment. Determination of the limits of the machinery must consider all phases of machinery life (transport; assembly, installation, and commissioning; teaching, programming; operation; maintenance, cleaning; troubleshooting; and dismantling) related to people, other machines, and the environment of the machines being evaluated. The following must be identified – use limits, space limits, time limits, and other limits.

Use limits include both intended use and reasonably foreseeable misuse:

- Various machine operating modes
- Different intervention procedures, including interventions required due to malfunctions
- The use of machinery by persons identified by personal characteristics (e.g. age limits)
- The anticipated level of training, experience, or ability of users

Space limits include range of movement for the machinery; space requirements for persons using the machinery, including operations and maintenance; human–machine interface; and machine–power supply interface.

Time limits include the lifespan of components and/or the machinery and recommended service intervals.

Examples of other limits are material properties, housekeeping requirements, and environmental factors influencing the machinery (e.g. temperature, dust, humidity).

Industrial plants are often very complex and consist of many different machines and machinery assemblies. When purchasing new or modifying existing machinery, the limits of each machine or machinery assembly have to be clearly defined. This is required in order to define the responsibility for the Declaration of Conformity. This responsibility usually lies with the machine manufacturer. This principle also applies when the machine is connected to the plant's process-control system. It sometimes happens, however, that the organisation responsible for overall design and procurement, which may be plant organisation, assembles different machines into an integral whole. In this case, the plant is responsible for issuing the Declaration of Conformity.

The definition of the limits of machines and machinery assemblies is critical. There could be conflicting interests between the supplier and the user that have to be addressed. From the user's point of view, it is important that all interfaces among individual machines are analysed in case they may give rise to new hazards and that this is done by the supplier of the total system. A supplier of industrial systems, on the other hand, is interested in a breakdown of responsibilities to the next level (i.e. to the suppliers of the individual machines).

Example: A contractor was responsible for a turnkey delivery of a new wire rolling mill to an aluminium plant. This consisted of the following parts:

- A casting wheel
- Two sets of rolling mills
- Reels for the collection of wires
- Material-handling equipment

The contractor referred to the sub-suppliers of these parts as being responsible for documentation of machinery safety and for issuing Declarations of Conformity. The client did not accept this answer and argued that the different parts interacted to serve a common purpose and had overlapping danger zones. In negotiations, it was decided to consider the whole rolling mill to be an assembly of machinery. Material-handling equipment was defined as machines, and the supplier was responsible for the risk assessments. The other parts were sub-assemblies and were documented by suppliers' declarations. The contractor was responsible for the risk assessment of the rolling mill as an integral whole.

Following are some guidelines for determining the limits of machines and machine assemblies:

1. Different machines are considered as being part of an assembly of machinery when:
   a. They interact to serve a common purpose.
   b. They are connected through a common control system.
   c. The danger zones overlap or the activity of one machine will automatically initiate hazardous movements in the interconnected machine.
2. A machine is considered to be an individual unit and not part of an assembly of machinery when:
   a. It is controlled manually by an operator.
   b. It has a separate control system.
   c. The danger zone of the machine is separated from the danger zone of other machines.

23.2.2   Hazard identification

Several methods are available for hazard identification, such as preliminary hazard analysis (PHA), what-if analysis, failure mode effect analysis (FMEA), and fault tree analysis (FTA). ISO 12100 does not suggest any methods to be applied for the identification of hazards but states that designers should identify hazards by considering task identification, possible states of the machine, and unintended behaviour by users or reasonable misuse of the machine.

The first step in the hazard identification is to note the tasks for each phase in the machinery life cycle (see checklist in Table 23.1). The next step is to identify hazards associated with the identified tasks (see checklist in Table 23.2). Table 23.3 shows an example of task and hazard identification for a new wire rolling mill at an aluminium plant.

In addition to the examples of hazards in the table, other reasonably foreseeable hazards not directly related to the tasks should be identified. First, possible states of the machine must be considered. Generally, there are two states: (1) the machine functions as intended and (2) the machine does not function as intended (e.g. due to failure of part[s], external disturbances, software errors, power supply disturbances, surrounding conditions). Second, unintended behaviour by users or foreseeable misuse of the machine must be identified.

23.2.3   Risk estimation and evaluation

The estimation and evaluation of risk for a machine follows the same steps as described for risk assessment in Chapter 20 and for coarse analysis

Table 23.1 Task identification checklist for all phases of the machinery life

Life phase	Example of tasks
Transport	Lifting, loading
	Transportation
	Unloading, unpacking
Assembly and installation	Assembly of the machine
	Connection to systems (e.g. power supply)
	Testing and trials
Hand-over	Setting
	Teaching/programming
	Process/tool changeover
Operation	Start-up
	Operation, all modes
	Feeding, filling, or loading the machine
	Removal of product from the machine
	Stopping the machine
	Minor adjustments or interventions
Maintenance and cleaning	Cleaning and housekeeping
	Preventive maintenance
	Corrective maintenance
	Removal/dismantling of parts
	Adjustments
Trouble-shooting/fault-finding	Removal/dismantling of parts
	Repairing and replacement of parts
	Restart after unscheduled stop
	Recovery of operation from jam or blockage
Dismantling/disabling	Disconnection
	Lifting, loading
	Transportation
	Unloading

Source: ISO, *Safety of Machinery. General Principles for Design. Risk Assessment and Risk Reduction*, International Standard ISO 12100:2010, International Organization for Standardization, Geneva, 2010.

in Chapter 21. Table 23.3 shows an example of a risk assessment for a new wire rolling mill at an aluminium plant. The result of the risk estimation is documented in the sixth column of the sheet. The matrix on individual risk usually applies in this case (see Table 21.4).

ISO 12100 states that four factors must be considered when reducing risk: (1) the safety of the machine, (2) the ability of the machine to perform

Table 23.2 Checklist of hazards for tasks associated with machines

Type of hazard	Example of origin	Example of potential consequences
Mechanical hazards	• Acceleration, deceleration • Cutting parts • Falling objects • Height from the ground • High pressure • Instability • Kinetic energy • Moving/rotating elements • Rough/slippery surface • Sharp edges • Stored energy	• Being run over • Being thrown • Crushing • Cutting • Drawing-in/trapping • Impact • Shearing • Slipping, tripping, falling • Stabbing, puncture, injections • Suffocation
Electric hazards	• Arc • Electromagnetic/electrostatic phenomena • Not enough distance to live parts under high voltage • Overload • Short-circuit	• Burn • Electrocution • Falling, being thrown • Fire • Shock
Thermal hazards	• Explosion • Flame • Heat radiation • Objects with high or low temperature	• Burn • Discomfort • Frostbite • Scald
Noise hazards	• Exhaust systems • Manufacturing process • Moving parts • Scraping surfaces • Rotating parts • Whistling pneumatics • Worn parts	• Loss of awareness • Permanent hearing loss • Stress • Tiredness
Vibration hazards	• Misalignment of moving parts • Mobile equipment • Scraping surfaces • Rotating parts • Vibrating equipment • Worn parts	• Discomfort • Medical disorders
Radiation hazards	• Ionising radiation source • Low-frequency electromagnetic radiation • Optical radiation	• Burn • Damage to eyes and skin • Other medical disorders

(Continued)

Table 23.2 (Continued) Checklist of hazards for tasks associated with machines

Type of hazard	Example of origin	Example of potential consequences
Material/ substance hazards	• Aerosol • Biological/microbiological agents • Dust • Flammable/explosive • Fluid • Gas	• Breathing difficulties, suffocation • Explosion • Infection • Poisoning • Other medical disorders
Ergonomic hazards	• Access • Design or location of displays and control devices • Mental overload/underload • Posture • Repetitive activity	• Fatigue • Musculoskeletal disorder • Stress
Hazards identified with the environment in which the machine is used	• Dust and fog • Lack of oxygen • Lightning • Pollution • Snow • Temperature • Water • Wind	• Slipping, falling • Suffocation • Machine failure/ malfunction leading to other hazards/origins

Source: ISO, *Safety of Machinery. General Principles for Design. Risk Assessment and Risk Reduction*, International Standard ISO 12100:2010, International Organization for Standardization, Geneva, 2010.

its function, (3) the usability of the machine, and (4) the manufacturing, operational, and dismantling costs of the machine. The as low as reasonably practicable (ALARP) principle thus is foundational for risk evaluation of machinery safety.

When high and medium risks are identified, measures have to be evaluated. These are documented in the seventh column. The team also has to consider the effects of the measures on the usability of the machinery.

In some cases, more detailed risk assessment of the machinery might be necessary (e.g. if the reliability of the safety system is a concern). For reliability assessment, FTA or FMEA may be appropriate methods.

Table 23.3 Record sheet for documentation of the detailed risk assessment for one of the rolling mills (example shows simplified results)

Life phase (Table 23.1)	No.	Task (Table 23.1)	Task description	Hazard and cause (Table 23.2)	Risk (Table 21.4)			Design measures/ safeguarding	Assumed precautions by user	Remaining risk		
					C*	F*	R*			C	F	R
Operation	1.1	Feeding, filling, or loading the machine	Manual guiding of bar from feeding wheel to first pair of rollers when automatic system is out of order	Mechanical hazard: Moving/ rotating elements Hand squeezed between rolls	C	A	Low	Not required because this is a low-frequency event	Use of personal protection	B	A	Low
	1.2	Operation, all modes		Noise hazard: Rotating/ moving parts	C	B	Medium	Replacement of gear boxes with low-noise emission types	Use of personal protection	B	B	Low

(Continued)

Table 23.3 (Continued) Record sheet for documentation of the detailed risk assessment for one of the rolling mills (example shows simplified results)

Life phase (Table 23.1)	No.	Task (Table 23.1)	Task description	Hazard and cause (Table 23.2)	Risk (Table 21.4) C*	F*	R*	Design measures/ safeguarding	Assumed precautions by user	Remaining risk C	F	R
Operation (continued)	1.3	Operation, all modes		Mechanical hazard: Falling objects Bar breaks and hits operator								
Maintenance and cleaning	2.1	Cleaning and housekeeping	Cleaning of rolls	Mechanical hazard: Moving/ rotating elements Caught between rolls moving in suction direction	C	C	High	Interlock preventing use of suction mode during cleaning		B	A	Low
				Material hazard: Fluid, aerosol	A	C	Low	Arrange for steam		A	C	Low

(Continued)

Table 23.3 (Continued) Record sheet for documentation of the detailed risk assessment for one of the rolling mills (example shows simplified results)

Life phase (Table 23.1)	No.	Task (Table 23.1)	Task description	Hazard and cause (Table 23.2)	Risk (Table 21.4)			Design measures/ safeguarding	Assumed precautions by user	Remaining risk		
					C*	F*	R*			C	F	R
Maintenance and cleaning (continued)	2.2	Corrective maintenance	Shifting rolls	Ergonomic hazard Strain injury due to heavy workload	B	B	Low	–	Lifting instructions	B	B	Low
Trouble-shooting	3.1	Removal/ dismantling parts	Removal of bars that have gotten stuck	Mechanical hazard: Moving/ rotating elements Caught between rolls that start to move: Incorrect operation of pendant control	C	B	Medium	Interlock to prevent incorrect operation	User instructions for removal of stuck bars	C	A	Low

*C = consequence, F = Frequency, R = risk. See risk matrix in Table 21.4.

chapter twenty-four

Comparison risk assessment

Comparison risk assessment fills a gap among existing methods on the assessment of the risk of occupational accidents (Kjellén 1995). It uses accident statistics from a reference system to assess the risk in an analysis object, which is a similar system under development. In the example described in this chapter, comparison risk assessment is used during the design of a new plant to predict the occupational accident-frequency rate for the plant after start-up (i.e. during operation and maintenance). Results are expressed as relative changes in the accident-frequency rate in relation to the experienced rate of the 'reference plant' that has been in operation for some years. In Chapter 27, we will use comparison risk assessment to assess the risk of fatal accidents during occupational road transportation. The aim in this case is to evaluate the effect of different safety measures directed at the driver, the vehicle, and the traffic environment, including the road standard.

As discussed in Section 20.2, occupational accidents occur in what Rasmussen (1997) characterised as loosely coupled systems with a relatively high accident frequency but a low magnitude of loss. This also applies to road transportation systems. These types of work systems are suited for an empirical risk management strategy based on statistical analysis, a condition that is utilised in comparison risk assessment.

24.1 Application of risk acceptance criteria in comparison analysis

Historical data on accident-frequency rates for a particular plant are utilised in some applications of comparison risk assessment. They often show a downward trend. A trend of this type can be explained by operational conditions, such as improved experience and HSE management. In the execution of comparison risk assessment during design, the applied risk acceptance criteria should not be sensitive to such trends. Rather than defining the risk-acceptance criterion as a fixed accident-frequency rate, the following criterion is suggested: 'The plant should be designed and the operation of it planned in such a way that operation's goals concerning the accident-frequency rate are not more difficult to meet than for existing plants'.

This means that there must not be a deterioration of the designed safety level for new plants as compared to existing plants.

Example: This criterion was implemented in the HSE program of an offshore project for the design and construction of a new floating production and drilling installation. The selected concept involved a number of challenges concerning the risk of occupational accidents:

- Floater movements
- Large open areas, where the personnel are exposed to wind and rainfall
- Increased handling of heavy drilling equipment
- Reduced crew

All these factors were expected to increase the accident-frequency rate. The acceptance criterion implied that compensatory actions had to be taken in design to meet the criterion. These included actions to reduce the exposure of the personnel to hazards and actions to reduce the probability of accidents.

Some chemical industries have defined individual acceptance criteria for the risk of fatalities, which is expressed as a tolerance limit for the likelihood of a fatal accident at work per year. A typical upper tolerance limit is a probability of one in one thousand (Schmidt 2007). It does not apply as an average but to the individual or group of workers at highest risk. This acceptance criterion may be expressed as follows: 'To avoid excessive risk to any individual, a criterion for the probability of fatal accidents of one in a thousand per annum in any function in work for the company shall represent a risk that shall not be tolerated'. To use this criterion in comparison risk assessment will require access to a reference system with accident statistics based on a large exposure volume because fatal accidents are relatively rare. This condition is fulfilled for a few types of work systems, such as road transportation.

24.2   Risk-assessment model

Comparison assessment is based on a model for the assessment of the accident risk in an organisation. This is determined by (1) the exposure (i.e. the frequency of hazardous activities that are performed by the organisation), (2) the probability of accidents when a hazardous activity is performed, and (3) the consequences of the accidents (Kjellén 1995). The activities are defined by a discrete set of types of activities, $\{A_i\}$. Associated with each type of activity is the possibility of the occurrence of injuries. We will analyse the risk of recordable injuries, that is, either lost-time injury (LTI) or total recordable injury (TRI). The injuries belong to the discrete set of types of injuries, $\{I_{i\alpha}\}$.

There is a probability $(p_{i\alpha})$ of the occurrence of an injury of type $I_{i\alpha}$ during the performance of the activity A_i. The probability (p_i) of the occurrence of any of the injuries in the set $\{I_{i\alpha}\}$ during the performance of the activity A_i is:

$$p_i = \sum_\alpha p_{i\alpha}$$

During a given time period (e.g. one year), the activity A_i is, on average, carried out N_i times. If Z denotes the number of injuries, the expected number of injuries of type $I_{i\alpha}$ in the activity A_i during this period equals:

$$E(Z_{i\alpha}) = P_{i\alpha} \times N_i$$

Similarly, the expected number of injuries in the activity A_i during the period equals:

$$E(Z_i) = P_i \times N_i$$

where $Z_i = \sum_\alpha Z_{i\alpha}$

Let e_i denote the mean number of working hours in the activity A_i during the time period in question. The injury frequency rate (λ) during this period equals:

$$\lambda = \frac{\sum_i E(Z_i)}{\sum_i e_i} \times 10^6$$

We assume that there are two workplaces, the 'new workplace' and the 'reference workplace'. The types of activities of the two workplaces belong to the same set $\{A_i\}$. The injury frequency rate and the distribution of accidents by activity and type are known as the reference workplace. We introduce the following relationships:

$$(E(Z_{i\alpha}))_{new} = k_{i\alpha} \times (E(Z_{i\alpha}))_{ref}$$

where

$$k_{i\alpha} = \frac{(p_{i\alpha})_{new}}{(p_{i\alpha})_{ref}} \times \frac{(N_i)_{new}}{(N_i)_{ref}}$$

The 'correction factor', $k_{i\alpha}$, is estimated for each type of activity and injury. This estimate is made up of two parts. First, the team evaluates the effects of changes between the new workplace and the reference workplace on the number of times an activity is carried out (exposure). After that, the team evaluates the effects of these changes on the probability of the occurrence of the injury. The expected number of injuries in activity A_i during the time period in question for the new workplace is calculated from the formula:

$$(E(Z_i))_{new} = \sum_\alpha k_{i\alpha} \times (E(Z_{i\alpha}))_{ref}$$

It is assumed that the total number of working hours during the time period in question (Σe_i) is known for both the new and the reference workplaces. The expected injury frequency rate for the new workplace is given by the equation:

$$\frac{\lambda_{\widehat{new}}}{\lambda_{\widehat{ref}}} = \frac{\left(\sum_i (Z_i)\right)_{new}}{\left(\sum_i (Z_i)\right)_{ref}} \times \frac{\left(\sum_i e_i\right)_{ref}}{\left(\sum_i e_i\right)_{new}}$$

24.2.1   Assumptions

An expert panel estimates the effect of the differences between the new workplace and the reference workplace on the risk of injuries. The panel employs a number of assumptions in these judgements when they evaluate the effects of design and operational conditions on the risk of injury. Of the different operational factors that affect this risk, only changes in manning level, in activity plans, and in specific work procedures are considered. No credit is given to other planned operational factors such as changes in the HSE management system or in safety training or information.

Factors in design or operation of the new workplace that have not been determined at the time of the analysis are assumed to be equal to those of the reference workplace.

Table 24.1 Decision rules in risk estimation

Relative change in probability	Evaluation
0	The risk of injury has been completely eliminated.
0.25	Major reduction in the probability of injury.
0.5	Large reduction in the probability of injury.
0.75	Moderate reduction in the probability of injury.
0.9	Small reduction in the probability of injury.
1	No change in the probability of injury.
1.1	Small increase in the probability of injury.
1.25	Moderate increase in the probability of injury.
1.5	Large increase in the probability of injury.
2	Major increase in the probability of injury.

The probability of the occurrence of an injury when an activity is carried out is assumed to be independent of the number of times that the activity is carried out. Effects of experience are thus disregarded. It is further assumed that the level of manning does not affect the expected probability of injuries in an activity. Manning is only considered when calculating the LTI rate.

The expert panel's judgements are of two types: assessments of changes in exposure and of changes in probability of injuries between the new and the reference workplace. In assessing changes in exposure, $(N_i)_{new}/(N_i)_{ref}$, the activity plans for the reference workplace and the new workplace are used as input. Mechanisation that results in a removal of the operator from the danger zone is also considered. These assessments usually are rather uncomplicated. If an activity has been completely eliminated at the new workplace, for example through mechanisation, this ratio takes the value 0.

Assessments of changes in the probability of injuries, $((p_{i\alpha})_{new}/(p_{i\alpha})_{ref})$, are more complex because they have to consider a number of different factors related to design and operation. To simplify these assessments, the team uses the decision rules according to Table 24.1.

24.3   The steps of the assessment

The assessment involves the establishment of a database of injuries for the reference workplace. This is manipulated in order to 'simulate' a database of expected injuries in the new workplace. It is performed in four steps, as shown in Figure 24.1.

In Step 1, the database for the reference workplace is established. A database on injuries is used in the analysis. The injuries in this database

INPUT

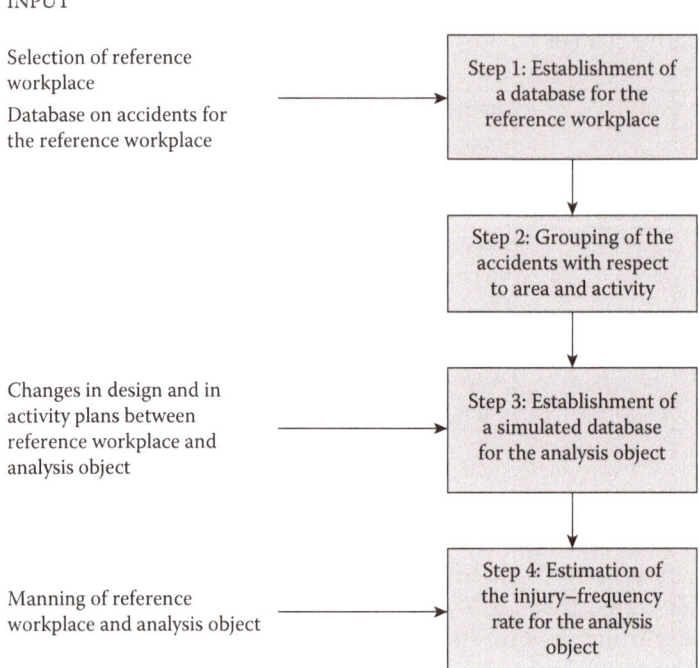

Figure 24.1 The steps of a comparison risk assessment.

have to be categorised by type of system, type of activity, and type of accidental event. From this database, a subset of data for the reference workplace is established. Two different sets of criteria are used in defining this
database. First, the reference workplace must be reasonably similar to the
new workplace with regard to technology and activity plans. Second, the
reference workplace database must be of an adequate size. The size determines the degree of resolution in the subsequent analysis.

Example (continued): An analysis team carried out a comparison
assessment of the floating production and drilling installation mentioned in Section 25.2. For this purpose, a database of about 100 injuries
from two existing platforms of a similar type was set up. The database
included so-called total recordable injuries (i.e. those resulting in medical treatment and transfer to another job in addition to the lost-time
injuries).

The purpose of Step 2 is to establish the distributions of injuries for
the reference workplace by type of activity and injury, $(E(Z_{i\alpha}))_{ref}$. Injuries in
the 'reference database' are grouped into activities. A hierarchical breakdown of the activities is used in this analysis. For each activity, the injuries
are grouped by type of accidental event. The analysis is truncated when

the breakdown has reached a meaningful level of detail for each type of activity and event.

Step 3 involves the establishment of the simulated new workplace database. First, the correction factor, $k_{i\alpha}$, is assessed for each combination of activity and event type. This is done by an expert panel that consists of personnel with operational experience from the reference workplace (subject-matter experts), designers from the project, and experts on risk assessment. The panel's task is to assess the effects of differences in design and operational plans between the new workplace and the reference workplace on:

1. Changes in the probability of injury each time an activity is carried out $((p_{i\alpha})_{new}/(p_{i\alpha})_{ref})$
2. Changes in job frequency $((N_i)_{new}/(N_i)_{ref})$, that is, the number of times each activity (A_i) is carried out each year

A checklist of different factors that affect the risk of accidents is used in this evaluation together with the decision rules according to Table 24.1. An example of a checklist for an offshore platform is shown in Table 24.2.

The mean number of injuries by type of activity and event for the new workplace $((E(Z_{i\alpha}))_{new})$ is then established on the basis of these assessments. The database for the reference workplace does not contain data on new types of activities in the new workplace. In this case, the expected number of injuries is determined by using injury data from activities that are as similar as possible at the reference workplace. The same applies to activities that have been redesigned such that new risks are introduced. Detailed job safety analyses may be required to identify and assess new risks.

The team documents its judgements on a record sheet according to Table 24.3. It is important that all judgements are carefully recorded.

Table 24.2 Checklist of factors that affect the exposure to accident risks and probability of accidents on an offshore platform

Design	Organisation
Factors affecting job frequency	
Walking distances	Manning
Production regularity	Activity planning
Factors affecting probability of injury	
Layout, gangways	Job experience
Lifting equipment	
Access, ergonomics	
Physical working environment (lighting, noise)	
Outdoor operations	
Safety measures	
Guarding	

Table 24.3 Extracts of results from a comparison risk assessment of kitchen work

Activity	N	Historical data for reference installation — Type of injury	Predictions for new installation — Change in job frequency	Change in probability	N	Comments
Cutting of meat, vegetables	6	Cut injury	0.7	0.9	3.8	Reduced cutting work; better layout and lighting; floater movements
Handling of hot pots, pans, and so on	2	Burns	1	1.25	2.5	Floater movements

Because expert judgements play such an important role, the analysis has to be transparent and allow for later review and verification.

Example (continued): The assessment of the floating platform covered all areas of the platform. Table 24.3 shows an example of the results for the kitchen. This part of the assessment was carried out by a team consisting of a supervisor and a catering worker from one of the reference platforms, an architect responsible for the design of the living quarters of the new platform, and a working environment specialist.

In Step 4, the injury-frequency rate is estimated for the new workplace. The number of injuries is summarised in order to establish the overall mean number of injuries for the new workplace and the reference workplace ($(\Sigma E(Z_i))_{new}/(\Sigma E(Z_i))_{ref}$). Next, the known manning level at the reference workplace and the planned manning level at the new workplace are used to establish $(\Sigma e_i)_{ref}$ and $(\Sigma e_i)_{new}$; λ_{new} is then calculated.

Example (continued): The total assessment predicted the occurrence of on average 16 injuries per year at the new installation, Table 24.4. There was a 2% reduction in relation to the historical data from the reference installations. Maintenance accounted for the largest reduction, and this was explained by reduced maintenance due to improved planning and improved access to equipment during inspections and material handling. Increases were expected in drilling due to a more extensive drilling program and more handling of heavy materials.

Table 24.4 Results of a comparison risk assessment of an offshore
floating platform

Activity	Expected number of accidents per year for the new installation	Historical data on the number of accidents per year for reference installations
Catering	4.2	4.5
Operation	1.8	1.5
Drilling	4.2	3.8
Maintenance	3.8	4.7
Walking	1.7	1.7
Others, unknown	0.4	0.3
Total	16.1	16.5

Calculations of the total recordable injury frequency rates for the new and the reference platforms showed a small decrease for the new platform. It was concluded that the acceptance criterion was met. The analysis, however, revealed a number of activities where an increase in the injury frequency of more than 20% was expected. The as low as reasonably practicable (ALARP) principle called for actions to reduce the risk of accidents in these activities in particular. They included:

• Handling of hot objects in the kitchen (floater movements)
• Cleaning of equipment in utility areas (more hydraulic equipment)
• Crane operations (reduced view from crane cabin, floater movements)
• Pipe handling in drilling (increased job frequency, floater movements)

section six

Putting the pieces together

Section VI illustrates applications of the methods and tools presented in Sections III to V for the management of safety in three different industries. Chapter 25 is built around a case study of the Ymer offshore platform with Norskoil as the operator. The case is fictive, but it draws experiences from a number of real field developments on the Norwegian continental shelf. We will look into how experience feedback is accomplished in different phases of a platform's life cycle (i.e. design, construction, and operation). We will illustrate both the prevention of occupational accidents and the prevention of major accidents.

Chapter 26 is organised in a similar way to Chapter 25 and illustrates the application of various methods for a large infrastructure project. Another fictive case study, of the David hydropower project, is brought through the different project phases into operation.

In Chapter 27, we will move to the field of road transportation safety. We will illustrate the application of different principles and methods of experience feedback in this new setting. This is done from the perspective of a company responsible for road transportation safety of their own employees and contractors.

chapter twenty-five

The oil and gas industry

25.1 Accidents in offshore oil and gas production

There are major accident risks involved in the development and operation of offshore oil and gas fields. The capsizing of the Alexander L. Kielland platform in 1980, the fire and explosion that destroyed the Piper Alpha platform in 1988, and the Deepwater Horizon drilling rig blowout and subsequent explosion and fire in 2010 remind us of this fact. The Deepwater Horizon catastrophe demonstrates that a single accident may cause a high number of fatalities, extensive environmental damage, and monetary losses. The explosion that followed the blowout killed 11 workers and injured 16 others. It resulted in a massive oil spill in the Gulf of Mexico, which was characterised as the largest environmental disaster in U.S. history (Deepwater Horizon Study Group 2011).

Ordinary occupational accidents also are a concern in this industry. There was one occupational fatality on the Norwegian continental shelf in the period 2008–2013, which results in a fatal accident rate (FAR) value of 0.4 (PSA 2014). The corresponding figure for the UK continental shelf was two fatalities and a FAR value of 0.6. The members of the International Oil and Gas Producers report a FAR value of 2.0 for 2013, based on 40 fatalities in all oil- and gas-related activities (OGP 2014a). This figure represents a positive development during the years from 2004, when the FAR value was 2.5 times higher.

The TRI rate for fixed installations was 7.3 on the Norwegian continental shelf in 2014 (PSA 2015a). Seventy-three per cent of the accidents occurred in construction (TRI rate = 11.6). The corresponding TRI rate for mobile facilities was 5.3.

An offshore installation has several functions (e.g. drilling, production, transportation, catering, maintenance, logistics, and so on), and the various functions/activities may lead to different types of accidents. In drilling, for example, contact with objects or machinery in motion is a predominant accident type. Accidents in manual handling such as cutting predominate in catering.

Safety is a central issue in all activities in connection with the development and operation of offshore fields. We will demonstrate how the different tools described in Sections III to V may be put for use in safety practice during the different phases of an offshore platform's life cycle. We will concentrate on design, construction, and operation. The selected case study, the Ymer field with Norskoil as the operator, is fictive. It draws experience from real installations in operation and from project work.

25.2   The Ymer platform

25.2.1   Design

Ymer is a semi-submersible production, drilling, and living quarters (PDQ) platform as shown in Figure 25.1. The platform is located directly above the subsea completed wells. They are connected to the platform via flexible risers. The oil is exported by pipeline.

The platform deck is built as an integrated structure located on the top of a substructure, consisting of four columns and a rectangular pontoon. The platform is fixed to the seabed by 16 anchor chains.

The drill floor and the derrick (drilling tower), with pipe-handling systems and the flare tower on the top, are located in the centre of the platform. In this area, we find the remaining systems for drilling and mud treatment. Subsea equipment such as valve assemblies (X-trees) and a blowout preventer (BOP) are stored and handled in the moon-pool area below the drill floor.

The aft of the integrated deck is dedicated to the process systems. There is an oil and gas-processing train with three-stage separation for

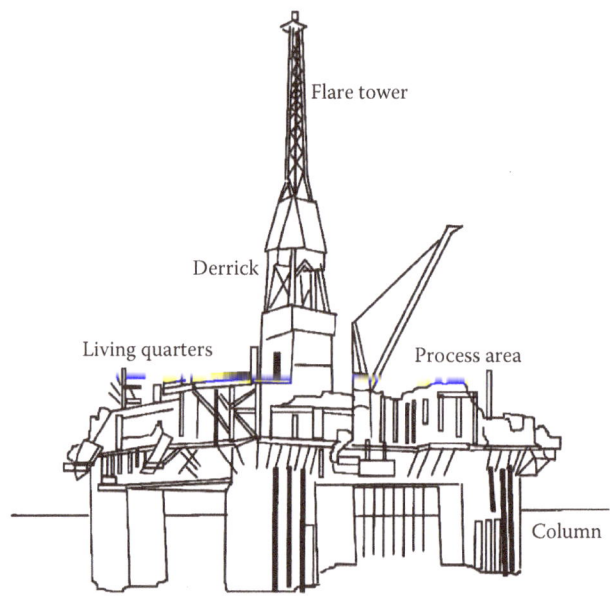

Figure 25.1 The Ymer PDQ platform.

production of 15,000 Sm^3 stabilised crude oil per day. Gas is re-compressed and re-injected during the initial production phase.

The living quarters are built as a separate module at the front end of the platform. Here, we find the accommodation areas and the offices. Most of the utility systems for electricity, water, air, and so on are located in the integrated deck below the living quarters. There are two main generators adjacent to the derrick.

25.2.2   Organisation and manning

The manning level of the Ymer platform is lower than on other comparable platforms. This is obtained by simple and functional design and by letting machinery suppliers take more responsibility for maintenance of their own delivered equipment. During the drilling period, Ymer is manned by 86 persons, 56 of whom work the day shift and 30 at night.

Work similar to that carried out on Ymer is found in land-based industry as well. The operation and maintenance crew (23 persons) has similar responsibilities to those of, for example, oil refinery personnel. Drilling involves the handling of heavy tools and equipment and is to some extent comparable to construction work. The catering crew's responsibilities are very similar to those of the employees of an ordinary hotel.

The main differences when compared to land-based industry are the location of the workplace, which is about 100 km out in the North Sea, and the shift schedule. All employees work offshore for two weeks in 12-hour shifts and then they have four weeks off from duty. Limited space is also a significant factor that differentiates offshore industry from land-based industry.

25.3   Prevention of accidents in design

According to basic safety engineering principles, safety by design is the first choice in the prevention of accidents. A safe design allows platform management to reach safety goals without undue efforts. An unsafe design, on the other hand, has to be compensated for through operational safety measures that may have a negative impact on production and on the quality of the employees' working life.

25.3.1   The phase model for the planning and execution of investment projects

In order to understand how safety is managed in an offshore project, we first have to look at how the total project period is split into separate phases. Figure 25.2 shows the phase model applied by Norskoil in the Ymer project. In order to continue from one phase to another, the project

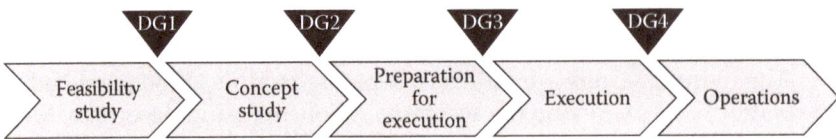

Decision gate 1: Approval to start concept planning
Decision gate 2: Project pre-sanction, approval to prepare for execution
Decision gate 3: Project sanction, approval to start execution
Decision gate 4: Approval to start operations

Figure 25.2 Phase model for planning and execution of investment projects.

has to pass so-called decision gates (DGs). This involves meeting certain criteria, which are different for each decision gate.

A project division within Norskoil carries out project work. Before DG1, the market organisation of Norskoil (i.e. the client) commissions the project division to begin work. The client has made the necessary initial evaluations of the potential resource basis of oil and gas and of the market beforehand. The scope of work for the *feasibility study* also has to be defined. This phase aims at demonstrating whether it is technically and economically feasible to develop the field. The project team establishes detailed reservoir models in order to assess the oil and gas reserves and to establish production rates over the field's expected lifespan (production profiles). In parallel, the project evaluates different field-development concepts such as fixed platforms, floating platforms, and subsea developments for tie-in to existing platforms in the area. This evaluation is based on information about existing infrastructure, water depth, reservoir models, and so on. The aim of the safety activities in this phase is to assess and evaluate whether it is feasible from a safety perspective to develop the field within the conditions defined by the authorities and by Norskoil's internal health, safety, and environment (HSE) policy and requirements. It is not necessary to choose the best concept from a safety point of view, but the rationale for making the selection has to be documented.

Provided that field development is judged to be feasible from a technical, economic, and safety point of view, the project gets approval to start the *concept study* (DG1). One development alternative is selected and further matured in this phase. The aim is to provide an adequate decision basis for investment approval to be given by the board of directors.

Norskoil decides to select a semi-submersible platform concept for further development of the Ymer field. At the second decision gate (DG2), Norskoil's board of directors makes the project 'go' (pre-sanction) decision, implying that the necessary capital for investments is reserved and that the project will be realised unless significant new information (e.g. on capital costs or the market situation) will emerge in the next phase. To pass DG2, the investment plan has to show adequate profitability. This implies that the capital and operational expenditures, as well as the revenues from production of oil and gas, are known with adequate certainty. Here, it is necessary that the project's scope has been adequately defined. The platform concept and the subsea and pipeline installations must be defined at an adequate level of detail.

The basic aim of *preparation for execution* is to develop the necessary basis for government approval and for the main contracts. The project matures the concept further and establishes a contractual strategy, technical specifications, and other tender documents. A plan for the development and operation (PDO) of the petroleum deposit and an environmental impact statement are developed and submitted to the authorities. They are based on input from different safety studies, including a coarse Risk and Emergency Preparedness Assessment, or REPA (Standards Norway 2010). A primary aim is to provide input to the documentation required by the authorities. The different safety analyses and studies are also important tools in concept optimisation and serve as input for technical specifications.

The procurement department establishes a list of qualified tenderers for engineering, procurement, and construction (EPC). They must be able to offer an adequate standard semi-submersible platform concept and must have the ability to mobilise a project organisation with adequate qualifications and experience. Tender documents are developed, and an invitation to tender is sent out. Bids are received and evaluated. An EPC contractor is selected from among the tenderers.

At DG3, Norskoil's board of directors makes the final appropriation of the terms of the main project contracts. *Project execution* consists of engineering (including procurement), fabrication, commissioning, and start-up of oil and gas production. The contract governs the relations between Norskoil and the contractor. The technical requirements for platform design are specified in the contract. They include a design basis, technical specifications and drawings, and so on, and serve as input for engineering. The contractor is responsible for the design of the platform and for procurement of equipment. Norskoil's project team follows up on the quality of the work, budget, and schedule. This team also handles design changes. An important task in the follow-up is to ensure that regulatory and company safety requirements are complied with in the design. The output from engineering consists of detailed drawings, purchase orders for equipment, and procedures for mechanical completion checks and commissioning of the different systems and parts of the installation. Mechanical completion (MC) and commissioning are activities that aim to ensure that the delivered modules and systems of the installation meet the specified requirements. By and large, MC refers to visual inspections of the platform by use of checklists, and commissioning refers to dynamic testing of equipment and systems during operational conditions.

Construction of the different modules and structures takes place at different building yards. Equipment is delivered from the fabrication sites and is installed, and the different modules are assembled together with the substructure. The platform then is mechanically completed and commissioned.

Next, the platform is installed and commissioned together with the subsea systems at its location in the North Sea. As the operator, Norskoil has the main responsibility for safety during all the offshore activities.

DG4 involves the decision to begin production of oil and gas. Norskoil's operations organisation is responsible for this decision. All essential processes, drilling, utility, and safety systems must be ready as well as the organisation and procedures for the operation of the platform. *Production* also involves maintenance and modifications of the installation. Again, as the operator, Norskoil has the main responsibility for safety during all of these activities.

25.3.2 Safety management principles

Norskoil has the overall safety responsibility in all phases of the platform's life cycle. The company's overall HSE policy states that Norskoil must comply with all regulatory requirements. Beyond that, Norskoil must be one of the foremost operators on the Norwegian continental shelf in relation to HSE work and proven results of such work. Management states that a high safety standard is a prerequisite for Norskoil to maintain a competitive business operation.

Norskoil has broken down its policy into HSE goals and action plans for the different project phases. They relate to:

- *Technical safety and barrier management:* Protection and emergency preparedness against major accidents due to fire, explosion, blowout, falling objects, ship collisions, and so on
- *Working environment:* Work-related factors that affect the health and well-being of the personnel, such as accident risks, noise, chemical substances, ergonomics, and psycho-social conditions
- *Environmental care:* Prevention of toxic releases into the air, sea, and ground
- *Production regularity:* Ensuring availability of oil and gas production systems

Norskoil applies common HSE management principles throughout the project phases in order to control and manage the different HSE aspects. The control loops of Figure 25.3 illustrate these principles.

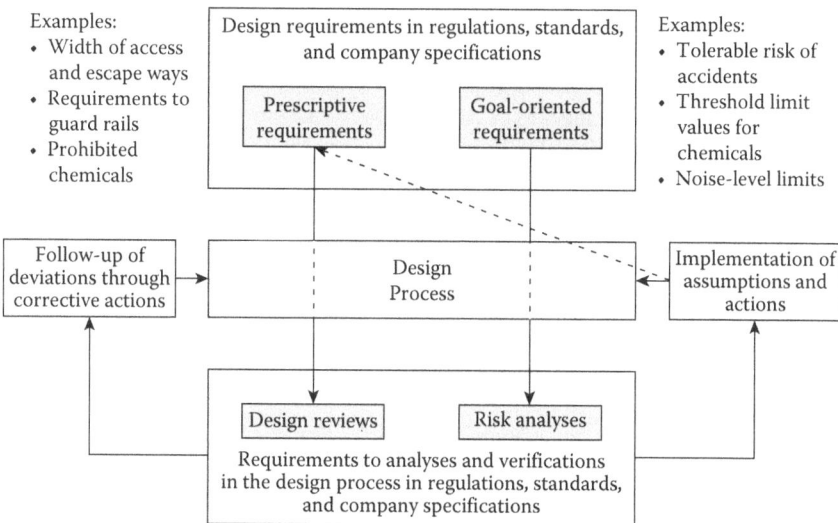

Figure 25.3 Different analysis and verification activities and their role in ensuring compliance with prescriptive and goal-oriented safety requirements of design.

The safety requirements, as stated in the regulations and in the contract, are our next focus. Norskoil has to translate the company's HSE policy and goals in relation to design into binding contractual requirements for Ymer. There are two types of safety requirements: goal-oriented and prescriptive.

The *goal-oriented* or *functional requirements* aim at defining satisfactory protection against hazards, without being specific about the detailed technical solution. Regulatory requirements, especially for the offshore industry, are often of this type. It is up to the contractor to choose any solution that meets the goal-oriented requirements, but the contractor has to document that the selected solution is in compliance with the requirements. This is done, for example, through the application of risk analyses. The results from these risk analyses serve as input to design.

The *prescriptive requirements* specify the detailed design solutions. Design reviews are applied to verify that these requirements are met.

The most important 'safety requirements' as to design are:

- Regulatory requirements
- Norskoil's overall HSE policy and goals
- Acceptance criteria for the risk of losses due to accidents
- NORSOK standards (Standards Norway 2004, 2008a, 2010) and European and international standards
- Norskoil's technical specifications

25.3.2.1   Acceptance criteria for the risk of losses due to accidents

According to management regulations (PSA 2015b), oil companies have to develop acceptance criteria for the risk of losses due to accidents. Risk assessment should be used to verify that new design solutions meet these criteria. Norskoil applies the following acceptance criteria for the Ymer platform:

1. *Frequency of fatalities:* The FAR must not exceed 10 fatalities per 10^8 work hours.
2. *Loss of barriers:*
   a. *Frequency of escalation:* The annual frequency of escalation due to impairment of explosion barriers among areas must be less than 1×10^{-4}.
   b. *Loss of evacuation possibilities:* There must be an upper limit on the frequency of loss of means for evacuation. This applies to evacuation from areas outside the area immediately affected by the accident. The average number of personnel in the area determines this limit (see Figure 25.4).
3. *Frequency of occupational accidents:* The frequency of occupational accidents must not be higher than for comparable platforms in operation. Norskoil also uses a risk matrix similar to the one shown in Table 21.4 in determining the acceptability of occupational accidents.
4. *Environmental risk:* The frequency of accidents that impose serious damage on the ecological system must not exceed 10^{-4} per year.

The different acceptance criteria apply to losses to personnel and the environment. There are no acceptance criteria concerning losses of material assets. Here, the safety measures are implemented on the basis of cost–benefit considerations. Two of Norskoil's acceptance criteria for risks to personnel use losses of barriers as the undesired outcome rather than injuries or fatalities.

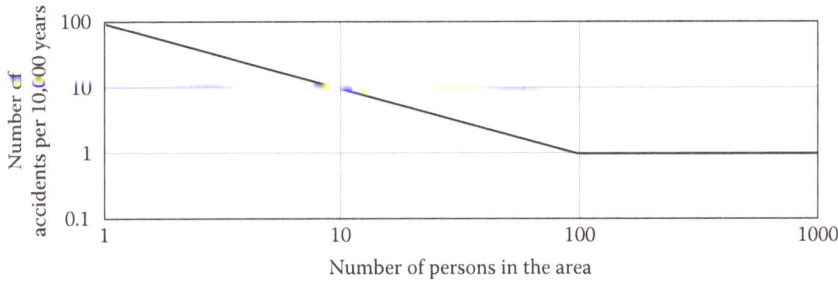

Figure 25.4 Acceptance criterion for the maximum frequency of loss of evacuation possibility due to major accidents.

The aim here is to guide the risk assessment towards the identification and evaluation of design aspects that have a known relation to risk.

In addition to meeting the risk acceptance criteria, Norskoil strives at continuous improvements in the level of safety through the application of the as low as reasonably practicable (ALARP) principle. This means that Norskoil accepts additional expenditures beyond those required to meet the acceptance criteria, provided that there is no gross disproportion between the costs and benefits of introducing further risk reducing measures.

25.3.2.2   *Experience transfer between operation and design*

When designing new production systems, a careful balance has to be maintained between safety measures in design and operation. In practice, we rarely are able to eliminate or guard against all hazards by design. Instead, we must rely on barriers that require constant vigilance by operations and maintenance management for operators to be effective (compare Ashby's law of requisite variety in Section 7.4.1), and the task of controlling these barriers must not be too complex and demanding.

This applies to all barriers requiring regular inspection, testing, and maintenance. It also applies to active barriers, where the barrier function is fulfilled through the interaction of technical, operational, and organisational elements. The operator (human barrier element) controls the technical barrier elements through procedural measures (organisational barrier elements).

Experience transfer between the design/project and operations personnel must help find an adequate balance between measures in design and operation. It also must help in:

1. Reduction of the complexity of the operational control tasks through:
   a. Identification and elimination or reduction of hazards
   b. Design of barriers between the hazard source and the operations personnel that require an acceptable effort to be established and maintained
   c. Design of feedback mechanisms to ensure that the barrier efficiency is maintained
2. Information to operations on how to maintain the barriers for which they are responsible

There are different experience transfer mechanisms between operations and projects; see Figure 25.5. These mechanisms are formal, involving documentation on paper or in databases, and informal, involving person-to-person communication.

In the formal transfer of experience, the *experience carriers* play an important role. They include Norskoil's technical specifications and

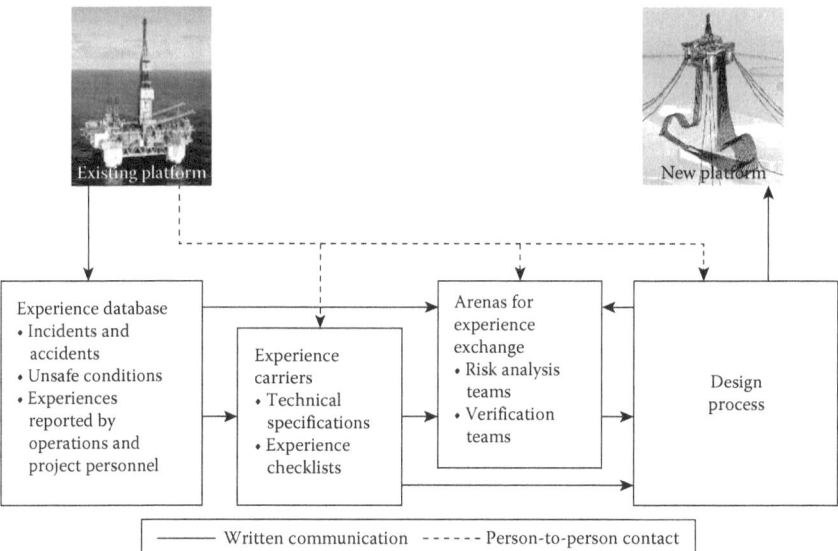

Figure 25.5 Experience flow from operations to projects to support the design process. (Adapted from Wendel, E., *SHE Experience Transfer from Operations to Projects*, Master's Thesis, Norwegian University of Science and Technology, Trondheim, 1998.)

experience checklists. The latter are less binding to the contractor. The experience carriers express operations experience in a language and format that is comprehensible to the design personnel. The process by which specific and detailed experience data are extracted, summarised, and evaluated for implementation by the experience carriers is crucial. This does not take place automatically. Rather, the organisation has to use a structured approach by establishing procedures for regular reviews of the experience databases.

The *experience exchange arenas* represent another means of experience transfer. The individual experiences are challenged in group discussions and combined in ways that promote the development of joint understanding and qualified experience.

For this project, an incident and accident database can be found among the *experience databases*. Although the project personnel have access to this database, there are few spontaneous queries of it. The experiences are considered too incomprehensible, and there are concerns about the data quality. A pre-condition for efficient use of an incident database in experience transfer is the existence of a community of specialists in practice with a good understanding of the database and clear ownership to it. The database has then proven valuable when used in black-spot analyses to establish experience checklists of needs for improvement in the new design.

The informal communication, through integration of operations personnel in the project organisation, is also shown. A relevant question is whether the experiences that are transferred through these person-to-person contacts really represent the operation organisation's experience or merely personal preferences of the involved operations personnel.

25.3.2.3   Control and verification activities

The framework regulations for the Norwegian offshore industry state that oil companies are responsible for ensuring that the regulatory requirements as to HSE are met; these are the so-called internal control requirements. Other regulations, such as the management, facilities, activities, and working environment regulations, detail the internal control requirements (PSA 2015b). Norskoil has an interest in controlling and verifying the company's HSE policy and goals.

In design, the EPC contractor must develop HSE management systems to control and verify compliance with all contractual requirements. They shall be based on the principles and requirements in the quality management standard ISO 9001 (ISO 2015). Design reviews, risk and reliability analyses, inspections, and tests are examples of control and verification activities. Norskoil's project team participates in some of the activities and carries out independent control and verification activities as well.

The NORSOK Standard S-002 on working environment reflects these basic safety management principles (Standards Norway 2004). Chapter 4 of the standard lists the various types of control and verification activities that must be performed by the contractor to cover all essential working environment factors. Here, we find requirements for coarse and detailed job safety analysis (JSA) and for quantitative risk analysis. In Chapter 5, we find the working environment requirements for design.

The project's and contractor's HSE management systems are documented in the respective organisation's *HSE program*. This is a document that describes:

- HSE objective and goals and acceptance criteria for the risk of losses due to accidents
- HSE requirements and applicable procedures for control and verification
- Organisation and responsibilities for implementation of HSE requirements in design and for control and verification activities
- Plans for control and verification activities, deliverables included

The project's HSE manager is responsible for the HSE program and has to update it for each new phase. It must reflect the criteria that have to be met at each decision gate. The initial HSE program, which is established at the start of the feasibility study phase, must have addressed the following questions at the end of this phase (DG1):

- Has the feasibility study concept been adequately documented through safety risk assessments? Are there specific safety challenges as to sensitive environmental resources, climatic conditions, water depth, and so on that have been identified and are the mitigation plans realistic to address these within the established investment plan?
- Have the applicable safety requirements and other steering documents (regulations, policy, goals, acceptance criteria, technical specifications, and so on) been identified and are they of adequate quality?
- Have the necessary authority contacts been established?
- Is the project organisation adequate, and is it manned with HSE personnel with adequate qualifications?

At the next decision gate (DG2), the project must be able to show that there are satisfactory answers to the following questions:

- Have all significant safety hazards in design and construction been adequately identified and addressed?
- Has it been documented that the selected concept is feasible with respect to safety and that all significant safety hazards have been resolved with adequate certainty with respect to cost and schedule?
- Have the plans for authority contacts been adequately defined and initiated?

The HSE program for preparation for execution will address the need to mature the development concept further with respect to safety and to develop the necessary tender and contract documents. At the end of this phase, the project must be able to give adequate answers to the following questions in order to pass DG3:

- Has the selected concept been adequately documented with respect to safety to meet Norskoil and regulatory requirements? Does the concept comply with the risk acceptance criteria? Have the environmental impact and the mitigation plan been adequately documented?
- Have the necessary authorities' approvals been received before proceeding with binding contracts?
- Have the safety requirements as to design and control and verification activities been adequately defined in the contracts?

- Are the contractors adequately qualified with respect to safety in order to perform the work?
- Has Norskoil established an adequate project organisation for follow-up of the contracts? Is adequate experience transfer from Operations ensured?

The HSE program for project execution will include the necessary activities for follow-up and verification of contract work. At the end of this phase, the project must be able to hand over adequate documentation to the operations organisation to meet the criteria for DG4. In all, operations have to be able to give positive answers to the following questions before the platform is ready for start-up and the introduction of oil and gas into the process systems:

- Are the technical systems, the operational procedures, and the organisation and personnel ready for start-up with respect to safety?
- Have the necessary authority approvals been received?

25.3.3   Prevention of major accidents

25.3.3.1   Concept selection and definition

In the concept selection process, different pros and cons of alternative concepts are evaluated against each other – primarily by use of economic criteria. It is not necessarily the best concept from a safety point of view that is selected but the concept that, from an overall evaluation, has the greater merit. The basic safety requirements and risk acceptance criteria have to be met by the selected concept.

Major incidents involve the accidental release of considerable energy and the potential for multiple fatalities and major environmental and material damage. In offshore oil and gas production, there are a few types of hazards or energies associated with the risk of major accidents. For a floating PDQ platform, these typically include fires and explosions due to leakage from process containment or blowout, dropped objects, maritime accidents (hull puncture due to ship collisions, extreme weather, or ballasting errors), and helicopter accidents. Here, we will focus on the prevention of fires and explosions because they dominate the risk picture. Figure 9.5 illustrated how fires and explosions are prevented through the implementation of independent barriers.

The project screens different development concepts. Risks that may make a concept unfeasible or result in high extra costs are focused on in special studies. An independent consultant carries out coarse REPA for the two development alternatives that are considered in accordance with NORSOK Standard Z-013 (Standards Norway 2010). They are a standard production ship and a standard semi-submersible PDQ platform. The aim is to obtain a total overview of the risks involved for concept selection and to check compliance with the acceptance criteria. It is important to remember that the REPA and the acceptance criteria evaluate the integrity of passive and embedded barriers in particular. These are to a large extent determined in the early project phase.

A REPA consists of two parts: a quantitative risk assessment and an emergency preparedness assessment. The risk assessment involves the following activities:

1. System description, including description of technical systems and relevant operational activities, part of life cycle to which the analysis applies, and personnel and environmental resources exposed to risk.
2. Identification of hazards and listing of initial events such as gas leakage. This work is based on generic information about the type of concept, and operational experience plays a minor role.
3. Accident modelling, consequence evaluation, and assessments of probabilities. Here, platform design and operational modes serve as important input. Probability assessments are based on generic accident statistics, averaging over different types of platform design, barrier types, and HSE management systems.
4. Evaluation of risk and comparison with the risk acceptance criteria.

The quantitative risk assessment is used to identify dimensioning accidental loads (Vinnem 2013). These are the accidental loads from major accidents such as gas explosions, hydrocarbon fires, or ship collisions that generate the most severe accidental loads that the safety barriers must be able to withstand. The dimensioning loads are used in the design of the barriers and thus translate Norskoil's risk acceptance criteria (goal-oriented requirements) into prescriptive design criteria; see Figure 25.3.

The risk analysis also gives input to the emergency preparedness assessment. It identifies the so-called defined situations of hazards and accidents (DSHA). These are major accidents (such as fires, explosions, and ship collisions) used in the planning of emergency responses. The aim of the emergency preparedness assessment is to verify that Norskoil's emergency preparedness requirements are satisfied for all DSHAs. The emergency preparedness assessment also considers potential less-serious accidents not defined by the REPA that the emergency response system needs to be dimensioned to manage (e.g. man overboard accidents).

Table 25.1 shows the results of the REPA for the two development alternatives. The REPA gives the ship alternative the best rating. In this

Table 25.1 Examples of results from the coarse REPA of two development alternatives for Ymer

Risk aspect	Norskoil's acceptance criterion	Risk level assessed for ship	Risk level assessed for semi-submersible with drilling
Loss of evacuation possibilities from the platform	1×10^{-4}	3.3×10^{-4}	6.4×10^{-4}
Frequency of escalation	1×10^{-4}	5.7×10^{-5}	6.5×10^{-5}
Fatal accident rate	10	6.1	6.9
Serious environmental damage	1×10^{-4}	$<1 \times 10^{-6}$	$<1 \times 10^{-6}$

alternative, an independent rig carries out the hazardous drilling, well completion, and maintenance activities far away from the rig. Risks on the rigs are excluded from the analysis, which gives the ship alternative an advantage. A production ship also has a safety advantage through its intrinsic design. There is good separation between safe and hazardous areas, and the hazardous areas are always located downwind.

The main safety concerns in the ship design involve:

- The turret through which oil and gas from the subsea installations are brought up to the ship. There are high risks of oil and gas leakage in the swivel inside the turret. Ignited gas leaks may cut off escape ways from the rear of the ship to the living quarters.
- Risers that, in some positions on the ship, are located beneath the living quarters. Ignited leaks will expose the muster area and lifeboats.
- Ship collisions in general and especially the shuttle-tanker activities close to the stern of the ship.

There are existing ship concepts that are considered feasible, provided that a collision warning system is introduced. It is also considered realistic that further design development of the turret will result in a feasible solution.

Major safety concerns with the semi-submersible are:

- The location of some subsea wells underneath the platform, which means that oil and gas will reach the sea surface under the platform in case of a blowout.
- Burning riser leaks that will expose the platform substructure and may cause it to collapse.
- Separation of safe and hazardous areas – the barrier between the drilling areas and the safe haven (living quarters) is especially critical.

According to the results of the REPA, the anchor-release system of the semi-submersible has to be designed such that the platform can withdraw to a safe location outside a burning pool fire on the sea surface in case of blowout. The risers and substructure have to be protected. The derrick must be able to withstand a burning blowout during the time it takes to relocate the platform to a safe position outside the pool fire on the sea. With these measures, the semi-submersible is considered a feasible solution.

The coarse REPA for the semi-submersible alternative is handed over to the authorities as an attachment to the plan for development and operation (PDO). When a contractor has been selected, the contractor's standard semi-submersible concept is subject to a new REPA.

Although operations are involved in evaluating the results of the REPA, their experience has a limited effect on the results. The methods utilised in the REPA are not impacted by operational experience because the REPA utilises generic failure data that average among different platforms with different operational philosophies.

During the concept study, the platform concept and subsea systems are further matured and optimised. A major safety challenge is to contribute to cost-efficient solutions within critical safety areas.

The EPC contractor updates the REPA to include the results of concept optimisation. Norskoil's HSE manager supervises this work. It is shown that the concept meets all applicable acceptance criteria, provided that a number of assumptions are met. These are used to define the accidental loads from fire, explosion, dropped objects, and ship collisions that the semi-submersible must be able to withstand. Another set of assumptions defines the safety and communication systems that have to be operable during an accident. There are also assumptions regarding the collision warning system and the withdrawal of the platform from the subsea wells in case of blowout. The REPA also results in a number of recommendations based on the ALARP principle. The assumptions and recommendations serve as input to design development and to specifications, and they need to be followed up in later project phases.

The REPA is used as a formal tool in defining the EPC contractor's obligations. Consequently, there is a need to clarify its limitations. It has been recognised that the REPA does not cover aspects in the interaction between design and operations, simply because its resolution is too low.

25.3.3.2   Project execution

The contractor is responsible for engineering. The Ymer project has a technical team in the contractor's offices for follow-up of the work. A main task is to ensure that the contractor implements the regulatory (and Norskoil's) safety requirements and risk acceptance criteria in the design in a satisfactory way. It follows that the assumptions from the REPA must also be implemented. It cannot be expected that all safety issues and questions have been resolved during the concept study of preparation for the execution phase. The project thus has to handle engineering-modification requests (EMRs) from the contractor and deviations from regulatory and company requirements (nonconformities). In resolving the EMRs and nonconformities, the main safety philosophy for the concept must not be violated.

Norskoil's technical team also issues variation orders (VOs), that is, changes in relation to what was agreed upon in the contract. The cost consequences of these changes are negotiated with the contractor. It is the technical team's duty to represent Norskoil's understanding of the regulatory and their own safety requirements and acceptance criteria and to be able to interpret these in discussions with the contractor.

Norskoil's HSE manager monitors and follows up on all assumptions, recommendations, and observations from the REPA and other related studies and evaluations. Nonconformities in design are evaluated and documented by the responsible discipline. They are checked and approved by the HSE manager. It is important to ensure that there are no violations of important safety philosophies and assumptions. For example, a layout change in the process area involving relocation of equipment must be checked against assumptions regarding explosion pressure build-up and ventilation.

The REPA is updated by the contractor to take design development into account and to evaluate the effects of changes. The results are compared to the risk acceptance criteria. It is a concern that the resolution of the REPA is often not sufficient to be able to discriminate between good and inadequate solutions to active safety barriers such as fire and gas detection. This has to do with the fact that the REPA uses statistics on failure rates that are averaged over a large number of installations with different characteristics.

It is important that the HSE management together with the technical team arrives at a consistent understanding of how to implement the general results of the analysis.

Example: Cable trays of aluminium are used in the design of process areas. During a fire, these trays will collapse earlier than steel trays. It cannot be derived directly from the analysis whether this is acceptable or not.

Other important issues that are brought up as a result of the updated REPA are:

- Requirements as to fire protection of the derrick to avoid collapse. A minimum of 20 minutes exposure of the derrick to fire without collapse is specified.

- It must be possible to withdraw the semi-submersible from outside a pool fire on the sea in case of blowout. It is concluded that the anchor system must be designed for a withdrawal of at least 135 m from the location of any well.
- Fire protection of risers to avoid escalation of a jet fire from one riser to neighbouring risers.

As a result of this update of the analysis, the specifications concerning accidental loads and safety requirements to systems during accidents are revised. Engineering has to assess the cost consequences of the revised requirements and the modifications of design that follow before they are implemented.

Due to the limitations of the REPA, it is supplemented with a barrier management activity, which includes the establishment and updating of a barrier strategy that is in parallel and integrated with the management of risk. There is also considerable work put into the design of electronic safety related systems such as those for smoke detection and for the activation of a fire suppression system (Norwegian Oil and Gas 2004). The aim is to ensure that these systems achieve a tolerably low frequency of dangerous failures or failure to respond on demand. Verification of compliance is based on reliability calculations utilising input from the equipment vendors and from operations. This will generate requirements as to equipment reliability to be met by vendors and requirements for inspection and testing intervals during operation.

Other risk analyses are carried out that cover operational aspects. A hazard and operability study (HAZOP) is an important tool in this respect (Rausand and Høyland 2004). It focuses on the process arrangements as well as the process control and shutdown systems.

Barrier availability is a concern due to insufficient feedback on barrier failures (Rasmussen 1993). The availability is affected by such design factors as equipment configuration, voting principles, and quality. Different operational factors such as inspections and testing procedures also affect availability. Norskoil has defined maximum safety unavailability for different safety systems based on industry standards (Norwegian Oil and Gas 2004):

- Process safety systems
- Fire and gas detection systems
- Emergency shutdown and gas relief systems
- Fire and evacuation alarms, emergency communication systems, and emergency lighting
- Emergency power
- Ballast systems

25.3.4 Prevention of occupational accidents

25.3.4.1 Concept definition and preparation for project execution

Occupational accidents are not a major concern in concept selection. The risk of this type of accidents is mainly decided by the detailed design solutions developed in later phases. Early on, the project focuses on some basic characteristics of the different development alternatives that have implications for the possibility of developing acceptable solutions in later phases. They include:

- Use of open solutions, where the personnel are exposed to wind and rainfall when operating process and utility systems
- Movements of the floater (ship or semi-submersible platform) due to heave
- General arrangements and their effects on walking and transportation

The project carries out coarse working environment evaluations including a *coarse analysis*. These serve as input to concept selection. Norskoil's operations department participates in these evaluations.

The tendency of the ship to weathervane is an advantage. It improves the climatic conditions (wind, rainfall) in naturally ventilated process areas and thus the risk of accidents. Access for inspection and maintenance in the turret is a concern due to very congested conditions.

Accident risks in drilling activities on the semi-submersible are a concern. A special challenge is to secure safe handling of heavy subsea equipment in the moon pool area in the centre of the platform. Advantages with the semi-submersible are the shorter distances to the different workplaces and less movement due to heave.

Figure 25.6 shows how the different analyses and evaluations of the risk of occupational accidents are coordinated with the design process (Kjellén 1990, 1998; Standards Norway 2004). They are part of a total evaluation of all safety aspects. Each activity has been defined and scheduled in Norskoil's and the contractor's HSE programs for the Ymer project. During conceptual design, the experience checklist is established and the coarse analysis is carried out.

Norskoil develops experience checklists for use in basic design during the concept study and preparation for execution. They are based on *accident-concentration analyses*, where accident data from similar existing installations in operation have been used. An accident database is used in the analysis. Norskoil's project team reviews the findings together with operations personnel from these two platforms. Jointly, they come up with recommendations on how to prevent the identified accident concentrations by design measures.

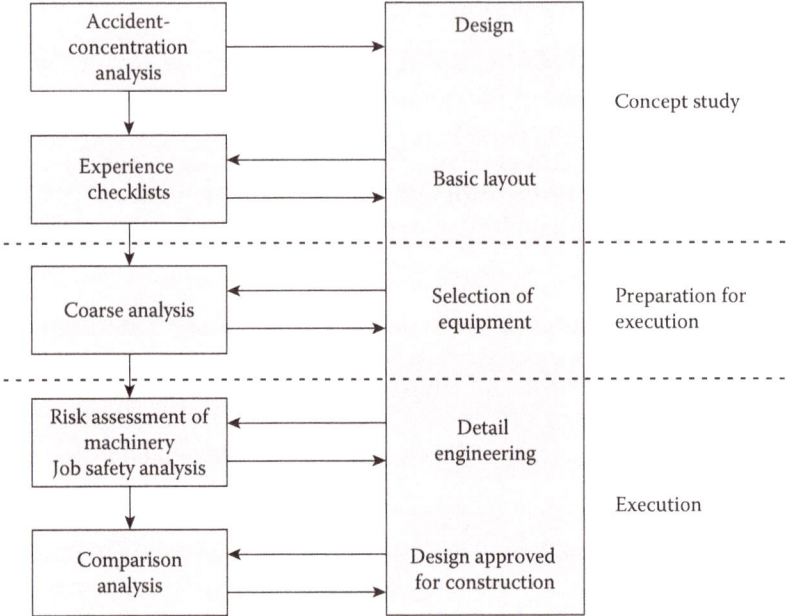

Figure 25.6 Use of different occupational risk analysis methods in design.

Example: There were 49 drill floor and derrick injuries in the database from two existing platforms. Nine of these involved moving heavy tools between the store and the drill floor. The experience checklist identifies hazards in connection with this activity and recommends that a continuous monorail is installed from the store to as close to the centre of the drill floor as possible. By using this monorail, manual handling of heavy tools will be made safer.

At the end of the concept study phase, layout drawings that describe the platform in sufficient detail are made available. The project performs coarse analyses of all areas of the platform (i.e. process, utility, drilling, and living quarters). They invite Norskoil's operations personnel to participate in the analyses. The analysis teams concentrate on expected problem areas related to layout, such as manual materials handling, work in open areas, access to equipment, location of noisy equipment, and solitary work. A number of layout changes are introduced as a result of this evaluation.

Example: A coarse analysis team reviews the handling of heavy tools between the store and the drill floor in the example mentioned earlier. They look specifically into the layout of the store and the location of the monorail. They also review the line of sight from winch control panels into the danger zone for material handling. The analysis team decides on the exact position of the monorail and the location of winches.

25.3.4.2   *Project execution*

The vendors are responsible for the *risk assessments of machinery*. The reports from these assessments are part of the required documentation to issue a declaration of conformity with the machinery regulations. The contractor defines the documentation on risk assessments that should be delivered by the vendors for review and verification. This includes the risk assessments for machinery and assemblies of machinery that are critical from a safety point of view; see, for example, NORSOK S-005 (Standards Norway 1999).

Example: The review by the contractor reveals inadequacies in the vendor's risk assessments for drilling equipment. There is a risk of getting caught in a rotating shaft of the remotely controlled manipulator arms for pipe handling during inspections. Machinery guarding has to be improved. The assessment also reveals insufficient provisions for isolation of hydraulic equipment from the power source. There is a risk of accidental start of the equipment during maintenance.

The EPC contractor is responsible for the *job safety analyses* (JSA). These cover the parts of the platform design that are the responsibility of the contractor, including interfaces with equipment. These analyses focus on tasks with a high risk of accidents (i.e. a high probability of accidents and/or severe consequences).

Example: Manual tasks in the derrick are a concern, especially those carried out outside work platforms. Here, a riding belt connected to a special winch has to be used. This operation is dangerous due to the risk of getting caught in or between pieces of equipment in the derrick. The JSA team lists all maintenance tasks in the derrick and reviews each task with a focus on the risk of getting caught by pipe-handling equipment. A primary safety measure is to eliminate the use of the riding belt. As a result of the analysis, the team decides to move lubrication points so that they are within reach from work platforms. Additional work platforms are also installed to improve access.

In a *comparison analysis,* the number of accidents per year of operation is estimated for the Ymer platform. Accident statistics for the two reference installations serve as a basis for these estimates. Three different teams representing the different areas of the platform make the estimates. All teams consist of project and operations personnel with thorough knowledge in this area. Results show that the total number of accidents per year is expected to decrease somewhat compared to the reference platforms. Reductions in the number of accidents in maintenance in particular contribute to this decrease. This has to do with improved equipment handling during maintenance, improved equipment reliability, and improved maintenance plans.

Example: The comparison analysis indicates that there will be a small increase in the frequency of drilling accidents from, on average, 5.8 accidents per platform year at the reference platforms to 5.9 at Ymer. This increase is mainly explained by increased drilling activities. However, the probability of accidents per activity has been reduced. Due to expected improvements in other areas, the acceptance criteria are met.

25.4   Construction site safety

There are two main safety concerns in this phase. One is to ensure that the platform itself has a satisfactory quality with respect to HSE. A second concern is to ensure that the construction work is carried out in a safe way. Here, we will focus on the latter aspect. It is in Norskoil's interest to avoid accidents during construction that:

- Have cost or schedule impact. A severe accident may delay start-up by up to one year, which will cause losses on the order of billions of Norwegian krones (NOK) for Norskoil and the other owners.
- Reduce the quality of the product (i.e. the platform).
- Result in injury to Norskoil's or the contractor's own personnel at the site.
- Result in loss of good will to Norskoil.

Also, bad safety statistics are regarded as a sign of sloppy management control at the site on the part of the construction contractor. Sloppy management of this type will also manifest itself in bad quality and delays (Grimaldi 1970; Levitt and Samelson 1993; Kjellén et al. 1997).

It is Norskoil's HSE policy to work with contractors in order to achieve satisfactory safety results. The aim is to avoid accidents resulting in serious harm to personnel or the environment or in extensive material damage or delay. Norskoil's safety goal for construction is a TRI rate of five or less. The company uses its HSE management system for contractors to accomplish this goal, as shown in Figure 25.7. Regulatory requirements, Norwegian and international standards for the oil and gas industry, together with Norskoil's contractor HSE policies and procedures define the requirements for the contractors (Standards Norway 2002; OGP 2010).

25.4.1   Step 1: Planning

In this step, Norskoil decides on the contract strategy; this is when it is decided to use an EPC contract for the floater. This is because the market for standard semi-submersible PDQ platforms is adequately competitive, and the design has been qualified in earlier projects. A *constructability risk assessment* is carried out that is similar to a coarse risk analysis of the different activities in construction and installation.

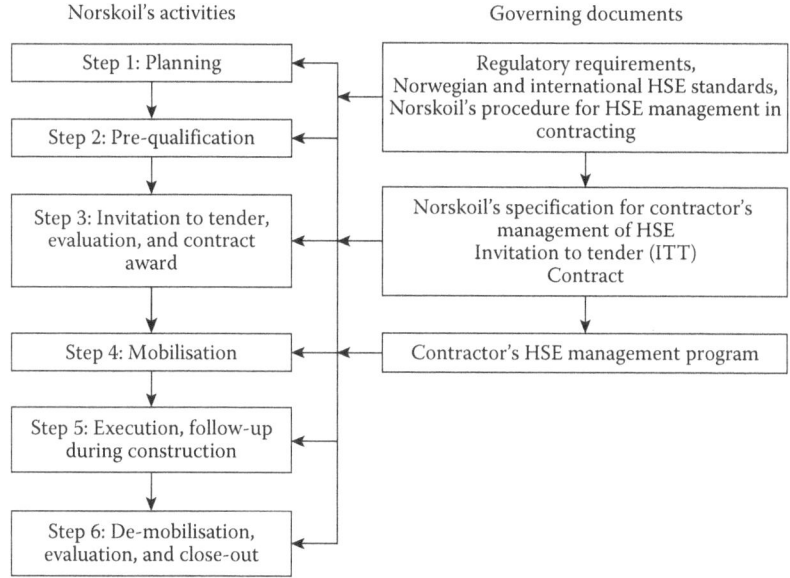

Norskoil's activities   Governing documents

Figure 25.7 Norskoil's principles for the management of HSE on construction sites.

25.4.2   Step 2: Pre-qualification

The basis for a successful follow-up on safety in construction is established during pre-qualification. The contractors who will be allowed to tender for the EPC contract are selected. The project team may select only highly qualified contractors with excellent existing safety results. If this strategy is chosen, the prospects for a safe execution of contract work will be good. This will, however, considerably reduce the number of bidders and thus reduce competition and increase costs. An alternative strategy is to accept contractors with lower safety qualifications and performance. In this case, Norskoil has to put more effort into follow-up of the contractors during the execution of the work in order to ensure an acceptable safety performance. Again, this strategy will have a cost impact through a more costly follow-up and also because of more accidents.

Norskoil has decided to set rather stringent safety requirements that the bidders must meet in order to be qualified. A pre-qualification questionnaire has been developed to be answered by the potential bidders. Finally, the short-listed bidders are *audited,* and this includes visits to fabrication sites. The questionnaire contains items regarding the potential bidders' HSE management system, resources and organisation, and accident statistics in previous work. Bidders with unacceptable results for any of the critical items in the questionnaire or audit are not allowed to bid.

Example: The Ymer contract is an EPC contract, where one contractor will be responsible for the design and fabrication of the complete platform. This contractor will in turn sub-contract parts of the work to other contractors. One of the potential bidders for the Ymer contract had allied itself with other yards to be able to tender for the complete contract. One of the yards in this alliance did not present well on HSE management and performance in the pre-qualification audit. The Ymer project management decided to accept this bidder alliance in order to achieve adequate competition for the contract.

25.4.3   Step 3: Invitation to tender, evaluation and contract award

In the next step, Norskoil's project team evaluates incoming bids. In addition to the economic evaluation of the bids, the project team makes a technical evaluation of the tenders. The proposed organisation, the competence of its members, and the plans for the work are the primary focus. With the tender, the bidders must submit a preliminary HSE program for the project as well as the names of their HSE manager and other key safety personnel.

Questions are discussed with the different bidders in bid-clarification meetings (BCMs). Any concerns about the safety performance or safety resources are brought up in these meetings. The BCMs also give Norskoil an opportunity to communicate its expectations to the bidders' leadership in the area of HSE and with respect to compliance with Norskoil's HSE policy and standards.

Example: Norskoil addressed the poor HSE program and performance of one bidder's sub-contractor with the bidder during the BCM. It was concluded that the poor safety performance (an LTI rate above 50) could partly be explained by the workmen's compensation system in the country of the sub-contractor. The workers actually received higher pay during sick leave due to accidents than they received if they were sick for other reasons. Deficiencies in the yard's HSE management system were also brought up. The bidder offered to provide additional personnel to follow up this particular sub-contractor with respect to safety. This offer was included in the tender.

Norskoil's specification for a contractor's management of safety in construction defines the company's requirements for a contractor's HSE program. This includes:

- Policy and goals
- Organisation and assignment of responsibilities for work on the project
- Activities for follow-up on safety, including workplace inspections, meetings, audits, and introduction of new employees

- Applicable safety rules for specified activities such as scaffolding, housekeeping, fire protection, and so on
- Emergency preparedness organisation and routines

After the contract is awarded, the EPC contractor and its sub-contractors have to submit their respective safety program for construction for Norskoil's review and acceptance. Norskoil's HSE manager for the Ymer project evaluates these draft programs. The EPC contractor has the overall responsibility for safety, and the sub-contractors' HSE programs are subordinated to this program. Norskoil's HSE manager will comment on items where the proposed program does not meet the project's (i.e. Norskoil's) requirements.

Example: Norskoil's HSE manager identifies unclear line responsibilities in the contractor's HSE program. It is stated that contractor's safety engineer is responsible for safety. The contractor is requested to clearly define safety as a line responsibility according to the requirements in Norskoil's specification, which is part of the contract. After the contractor has implemented the comments from Norskoil, the program is accepted by the project.

25.4.4   Step 4: Mobilisation

At this stage, the different HSE programs have been accepted by Norskoil. Norskoil's site personnel for follow-up of the contract start to mobilise at the fabrication yards. Teambuilding is arranged between the contractor and Norskoil's personnel to align the two parties with respect to Norskoil's expectations for the contractor's HSE culture and safety performance and how the contractor is going to meet these expectations. The teambuilding workshop results in mutual agreements on information and follow-up activities.

Norskoil also arranges a *mobilisation safety audit* to check that contractor's obligations in the contract and in the contractor's HSE program are implemented in practice.

Example: The mobilisation audit reveals that the contractor's documentation on lifting equipment at the yard to be used in contract work is incomplete. Certificates are lacking, and some of the equipment has not been subject to inspection and testing by qualified persons. The contractor is instructed to rectify the deficiencies as a condition for start-up of work.

25.4.5   Step 5: Execution, follow-up during construction

The EPC contractor establishes an organisation for follow-up of safety in construction. A site safety coordinator, reporting to the contractor's fabrication manager, does most of the practical work. Norskoil's onsite safety engineer monitors this work and intervenes if the actual safety standard does not meet the project requirements.

Norskoil's specification for a contractor's management of safety states detailed requirements concerning *reporting incidents and accidents* to the oil company. The project team meets regularly with the EPC contractor's management staff. Here, safety performance is brought up as a fixed item on the agenda, and areas of concern are highlighted. The EPC contractor has similar management meetings with sub-contractors, where safety matters are also discussed. These regular management meetings are the main occasions for decisions on remedial actions where the safety performance is inadequate.

There are regular *workplace inspections* at the fabrication yards. The frequency and scope of these inspections are defined in the different HSE programs and are based on risk assessments. The EPC contractor's site safety coordinator participates occasionally in the inspections at the sub-contractors' yards. Also, Norskoil is invited to participate but does so infrequently because the company only has permanent site representation at the EPC contractor's own yard.

Example: The EPC contractor's site safety coordinator identifies sloppy housekeeping conditions at one of the yards during the regular workplace inspection. He decides to monitor this yard more closely and participates in subsequent inspections. When the situation does not improve, this concern is brought up at the monthly management meeting. It is decided in the meeting to stop work for clean-up. The yard has to pay for the delay costs.

It is the responsibility of Norskoil's HSE manager to follow up on the *accident and incident statistics* from the yards. In this way, he monitors the safety performance and compares the results with the project's goals. Based on this information, Norskoil monitors the TRI rate and the RUO-rate, as shown in Figure 25.8.

Safety audits are the main tool of Norskoil's HSE manager for follow-up on safety in construction. The contractor's site safety coordinator also performs audits of this type at the sub-contractors' yards. The audit team checks actual performance against regulatory and contractual requirements and the yards' HSE programs. The audits usually consist of two parts (i.e. a system portion and a technical portion). In the system audit, the HSE manager checks safety documentation and practices in relation to the commitments in the HSE program. The technical audit is carried out as a workplace inspection. The audits follow a schedule and each audit covers a separate theme. Following are examples of audit themes:

- Reporting of accidents and incidents
- Emergency preparedness

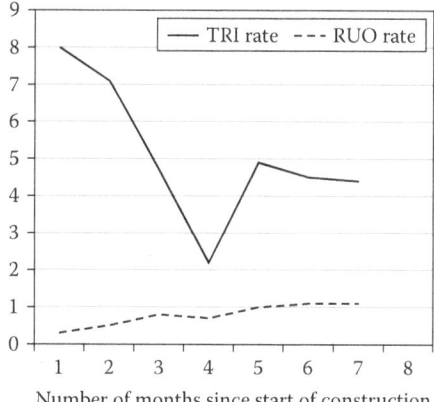

Figure 25.8 The TRI rate and RUO rate for the Ymer project from start of construction.

- Cranes and lifting safety
- Safe work practices, compliance with safety rules
- Housekeeping
- Fire precautions
- Handling of hazardous chemicals

Example: The safety audit of a yard identifies three nonconformities and seven observations. Poor housekeeping and unsecured work at height represent typical findings. The observations are sent over to the EPC contractor to be followed up with the yard for corrective action.

The yard proposes to improve housekeeping by introducing new containers and by defining responsibilities for regular housekeeping activities. The yard also proposes improved training of supervisors and workers in the management of risk in hazardous tasks such as work at height to achieve compliance in practice. The training will focus on a stepwise method for planning, execution, and evaluation of the work. Use of JSA, identifying and integration of relevant safety requirements, and team assessment in toolbox talks are central elements of the method. Norskoil accepts the corrective actions.

Norskoil also carries out ad hoc audits in cases of serious incidents or other negative experiences.

Example: A fire occurs in the air intake filter of one of the main generators while it is subject to preparation for the factory acceptance test (FAT) by the vendor. A welding spark ignites the filter. Norskoil's HSE manager performs a Safety Management and Organisation Review Technique (SMORT) investigation of the fire by referring to Norskoil's right to perform audits according to the contract. This reveals a serious misunder-standing and disagreements between the contractor and the vendor on safety responsibilities.

25.4.6   Step 6: De-mobilisation, evaluation and close-out

After completion of construction work at the yards, the HSE manager performs an evaluation of this phase and the resulting safety performance of the contractors. There are two aims:

1. To update Norskoil's steering documents, especially the internal procedure and the safety specification for contract work. In this way, the decontextualised experiences from the construction of the Ymer platform are brought forward to new projects.
2. To make information about experiences with the different contrac-tors available in the database of Norskoil's procurement department for use in future contracting work.

25.5   Safety during plant operation

Ymer's platform management faces a number of safety challenges, especially during the first period after start-up. The organisation is new and unfamiliar with the platform and the management system. Construction work on the platform has not been fully completed at start-up and con-tinues in parallel with production and drilling. It is of major importance to avoid construction work on systems that have not been de-energised. Here, the permit-to-work (PTW) system plays an important role.

After the construction team has been demobilised, the platform goes into a period of routine operation. It is important to maintain high alert-ness to safety challenges and to avoid complacency, although the risk is not as high as during the early phase after start-up.

25.5.1   HSE management principles

The safety management principles applied at Ymer during the operation phase are described and analysed in relation to the plan–do–check–act (PDCA) wheel or cycle (see Figure 7.5). We view safety versus occupa-tional accidents as a quality aspect that platform management has to satisfy.

Safety is a line responsibility. It starts from the top of the organisa-tion (i.e. the platform manager). As the operator, Norskoil also has overall responsibility for safety in work on the platform that is carried out by the drilling and construction contractors.

Norskoil has documented its HSE management system for the operation of the Ymer platform. It is not collected in a handbook but is integrated into the total platform-management system that is available via intranet. It covers the following elements:

- Top-management commitment and leadership
- Policy, goals, and acceptance criteria
- Organisation and responsibilities
- A list of references to applicable regulatory requirements
- Requirements for safety systems/barriers
- Procedures and work instructions for the identification and evaluation of safety risks and for handling changes and emergency responses
- Procedures and work instructions concerning safety performance monitoring and follow-up of safety activities
- Periodic action plans that detail the implementation of the overall HSE policy and goals
- Auditing and review activities

Let us relate these different items to the quality control loop shown in Figure 7.3. The overall safety requirements (the norm in Figure 7.3) are defined in the section on policy and goals. The list of applicable regulatory requirements also belongs in this part of the control loop together with the more detailed requirements at lower levels that are defined in technical requirements, procedures, work instructions, and so on. The goals are more operationally specified. Norskoil updates these on a regular basis. A long-term goal is to reduce TRI frequency rate to zero.

Implementation is ensured through the Ymer platform organisation as defined by its responsibilities and qualifications. There are yearly safety action plans that break down the policy and goals into tangible actions.

Control and verification are accomplished through hazard identification and assessment, safety performance monitoring, and auditing. Ymer applies some of the tools described in Sections III through V of this book for control and verification. This is followed by corrective actions. Norskoil applies the same system for follow up of these as for incident reporting and follow up (see Figure 25.9).

Ymer's emergency preparedness plan (including organisation, resources, and roles) is documented as a separate part of the HSE management system. This plan also includes the action plans for the different accident scenarios that have been defined in advance, such as the defined situations of hazards and accidents (DSHA).

The drilling and construction contractors have developed their own HSE management systems, but they also have to comply with the HSE management system for Ymer, established by Norskoil as the operator.

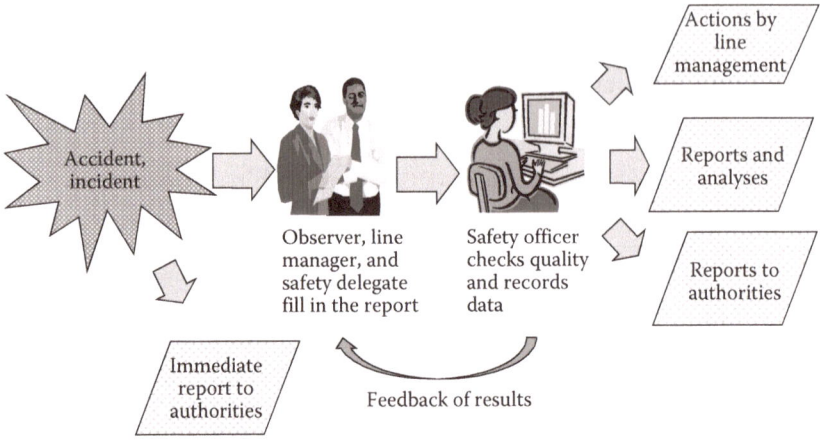

Figure 25.9 Investigation, registration, and follow up of accident and incident reports.

Contractors' HSE management systems focus on the control of hazards in special activities under their responsibility.

25.5.2   Policy and goals

Norskoil's overall HSE policy has been implemented in Ymer's HSE management system. In addition, Ymer's management has defined goals for the safety performance of the platform.

Example: Ymer has defined the following safety performance goals related to the first year after start up:

- TRI rate < 3
- RUO rate > 5
- Less than three events with potential for permanent disability or fatality (Grades 3–5 according to Table 5.5)

25.5.3   Planning and implementation

The safety action plans represent a cornerstone in the implementation of HSE policy and goals. Both Ymer's own organisations and contractors develop such plans. This is done jointly by management and the employees. Management is responsible for the allocation of the necessary resources. The actions are implemented and followed up through clear responsibilities and deadlines. Ymer's management applies the principles of the Deming circle in this work.

Example: The drilling contractor experiences problems with a new unit for injection of drilling mud. It consists of screw conveyors, a mill, tanks, and pumps. Personnel operating this unit are exposed to a high noise level, and levels of oil mist above the threshold limit value when remedying operational disturbances. There is also a risk of getting caught in the screw conveyor. The drilling employees have submitted unwanted occurrence reports on this hazard. Jointly, management and the employees decide to look into a redesign of the unit and list this as an item on the action plan. A problem-solving group is established for this purpose, and it receives a budget for its own work and for remedial actions. The driller is responsible for this group.

Norskoil has realised that the attitudes and behaviour of its management personnel are of basic importance to the successful implementation of HSE policy. Ymer management is trained in leadership through communication, coaching, and acting as a role model to ensure compliance in practice. This involves:

- Motivating the employees to take responsibility for safety. This is to be accomplished through education and training and through empowerment of the employees by their participation in job planning and evaluation where safety-related activities and decisions such as JSA play an important role.
- Marshalling correct behaviour through management's own vigour and example.
- Toolbox talks, other meetings, and communication on HSE policy, goals, and results.
- Incentivising a good culture for the reporting of unwanted occurrences through feedback of results and acknowledgment.
- Visibility and trustworthiness in matters relating to safety by participation in workplace inspections and investigations of serious accidents and by requesting accountability for safety-related decisions and results.

Example: The production manager observes a repairman working on a pipe rack 3 m above the deck without the required fall protection. On speaking to the repairman, the manager is told that this is a high-priority job and that the scaffolders have not been available. The production manager stops the job and oversees the erection of scaffolds so that the job can be completed in a safe way. He also agrees with the repairman that the health of personnel is important and must not be put in jeopardy because of high-priority work.

Regular toolbox talks with all personnel are important occasions for communication between management and the employees. Management uses these meetings for information on safety matters (new procedures, instructions, and so on) and for discussions on experiences and new goals and action plans. The employees bring up safety concerns and experiences in the meetings.

25.5.4   Control and verification

25.5.4.1   Reporting of accidents and unwanted occurrences

Ymer uses a computerised database for storing and retrieval of reports on accidents and unwanted occurrences (see Figure 25.9). Two procedures describe the routines: one on the routine reporting and follow-up of events and the second on the investigation routines in cases of serious incidents. The following types of events and conditions are reported, and information is stored in the database:

- Accidents resulting in injury to personnel, material damage, oil/ chemical spills, and production stops
- Unwanted occurrences
- Gas leaks
- Unsatisfactory conditions
- Safety nonconformities

Management emphasises the importance of reporting all potential problems. The reporting is promoted through simple routines and efficient follow-up. All reports are assigned actions for follow-up with a responsible person, and a deadline is set. The database is used for the follow-up actions.

One of the safety officer's tasks is to assign a risk score to each reported event. The supervisors formerly carried out this scoring, but this routine was abandoned due to a reluctance to use high-risk scores. High-risk events are followed up closely by top management. Results and experiences of a general value are fed into an experience database for use both in new projects and in operations on other platforms.

Problem-solving groups are utilised in the development of remedial actions. The aim is to use the knowledge of the personnel directly involved to solve safety problems.

Information is distributed through periodic summaries of accident statistics, status reports on safety measures, special reports, and so on. There are different uses for this information – such as input to decisions on remedial actions, evaluation of the effects of these actions, and monitoring of safety performance. The database also provides mandatory reports on accidents to insurance companies and the authorities.

The *periodic summaries* on accident statistics are distributed quarterly. They provide feedback on safety performance to line management and the safety organisation. Ymer's quarterly report contains control charts on the accumulated TRI rate since the start of the year, the RUO rate per month, and the distribution of accidents by accident type, injured part of the body, and so on. An edited summary of each accident case is also presented. The database helps in the production of standard reports and makes it possible to tailor these to the needs of each platform.

The database also gives feedback to decision-makers on the responsibilities, target dates, and *status of remedial actions*. This is a means of ensuring timely implementation.

Line managers, safety personnel (including safety representatives), and decision-makers responsible for design, procurement, training and education, and so on, have access to the database for *queries*. This possibility is utilised infrequently. When used, the majority of the questions are rather trivial. They mainly concern information on individual accident cases relating to a specific type of machine or job, type of hazard, and so on. The database is well suited for the retrieval and summarisation of relevant accident cases. Data on accidents are also used in special studies such as risk analyses and accident concentration analyses.

25.5.4.2   Workplace inspections
There are different types of scheduled workplace inspection activities. The department manager and the safety representative make biweekly inspections of their area. They follow a joint plan for the whole platform, which is based on a risk assessment, and they cover different items each time. Housekeeping is a mandatory item on the agenda. Most of the identified problems are remedied on the spot and are not recorded. Actions that are not executed immediately are documented for follow-up. Action status is reviewed at safety meetings.

Ymer practices one-on-one rounds at the different levels of the platform organisation. Approximately twice a year, Norskoil's assistant director for production makes a one-on-one round with the platform manager. They are not notified in advance, and the focus is on safe behaviour and housekeeping. The platform manager and the department managers make such rounds on a monthly basis.

The working-environment committee for the Ymer field also makes a workplace inspection once a year. The focus is on ergonomics and industrial hygiene, and experts covering these areas participate in the inspections.

25.5.4.3   Control of barrier availability
Norskoil has established a comprehensive program for inspection and testing of barriers in accordance with Norwegian and international standards (Norwegian Oil and Gas 2004; IEC 2010). Table 25.2 shows examples of barrier functions, systems, and elements that are subject to inspection and testing.

The inspection and testing activities are subject to detailed planning. Procedures define the criticality of the technical components and systems

Table 25.2 Examples of barrier functions, systems, and elements, and methods for control of availability

Barrier functions	Barrier systems and elements	Control method
Containment	Pipes and vessels	Periodic inspection, non-destructive testing, and measurement
Fire and gas detection, emergency shutdown, and pressure relief	Fire and gas detectors, Emergency shutdown valves, blow-down valves, pressure safety valves	Periodic testing Leak testing
Isolation of ignition sources	Explosion proof barriers on electrical equipment Emergency shutdown of equipment	Periodic inspection and testing of shutdown of non-classified equipment in exposed areas
Ventilation	Mechanical ventilation system	Periodic testing and inspection
Separation	Fire and blast walls	Periodic inspection and testing of tightness
Integrity of load-bearing structure	Load-bearing structure Passive fire protection	Periodic inspection
Active fire protection	Active fire protection system and equipment	Periodic inspection of firefighting equipment Periodic testing of fire pumps, sprinkler systems, and foam skids
Escape, evacuation	Evacuation alarm Emergency lighting Emergency communication system	Inspections Periodic testing Periodic evacuation drills
Common	Emergency power	Periodic testing

and the inspection and testing intervals. Based on these procedures, yearly inspection and testing programs are established. The results are followed up with immediate corrective actions where required and with more long-term measures. The status is reported to the authorities on a regular basis. The Petroleum Safety Authority Norway uses the data to analyse status and trends in the performance of barriers against major accidents (PSA 2014).

It is important that the tests are realistic and that they cover all sub-systems to check the availability of the system as a whole. The fire and gas detection system, for example, is tested all the way from the detector, which is activated by gas, light or smoke, to the display in the central control room.

25.5.4.4   Risk assessments

Ymer has implemented coarse and detailed JSAs as a systematic measure to control workplace hazards. All employees are involved in these activities. They participate in the listing of all activities on the platform. Each activity is assessed and documented with respect to:

- How often it is performed and number of personnel involved
- Possibility of injury/loss
- Risk estimation based on expected frequency and consequences
- Decisions regarding remedial actions and detailed JSA

The documentation is reviewed yearly to identify the need for updating. JSAs are always performed before the start of new jobs that may be critical from a safety point of view. The supervisor reviews the need for such analyses as a routine procedure in the PTW system. The following types of jobs have to be assessed in a JSA before they can start:

- Entering of tanks, where there is a risk of poisoning or suffocation
- Heavy lifting involving risks of dropped objects
- Non-routine maintenance of systems containing hydrocarbons

The REPA is also subject to updating with new incidents and in connection with modifications. The aim in the latter case is to check that the modifications do not violate any of the acceptance criteria for the risk of major accidents.

25.5.4.5   Audits

Safety auditing is used as a tool for independent examinations of Ymer's HSE management system. Norskoil's audit department carries out independent audits of all platforms where the company is responsible as the operator. They follow a yearly plan and cover different items. The audit team is led by an authorised auditor and includes experts in various fields.

Example: Ymer has to go through major modification work involving upgrading of one of the compressors. A contractor will be responsible for the work. Because this is a non-routine operation, the platform manager initiates an audit to check the suitability of the HSE management system in this new situation. The audit team consists of a lead auditor, a platform manager from a platform that recently has undergone a similar activity, and an HSE manager. They audit both Ymer's and the contractor's

HSE management systems by conducting interviews and reviewing documents.

The conclusions are rather serious. Ymer's management has not adequately foreseen the safety challenges involved in carrying out modifications in parallel with ongoing production. There is a need to upgrade the PTW system so that it can handle a much higher activity level. Systematic risk assessments must be performed on all modification work, and procedures need to be established on the basis of the results.

chapter twenty-six

The hydropower industry

26.1 Accident risks in hydropower development and operation

Hydropower is used to generate electricity by harnessing the energy when water flows to a lower altitude. Countries in an early phase of economic development tend to invest in large hydropower projects to provide electricity for the population and industry. These projects represent a challenge from a safety point of view because the construction industry in general, and particularly in developing countries, is characterised by a high risk of accidents. The International Labour Organization (ILO) estimates the fatal accident risk in construction to be five times the average fatal accident risk among employees worldwide (Murie 2007).

Here, we will present a case involving a so-called run-of-river hydropower project. In this type of project, the water is transferred from a river and brought by tunnels and pipes and discharged further down into the same or another water system. The drop between the inlet and outlet locations and the flow of water are two parameters affecting the generation of power. In the case presented here, the hydropower project is located in a high mountain range to exploit the large height differences. Construction involves considerable challenges in managing geological hazards such as rock falls and landslides and transportation risks such as road departures.

Dam projects utilise the water pressure that is built up by the water column in the dam. This type of project is often located in narrow canyons, where the physical environment involves steep slopes with falling rock and landslide risks. Dams located on the plains need to be large to generate enough flow to compensate for the lower drop. A dam break involving large volumes of water and a large population density downstream of the dam may have catastrophic consequences.

Table 26.1 gives an overview of typical accident risks for three activities at different development stages of a medium-sized hydropower project. The rough exposure estimate in man-hours applies to a medium-sized (ca. 150 MW) run-of-river project including the construction of a dam to regulate the flow of water on a short-term basis (24 hours). Site investigations are carried out over one to two years at an early stage of the project (e.g. feasibility study phase) with the purpose of collecting data as input for design and for assessments of economic and environmental viability. At this stage,

Table 26.1 Typical accident risks for three different activities in the development and operation of a hydropower project

Activity	Site investigations	Construction	Operation and maintenance
Rough exposure estimate	~15,000 man-hours (accumulated)	~16 million man-hours (accumulated)	~30,000 man-hours/year
Major accident risks	Transportation accident	Fire in tunnel Tunnel collapse Transportation accident Natural hazards (rock fall, landslide, avalanche, flooding)	Dam failure Flooding of powerhouse Fire, explosion in subsurface powerhouse
Occupational accident risks	Transportation accident Accident due to natural hazards Falling during movement and manual transportation of equipment Getting squeezed during rigging	Drowning Falling from height Hit by load (material handling) Rock fall, rock burst in tunneling Fire in tunnel Transportation accident (roll-over, hit by vehicle, driving off road)	Drowning Electrocution Accident with moving machinery parts and equipment
Third-party accident risks	Not applicable	Drowning Transportation accident (hit by project vehicle)	Drowning due to dam break or penstock rupture Falling into water-conveying canal or dam and drowning

Source: Kjellén, U.: In Alternative energy and shale gas encyclopedia. 413–422. 2016. Copyright Wiley. Reproduced with permission.

the quality of the infra–structure, such as the road network, in the area may be poor, and this represents a considerable safety challenge for the project. The figures in the table for the construction phase are based on a workforce of about 2000 construction workers employed for a period of four years.

26.1.1 *Major accidents with multiple fatalities*

Dam breaks, although rare, represent the single largest potential for major accidents in hydropower projects (ICOLD 2005). The potential for catastrophic consequences is determined by the size of the dam, the size of the area downstream that will be inundated in a dam collapse, and the size of the population at risk. One example of this type of major accident is the Banqiao dam failure in China in 1975, where 26,000 lives were lost in the subsequent flooding.

Dam failures are rare. Statistics by the International Commission on Large Dams (ICOLD 1995) show that about 0.5% of the dams built after 1950 had failed in the following period of 40 years.

Flooding of the powerhouse may also have major accident potential. The Sayano-Shushenskaya hydropower plant experienced a rupture of the casing of one of the turbines in 2009, and the subsequent flooding of the powerhouse resulted in 75 fatalities (International Water Power and Dam Construction 2010). The reason for the large number of fatalities was an increased manning of the powerhouse during maintenance work. The use of remotely controlled power plants, which are unmanned during normal operation, reduces the likelihood of severe consequences from this type of accident.

Subsurface powerhouses usually are equipped with large transformers. These may explode due to the transformer oil being ignited by an electric arc (Hansen et al. 2002). If the transformers have not been adequately separated from the rest of the powerhouse, the explosion and subsequent fire may escalate, resulting in smoke spreading throughout the power plant and trapping personnel. An example of this type of accident is the transformer explosion in 1973 that killed three persons in the Tronstad hydropower plant. The plant was located 750 m inside a mountain.

Natural hazards such as rock or landslides, avalanches, and flash floods may have catastrophic consequences, especially during the construction phase when several thousand people may be present at the site. Figure 26.1 shows an example of a rock slide impairing a section of a construction site access road.

Accidents involving misfire of explosives, tunnel fire, and rock fall during excavation of tunnels and caverns also have multiple fatality potential. Traditional drill-and-blast methods for tunnel excavation may expose as many as 30 people to these hazards when working at or close to the tunnel face. With the application of mechanised drill-and-blast technology, this number may be reduced to not more than four persons simultaneously at the face. This condition, combined with the fact that the workers spend most of the time operating machinery equipped with protected cabins, reduces the risk of major accidents considerably. Diesel-powered engines used in the tunnel are the main culprits of tunnel fires, and the resulting development of smoke may trap personnel inside the tunnel (Ingason et al. 2010). Rock fall and collapsing tunnels are caused by unpredicted geological conditions or by failure in tunnel support (Sidenfuss 2006).

Figure 26.1 Rock slide during construction of a hydropower plant.

Road departures during personnel transportation represent another type of accident with major loss potential (see further information in Chapter 27).

26.1.2 *Occupational accidents*

Here, we will focus on the construction phase, where the gross majority of the work exposure occurs. Hydropower construction – like infrastructure projects in general – consists of a large proportion of heavy civil construction. Examples are road construction, earth movement, tunnel and cavern excavation, and the construction of dams. The civil construction is followed by the installation of electrical and mechanical equipment in the power plant and by the installation of hydraulic steelworks – for example, dam gates and penstocks. Construction of transmission lines and substations is also part of a hydropower project's scope in this phase.

Figure 26.2 shows the distribution of fatalities in connection with an international renewable energy company's projects in south-eastern Europe, Asia, and South America. The statistics are built on experiences from nine Greenfield hydropower projects during a period of 10 calendar years and with an accumulated construction period of 30 project years. The statistics include one project in the Himalayas that experienced 10 fatalities alone.

Five of the projects in this sample experienced a fatal accident rate (FAR) of about 20 during the last 20 months of the 10 calendar years. The corresponding total recordable injury (TRI) rate was about four.

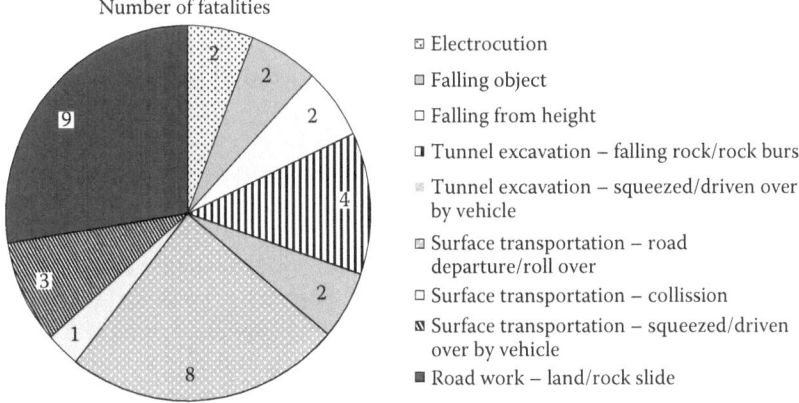

Number of fatalities

- ▨ Electrocution
- ▤ Falling object
- □ Falling from height
- ▯ Tunnel excavation – falling rock/rock burst
- ▦ Tunnel excavation – squeezed/driven over by vehicle
- ▧ Surface transportation – road departure/roll over
- □ Surface transportation – collission
- ▩ Surface transportation – squeezed/driven over by vehicle
- ■ Road work – land/rock slide

Figure 26.2 Fatal accident statistics from hydropower construction projects (N = 33).

The fatal accident statistics are dominated by road transportation (12 of 33 fatalities). Eight of the fatalities involved road departure and roll-over accidents. These accidents usually have occurred on roads characterised by steep inclination and vertical drop on one side of the road. There were three fatalities due to people being hit or driven over by a vehicle.

Road work is also associated with a large share of the fatalities (nine), all due to geophysical hazards. Three workers involved in scaling of terrain in connection with road work were killed in a single rock-slide accident in south-eastern Europe.

Tunnel excavation accounts for six of the fatalities – four due to falling rock or rock burst and two when a worker had been squeezed or driven over by heavy mobile machinery.

Accident statistics from the renewable energy company's 16 operating plants in Asia and South America, with a total installed capacity of 1500 MW and representing 60 plant-operation years, did not include any fatalities (Kjellén 2016). The TRI rate of the operating plants was about two. This also included work on transmission lines owned by the company. The TRI statistics were dominated by falls from height followed by motor vehicle accidents. There were no recordable injuries for work on high-voltage systems or in connection with waterways, which is a positive sign indicating that there are sufficient routines to manage the risks in these jobs.

26.2   The David hydropower plant

We shall use the fictive David hydropower project (HPP) as an example. It is located at an altitude of between 2880 and 3825 metres above sea level in the South American Andes. The installed capacity is 80 MW from one Pelton turbine. Figure 26.3 shows the layout of the project. The total length of the tunnels is 11 km. The project also includes the construction of 96 km of transmission lines.

The construction of the David HPP takes three years and employs in average of 1500 construction workers. There are three main contracts – one for civil works, one for electro-mechanical equipment and installations, and one for hydraulic steelwork.

During operations, the plant is manned continuously by two 12-hour shifts. The day shift consists of a total of 10 persons – the plant and maintenance managers, three operators, and five maintenance personnel. The night shift consists of one control room operator and one field operator. There also are contracted personnel at the plant (e.g. for security).

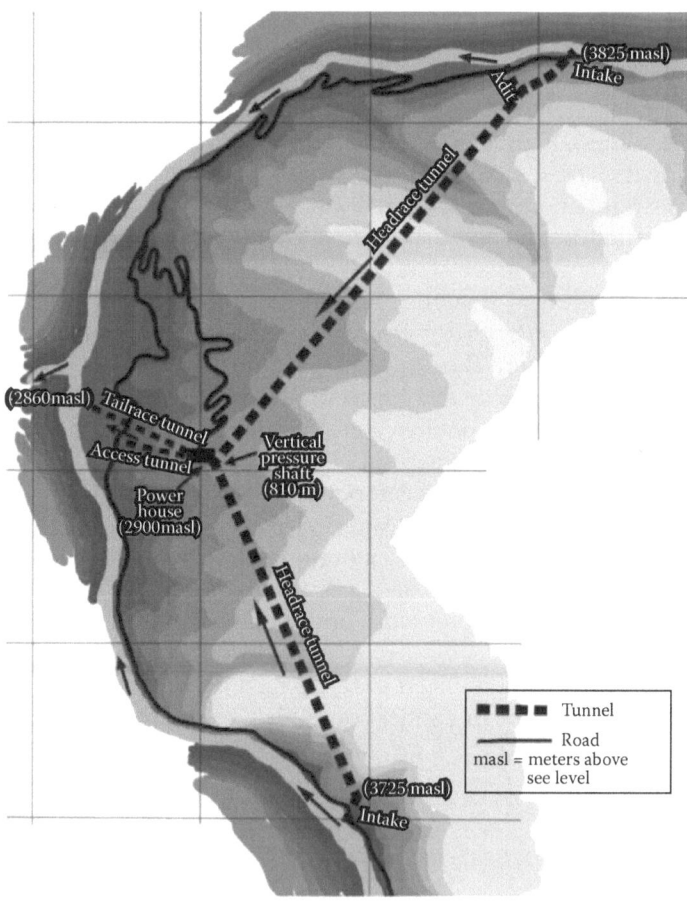

Figure 26.3 David hydropower project layout.

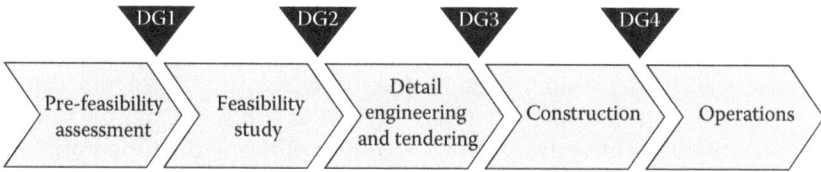

Decision gate 1: Approval to start the feasibility study and permitting
Decision gate 2: Project pre-sanction, approval to start detail engineering and tendering
Decision gate 3: Investment decision, approval to start construction
Decision gate 4: Approval to start operations

Figure 26.4 Typical phases in the value chain of an energy project.

26.2.1   The phase model for project management

The phase model used by the owner of David HPP is shown in Figure 26.4. It is similar to the phase model in the example from the oil and gas industry in Chapter 25. As for Norskoil, the owner of David HPP considers the second decision gate (DG2) to be the most critical in deciding on investments in the project. There are some differences, however. The David HPP feasibility study is made in two steps to avoid the significant expenditures for a full feasibility study for projects that are not viable. The concept study phase is critical in offshore oil and gas developments but of less significance in hydropower developments because the concept is decided mainly by the hydrology and topography in the area.

The management of HSE is an integral part of the owner's phase model for project management. It breaks down into two main areas:

1. *The management of HSE in design:* The aim is to ensure that the chosen design concept is adequately safe to construct, operate, and maintain.
2. *The management of HSE in contract work in field investigation (primarily during the feasibility study) and construction:* Here, we will focus on the latter phase. The aim is to ensure a safe execution of construction work and contractors' compliance with regulatory requirements and the owner's requirements as to HSE.

26.2.2   Management of accident risks in design

The owner has developed a specification for HSE in the design of hydropower plants similar to the NORSOK standards utilised by Norskoil (Chapter 25). The specification contains the owner's 'organisational memory' of past experiences related to the construction and operation of hydropower plants. It was originally established through discussions with senior project and operations personnel and is kept updated as a living document with new experiences.

The specification identifies international safety standards, such as National Fire Protection Association (NFPA) 850 on fire protection of electric generating plants, and defines how to implement the different guidelines in the owner's plants (NFPA 2015). A special part of the specifications defines the owner's requirements for documentation of the safety standard in design through risk analysis and verification activities. The specification is included in contracts with designers and construction companies.

Table 26.2 shows the different decision criteria used to check whether the design is adequately matured from an HSE point of view for the project to pass into the next phase. The project develops a preliminary design of the hydropower plant in the pre-feasibility study phase. This is subject to a hazard identification (HAZID), covering consequences of design for both construction (constructability HAZID) and operation and maintenance. Typically, the constructability HAZID addresses the location of the power plant and the availability of safe access roads for transportation of personnel and materials; the availability of medical

Table 26.2 Decision criteria related to HSE in design for the different decision gates

Decision gate	Decision criteria
DG1	• Pre-feasibility concept adequately documented as to HSE risks in construction and operation. • No conditions have been identified that make meeting the owner's HSE requirements for design and construction unfeasible.
DG2	• The feasibility study concept has been adequately documented from an HSE point of view. • The feasibility study design concept complies with the owner's HSE standard for construction (constructability) and operation and maintenance. Identified HSE issues in design are manageable within the allocated project budget and schedule.
DG3	• The design has been adequately matured and documented. Specified HSE design reviews and risk assessments have been performed, and all critical deviations from the owner's HSE design specifications and critical risks in the risk register have been resolved.
DG4	• All HSE-related checkpoints in the operational readiness review concerning personnel, facilities, and management systems and procedures have been reviewed, and there are no outstanding items classified as high risk (RED) (i.e. either closed or accepted with waiver).

emergency facilities; and natural hazards such as slope instability, earthquakes, flooding, and so on. The HAZID for safety consequences of the design to operations phase focuses on the layout of the plant and the design of dams and waterways. The first coarse fire and explosion risk assessment is done in this phase, checking the potential for escalation of, for example, a transformer explosion into manned areas and escape ways.

The design is matured in the feasibility study phase to a stage where it is approved for tendering. The design is subject to different risk assessment and verification activities such as:

- Update of the coarse HAZID from the previous phase
- Design review to verify implementation of the requirements in the HSE design specification
- Special studies on critical issues

Example 1: The constructability HAZID brought up a number of safety risks for the David HPP. Transportation on site roads was a critical issue due to the steep slopes and high geotechnical risk. The area is prone to natural hazards such as glacier lake outburst flood (GLOF), flash flooding, and landslides. It was concluded that the project site layout has to be verified showing that camps, construction sites, and roads are not within the danger zone of the identified hazards.

Another critical constructability issue for the David HPP is its location at a high altitude. This has a number of impacts, including increased health hazards during tunnel excavation. Table 26.3 shows the result of the constructability HAZID for this risk. The mitigation measures have cost consequences and must be implemented in the tender documents for the civil contract.

Example 2: The transformer cavern for the David HPP is located adjacent to the main access tunnel to the power plant. The access tunnel also serves as the main escape way in case of fires, explosions, or flooding. The HAZID identifies the possibility of a transformer explosion followed by a fire that blocks the main escape way. The owner has a risk acceptance criterion for this type of event, stating that 'the main escape way shall not be made unavailable due to an accident more often than once per 10,000 years'. Calculations of the dimensioning pressure for an explosion with a frequency of more than one in 10,000 years shows that it exceeds the design pressure for the transformer rooms. Further reinforcement of the rooms to meet the required strength to withstand the dimensioning explosion load would not be feasible. The issue was resolved by designing an additional fire and explosion wall between the transformer cavern and the main escape way.

Table 26.3 Extract from the constructability HAZID of the David HPP

Risk event	Description	Mitigation DG2	Mitigation DG3-4	Risk Current	Risk Residual
Increased health risk of pollution in tunnels (CO, NO$_x$, SiO$_2$, dust) due to the high altitude.	Less efficient combustion of diesel engines. Increased inhalation due to reduced oxygen concentration. Longer distances from portal to face then planned due to uneven progress at tunnel faces.	Establish threshold limit value (TLV) for toxic gases for actual height above sea level. Adapt capacity of ventilation system considering the high altitude. Plan for needs of increased ventilation capacity.	Contractor to submit calculation verifying ventilation design for project approval before start up. Include measures to reduce pollution in methods statements for excavation including avoidance of diesel engines, water spraying, regular monitoring, and stop criteria for high pollution levels.	High	Medium

The HSE design review is carried out at a stage in the feasibility study when the design drawings are adequately matured but still possible to change.

Example 3: A design review of the David HPP was carried out by the design consultant with participation of experts on safety and operations and maintenance from the owner. It consisted of a systematic review of the requirements in the owner's HSE specification in the design of relevance to this phase. Following are examples of issues identified in this review:

- The control room and the main workshops have to be moved to the portal of the main access tunnel to avoid permanent workplaces inside the subsurface powerhouse. Only local control panels are allowed in the powerhouse.
- The battery room has to be moved to above the critical water level for flooding of the plant.
- It is necessary to establish access ways from all areas of the powerhouse to the alternative escape tunnel. This measure is based on the requirements of two alternative escape ways from all areas of the powerhouse.
- A special study is required to verify safe escape in a fire scenario considering air flow directions and capacity.

The design was updated based on the input from the design review and was issued for the tender for detail design and construction of the power plant.

Design is further matured in the detail engineering and tendering phase based on detailed designs for the major equipment – such as turbines, generators, and valves. Safe access for operation and maintenance and maintenance handling of equipment are assessed in detailed job safety analysis and design reviews. The owner's project team will follow up on the HSE design requirements during fabrication and installation of equipment before the plant begins operation.

26.3   HSE management in construction

The owner has developed two central documents to manage HSE in construction:

1. A specification for contractors' management of HSE at construction sites
2. A procedure on the management of HSE in construction of the plant

The first document is directed at the contractors involved in the power plant construction, forming part of the contract; the second document is directed at the owner's own personnel responsible for planning and project follow-up. The specification is based on generally accepted

HSE management principles of international standards and Occupational Health and Safety Assessment Series (OHSAS) 18001 in particular (British Standards Institution 2007). It also includes relevant requirements in the performance standards issued by the International Finance Corporation (IFC 2012). It defines the owner's requirements for documentation of contractors' HSE management systems and for the various HSE management activities such as induction and training, investigation and reporting of accidents and unwanted occurrences, inspections, use of risk assessments, and the management of hazardous work. The owner also defines more detailed requirements for the HSE management of critical activities such as transportation and tunnel work. A main principle in the specification is that one contractor (usually the civil contractor) is appointed as 'principal contractor' and is responsible for coordination of safety work at the site and for site security and emergency preparedness.

The procedure defines the owner's requirements to own project organisation including responsibilities for HSE during the different project phases. It also describes the owner's instructions to the project team regarding its management of HSE in the different phases of the procurement process (see Figure 26.5).

Strategy work is carried out in the feasibility study phase, or shortly thereafter, before the start of tendering work. This gives the owner an overview of the local contracting market and the resources and capabilities of potential bidders for the project. It serves as input for the work involving contract strategy, where the number and types of contracts are decided. The ability of the contractors to execute the work in an adequately safe way is an important parameter in these evaluations.

The owner carries out pre-qualification through questionnaires and audits of potential bidders. Examples of HSE elements used in the evaluation are:

- Typical HSE programme for manufacturing and construction
- Resources and HSE organisation (including access to qualified personnel for HSE management and emergency preparedness, and emergency preparedness equipment)
- Safety statistics (fatalities, TRI rates, lost-time injury [LTI] rates) for the last five years
- HSE standard of sites where the supplier has ongoing activities
- Experiences with potential bidders from previous projects
- Follow-up by contractor's home office, including audit plans

In the tender preparation, the project's HSE manager provides input for the 'invitation to tender' (ITT) regarding instructions on documentation that should be included in the tender. The documentation list is similar to the issues brought up during pre-qualification. Tender evaluation

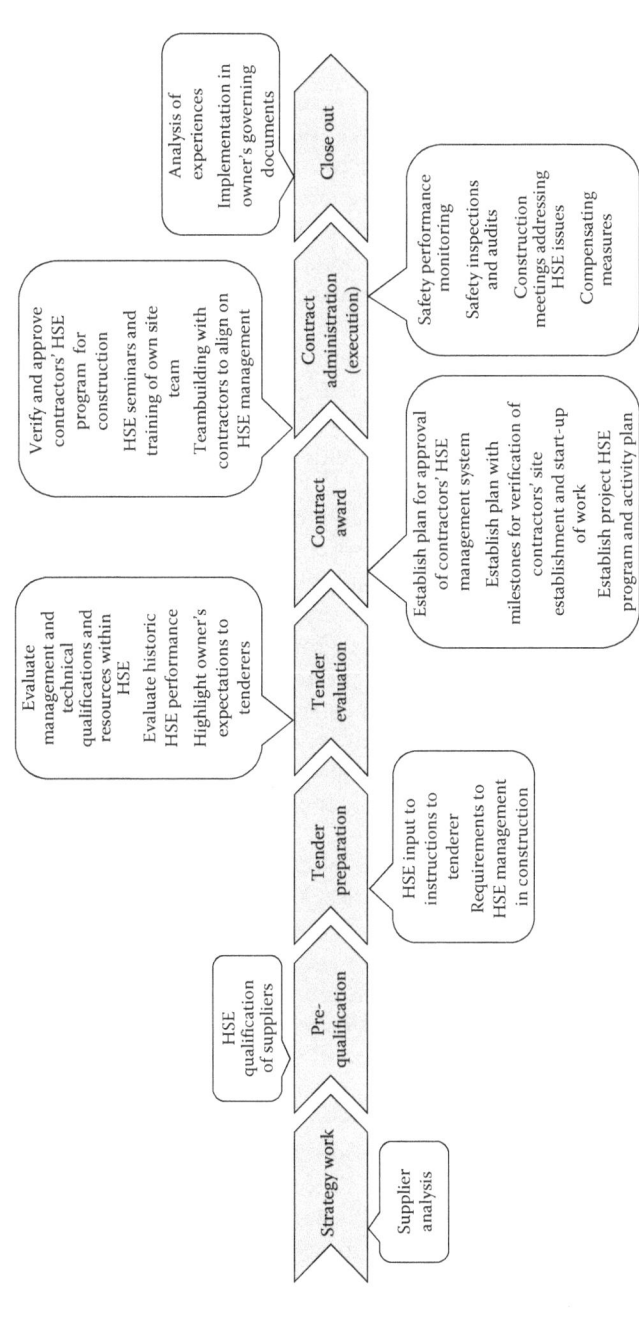

Figure 26.5 The contracting process and associated safety activities in a construction project. (Adapted from Kjellén, U., *Saf. Sci.*, 50, 1941–1951, 2012.) Similar principles apply to the contracting of field investigation work and maintenance work in an operating plant.

includes a technical and administrative portion (where HSE belongs) and a commercial part. Because all tenderers have been pre-qualified, HSE normally does not exclude any bidder from further evaluation. A more relevant question is how to compensate for an expected lower HSE management performance of a bidder that is considered for the contract award. Normally, this will require an increased effort by the owner for follow-up during construction, and the costs for this must be included in the total costs when evaluating the different bidders. This issue is especially critical for projects in remote locations, where there is insufficient competition for the contracts from well-reputed national or international contractors.

The bid clarification meetings are used primarily to explain critical issues in the bid, such as how a contractor is going to plan and execute hazardous work. These also give the project's HSE manager the opportunity to meet the contractor's appointed project, construction, and HSE managers to form an impression as to their qualifications. In addition, these meetings offer opportunities to highlight the owner's expectations as to a contractor's safety performance.

The first issue after the contract's awarding normally is to verify and approve contractors' HSE management programs. These will become binding for the contractor in a way that is similar to the HSE specification in the contract – subject to the owner's approval.

The owner's project team will plan for start-up activities including internal safety training and seminars and team-building workshops with contractors. These play a vital role in establishing a positive climate for cooperation regarding safety beginning from the start of construction.

The contract defines milestones in site establishment that the contractors have to meet before the start of construction work. These include:

- Emergency response facilities established (including medical and ambulance services, communication systems).
- Emergency response organisation established and trained.
- Site fenced in and security guards established at site.
- Safety induction program established.
- HSE management routines established (accident, unwanted occurrence reporting, site inspections, and toolbox meetings).
- Transportation risk assessment executed and results implemented in transportation safety management plan.
- Construction roads have met contract requirements as to safety standards.
- Storage for hazardous substances established.
- Waste-handling facilities and routines established.

The site team will monitor progress and verify that the milestones have been met. The site team may experience a dilemma in balancing tolerating delays in construction activities due to poor progress in meeting milestones or accepting substandard safety solutions to allow construction to start. The latter alternative will affect contractors' respect for contractual safety requirements and will be a poor solution in the long run as well.

The project will monitor the different contractors' safety performance during construction. The owner uses the number of fatalities, the TRI rate, and the rate of unwanted occurrences (RUO) in the monitoring of the safety performance.

The David HPP has established its own team of safety inspectors. There are inspectors with a general safety background and others focusing specifically on safety in the tunnels.

Example 1: An audit by the owner's home office reveals that the project's safety inspectors merely duplicate the inspections by the contractors' safety inspectors and that the observations are transmitted to the contractors for correction (i.e. first order of feedback according to Van Court Hare, see Chapter 7). The auditors recommend that a system of directed safety inspections and audits be introduced, giving priority to construction activities that are identified as high risk according to the project's risk register for construction work. A decision is made to focus on the following activities: Work in the danger zone of mobile machinery, falling rock hazards in tunnels, tunnel air quality and ventilation, work at height, and crane operations. The safety inspectors develop checklists for each of these areas to be used in the inspections. The checklist for crane safety includes:

- Crane operator, rigger, and signal man can show adequate documentation of qualifications.
- Cranes are certified and subject to preventive maintenance.
- Crane and lifting equipment are certified and regularly inspected and tested.
- Instruction manuals for the equipment are available in the local language.
- Crane operations maintain adequate safety distance to power lines.
- Cranes adequately secured from tip-over during lifting.
- Lifting operations documented in job safety analysis and brought up in toolbox meetings.
- Area for lifting operation (danger zone) is roped off and kept free from people.

The introduction of the new system reveals alarming deviations among subcontractors, resulting in work stoppage to rectify these conditions.

Contractors are required to report all recordable injuries and high potential (HIPO) incidents without delay and to submit a first (Level 1) report within one day. It should be followed by a Level 2 report within seven days. The owner has the right in accordance with the contract to appoint an observer to participate in the Level 2 investigations by Contractors.

The owner has decided that all fatalities occurring at the project's construction sites and 30% of the high-potential incidents shall be subject to Level 3 investigations. There are trained project personnel available to lead such investigations.

Example 2: The David HPP experienced an electrocution incident due to electrical shock during rock-support work in tunnel excavation. Electricity passed through the body of one worker while he was involved in transferring concrete mix from a mixer truck to a robot used in spraying the concrete on the tunnel wall. The investigation showed that the robot had been equipped with a 240 V halogen floodlight at the construction site to improve illumination of the area to be sprayed by shotcrete. The halogen lamp was not suited for the tunnel environment. The investigation also revealed a faulty earth connection for the lamp. An isolation fault caused an electric current to flow from the lamp to the robot through the body of the worker to the mixer truck and then to the ground.

The investigation further revealed that there were several indications of a faulty electrical system. The 240 V temporary electrical system in the tunnel was equipped with a residual current device used to prevent workers from receiving a fatal electric shock. It had tripped several times, but the electrician had not initiated any action other than to reset it.

The investigation identified a number of root causes. The contractor's system for the management of change was never applied. The maintenance department had modified the robot and, in doing so, had violated the intrinsically safe characteristics of the instrument power system of the robot without assessing the risk. Contractors' electricians responsible for installation and inspections of the temporary electrical systems were not adequately qualified and had developed a complacent attitude towards electrical safety. The project team at the site did not give the temporary electrical systems adequate attention but focused on the follow-up of the permanent electrical installations for operation of the plant.

The contractor responded to the investigation by confirming actions to improve the management of change and electrical safety. The project conducted an audit of the safety of temporary electrical systems shortly after the investigation. As a result, the project's inspection team at the David construction site was strengthened with an electrical safety inspector.

After the construction period, the David HPP was commissioned by a joint team consisting of the electro-mechanical contractor, personnel from owner operations, and equipment vendor personnel. A permit-to-work (PTW) system was introduced in this phase for all work on the energized power plant systems. This was managed by the electro-mechanical contractor during commissioning and by owner operations thereafter. When the civil contractor demobilised, the responsibility for the emergency response system was transferred to the owner's project organisation until the plant was handed over to the operations department.

26.4 HSE management in operation and maintenance

The owner has manned the David HPP with his own employees for the operation of the plant and for regular inspections and minor preventive maintenance and repair. All larger jobs are contracted.

HSE management is based on an industry standard according to OHSAS 18001 (British Standards Institution 2007). The David HPP applies an HSE management system for hydropower plants based on the owner's standard but that is adapted to site-specific circumstances as identified in the risk assessment of the plant.

That PTW system is used for all work in the plant except for pre-defined tasks such as routine operations tasks, tasks carried out in the maintenance workshop, and office work. The plant manager monitors the system, and the shift supervisor is responsible for implementation. He/she shall ensure that the hazards related to the plant are identified and that the necessary barriers are implemented before the start of work. Special permits are required for hazardous duties including work on high-voltage equipment, work on high-pressure water systems, entry into confined space, and work with radioactive materials and explosives. Operators have been trained as task safety supervisors for work on high-voltage systems and on main waterways, with the sole duty of following up to be certain that work is carried out in accordance with the PTW.

The plant has established an *emergency response plan* and has trained its personnel to fill the different roles according to this plan. The plant manager or his deputy will act as emergency manager, and the maintenance manager or his deputy as on-scene commander, should an emergency event occur. The fact that the David HPP is remotely located is of special concern, and the plant cannot assume there will be any external support during the first 60 minutes after an accident. Personnel are trained to provide essential first aid during this first period after a severe accident. They are also instructed to evacuate the plant in case of fire, rather than to try to fight the fire, if there is any doubt about the outcome.

The *management of contract work* in the plant is done according to a simplified version of the procedure for the HSE management of contractors during construction. A separation is made between work involving low risk – such as minor building maintenance, medium risk – such as preventive maintenance, and high risk – for example, confined space entry and hot work inside the plant. This classification is used to determine the resources needed for follow-up. All contracts include the safety rules and emergency routines for work inside the plant perimeter and the plant HSE management systems applicable to contractors. All work will require PTW. The contractor personnel have to go through a safety induction before starting work. This focuses on plant rules, emergency routines, and the PTW system.

The case involving flooding of a pressure tunnel during repair work presented in Section 13.6 illustrates the importance of adequate management of contractor safety and of a well-functioning PTW system.

chapter twenty-seven

Work-related road transportation

27.1 Accidents in road transportation

In road transportation, the driver controls large amounts of kinetic energy. From elementary physics, we know that there is a linear increase in kinetic energy with the weight of the vehicle and a quadratic increase with its velocity. These basic facts show up in the accident statistics. The World Health Organization (WHO) estimates that there are 1.2 million people killed yearly worldwide in road transportation accidents (WHO 2015). The fatalities are distributed about 50/50 between unprotected pedestrians, cyclists, and motorcyclists on the one hand and drivers and passengers of heavier vehicles on the other hand.

The risk of sustaining a fatality in road transportation varies considerably among different countries. In the UK, Germany, Norway, and Sweden, there are three to four fatalities in road transportation accidents per 100,000 inhabitants each year (WHO 2015). The corresponding figure for the United States is 11. These figures represent significant improvements in transportation safety during the last decades. In the United States, for example, the figure 11 represents an improvement of almost 65% since the late 1960s. The turnaround in these high-income countries came with the introduction of national policies on road safety management (IRAP, undated).

The situation is different in low- and medium-income countries, where the number of fatalities per 100,000 inhabitants per year in road transportation is considerably higher; the figure is around 15 for countries such as India and Peru and between 25 and 30 for sub-Saharan countries such as Kenya, Zambia, and South Africa. These figures do not take into account differences in the spread, distribution, and use of different types of motor vehicles among countries. A detailed study based on Indian national statistics from 2010 to 2012 showed that the number of fatalities per one million vehicle-kilometres was 0.08 for cars and light vehicles (Høye 2014). This number is about 30 times the corresponding rate for Norway.

In this chapter, we will focus on the management of safety in transportation from a company's perspective. Road transportation accidents represent a significant occupational safety issue. It is estimated that road accidents sustained while the victim is at work account for between 20%

and 40% of all occupational fatalities (Fort et al. 2010). In the United States, 35% of the work fatalities reported to the Bureau of Labor Statistics in 2000 to 2004 were associated with motor vehicles (Murray 2007). Two-thirds of the fatalities occurred due to vehicle crashes on highways. The remaining one-third was approximately equally distributed between vehicle crashes off highways or on industrial premises and pedestrians being hit by a vehicle.

A UK study from the 1990s showed that company car drivers were about 50% more likely to be involved in crashes while at work than the general public (Fort et al. 2010). A Norwegian study of road transportation fatalities in 2005 to 2010 showed that 36% of these fatalities involved people at work (Phillips and Frislid Meyer 2012). Heavy vehicles were involved in several of the fatal accidents, but the drivers of these vehicles were rarely severely injured.

27.2 Principles for the management of safety in road transportation

27.2.1 Measures of risk

In the previous section, we used different measures for the risk of transportation accidents. Let us review the definition of the risk for accidents from Section 15.1 and look closer into how it is applied in the field of transportation safety. We defined the risk of accidents as a combination of measurement of the probability or frequency of accidents involving losses per unit of exposure in a specified activity and the extent of the losses (consequences). In occupational safety, we use the number of man-hours as a measure of exposure. In calculating the risk of road transportation accidents, we usually use other exposure measures. When we want to assess the effects of transportation accidents on the *health of the general public*, exposure is measured as the number of inhabitants at risk during a specified period. We are also concerned with the risk of accidents in relation to the *transportation work* carried out. Here we use, for example, the number of vehicle-kilometres as an exposure measure. Table 27.1 summarises some common transportation-risk measures.

It follows from the definition of the risk of transportation accidents that there are three different means of reducing losses due to transportation accidents as seen from society's point of view. These include:

- Reducing exposure (i.e. the extent of road transportation)
- Reducing the probability of accidents per unit of exposure (e.g. vehicle-kilometre)
- Reducing the consequences of road transportation accidents

Table 27.1 Some common measures of the risk of road transportation accidents

Measure	Definition
Health risk, general	Number of fatalities (or injuries) per inhabitant and year
Health risk, transportation	Number of fatalities (or injuries) per million hours in transportation
Transportation-accident risk	Number of transportation accidents per million vehicle-kilometres
Transportation-injury risk	Number of injuries (including fatalities) per vehicle-kilometre
Crash-involvement rate	Number of vehicles involved in crashes per 100 million km
Fatal crash-involvement rate	Number of vehicles involved in fatal crashes per 100 million km

We will come back to these different means in the next section, when we discuss how the different 'components' of the transportation system affect the risk of accidents.

27.2.2 Points for intervention

We touched upon Haddon's phase model of transportation accidents in Section 4.3. It divides an accident sequence into three distinct phases: the pre-crash phase, crash phase, and post-crash phase (Haddon 1968; Murray et al. 2014). During the pre-crash phase, the road users interact in ways that normally follow well-controlled patterns but that occasionally result in transportation conflicts. Injury and damage occur during the crash phase. This is when there is an uncontrolled energy exchange among the involved vehicles and between vehicles and other road users and/or obstructions in the environment. Losses may be limited through actions in the post-crash phase, including first aid, firefighting, medical treatment, and so on. Table 27.2 illustrates the Haddon matrix by giving examples of factors related to the three phases and the 'components' of the road transportation system.

During the pre-crash phase, the actions of the road users have significant effects on whether an accident will occur or not and also on its consequences. We will use the model of the driver-vehicle-environment system according to Figure 27.1 to analyse the situation further. Here, the driver is regarded as one of the components of the transportation system, and the model focuses on the information processing of the driver. The driver receives and processes information mainly from the transportation environment. The driver acts on the information, and these actions affect the movements of the vehicle, which in turn affect

Table 27.2 The Haddon matrix with examples of factors that affect the risk of
road transportation accidents

Phase	Human (driver) factors	Vehicle/ equipment factors	Road/environment factors
Pre-crash	Age	Speed	Road design and
	Gender	Brakes, tires	layout, friction
	Health, vision	Road holding	Weather conditions,
	Driving training and	ability	precipitation
	experience	Active safety	Roadside slope
	Attitudes, personal	standard	stability
	traits	Visibility	Traffic environment,
	Fatigue	Maintenance	traffic density
	Alcohol, drug intake	standard	Traffic control
Crash	Age	Speed	Roadside barriers,
	Health	Size, weight	median dividers
	Use of seat belt	Crashworthiness	Roadside hazards
		Seat belts, airbags	Traffic density
		Integrity of fuel	Enforcement of speed
		system	limits
Post-crash	Age	First-aid	Communication
	Health	equipment	network
	Experience in	Communication	Distance to and quality
	emergency handling	equipment	of emergency care

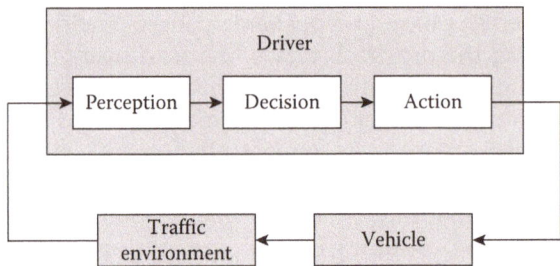

Figure 27.1 A model of the driver-vehicle-environment system. (Reprinted from
Englund, A., et al., *Trafiksäkerhet – En kunskapsöversikt*, Studentlitteratur, Lund,
In Swedish, Copyright 2006, with permission from Studentlitteratur.)

the transportation environment. The driver registers changes in the transportation environment and acts on these in an ongoing process.

27.2.2.1   The driver

The driver affects the accident risk in many ways. He has a crucial influence on road use and thus exposure. In professional driving, the decision about road use usually is made by the employer. Driving behaviour is a critical determinant of the pre-crash phase and hence on the probability of transportation accidents. Although the crash phase is to some extent controlled by the driver, previous driver actions to control speed and direction will affect the outcome.

Early research focused on the possibility of identifying accident-prone drivers (i.e. drivers who, due to certain personal characteristics, were more likely than others to be involved in transportation accidents). The aim was to improve transportation safety by identifying such individuals (e.g. by use of psychological tests) and excluding them from the transportation environment. Accident statistics showed that some drivers experienced more accidents than others. Although this so-called *accident-proneness theory* is often accepted as credible by the layman, it has been criticised by the research community for several reasons and today is of little practical significance (McKennan 1983). More recent studies show that drivers with behavioural traits associated with anger, aggression, and frustration are more frequently involved in collisions (Darby et al. 2009). Behavioural scores on psychological tests only account for a small portion (about 11%) of the variation in collision involvement. To summarise:

- Differences among drivers as to accident experience are to a large extent statistical artefacts that may be explained by pure chance.
- A small part of the differences in accident experience among drivers may be explained by personal characteristics. It is more challenging, however, to identify the specific characteristics of accident-prone drivers. This makes it difficult to develop valid and reliable tests.
- The use of psychological tests to exclude accident-prone drivers would involve many instances of wrong diagnoses due to an inadequate correlation between test results and accident risk. It is politically unacceptable to deny people the right to drive (including the right to work as professional drivers) on such a fragile basis.

Age and driving experience are temporary personal characteristics that, according to the accident statistics, are significant determinants of accident risk. Statistics show that the risk of accidents per kilometre driving follows what is known as a U-curve, with a minimum risk around the age of 50 (Loeb et al. 1994). Both young drivers (aged 18–24) and old drivers (aged above 75) experience an accident risk that is many times higher than that of middle-aged drivers. Increased experience has a greater impact on accident risk than increased age (Maycock 1997). Drivers who first experience transportation at a young age will start at a very high accident risk but will show rapid improvements during the first year of driving. Older inexperienced drivers will start at a lower risk level but will improve at a slower rate.

Fatigue (i.e. a gradual and cumulative loss of vigilance and alertness) also is a significant determinant of accident risk. A UK study suggests that fatigue is a contributing factor in as many as 20% of all road accidents (Jackson et al. 2011). The driver's experience of fatigue is related to time of day, driving schedule, eating habits, physical activities, and so on. The commercial driver has only limited control over these different factors and has been the focus of research into the effects of fatigue on the risk of accidents. Increases in the risk of accidents during certain periods of the 24-hour cycle are to a large extent explained by the circadian rhythm (i.e. 'body clock'). Studies of truck drivers show that accident risk doubles during the period between midnight and six o'clock in the morning, as compared to other parts of the day and night; see Jackson et al. (2011) for an overview. The risk of single accidents is higher during the night, and the consequences are more severe due to higher speed. Also, drivers are more susceptible to accidents after driving between 12 and 15 hours. The likelihood of accidents increases with driving time. Different studies show that accident risk increases by a factor of two or more for drivers working more than a 'normal' working day of 8–10 hours.

Fatigue as a causal factor has been studied in accident investigations. A distinction is made between falling asleep as a direct cause of the accident ('lapse hypotheses') and a reduction in the driver's general performance as a result of fatigue ('instability hypothesis'). Different studies of truck accidents estimate that the driver fell asleep in between 1% and 10% of the accidents (Englund et al. 2006). A study of fatal truck crashes concluded that fatigue was a causal or contributing factor in a third of the cases (Loeb et al. 1994).

There is clear evidence of the negative effects of *alcohol* on transportation safety, and the risk of accidents increases with the alcohol contents in a driver's blood. A Swedish study showed that accident risk increased by a factor of four at an alcohol blood concentration of 0.04% and a factor of 40 at a concentration of 0.15% (Englund et al. 2006; see also WHO 2015). Although the portion of drivers that are affected by alcohol is relatively low in Sweden (less than 0.2%), alcohol is a contributing factor in about a third of fatal transportation accidents. The significance of alcohol as a causal factor varies among countries, depending on alcohol consumption habits.

Alcohol and other types of drugs are also a concern in commercial driving. A U.S. study showed that about a third of fatally injured drivers had positive results in toxicological tests, where 13% of the drivers tested positive for alcohol, 13% for marijuana, and 9% for cocaine (Loeb et al. 1994).

27.2.2.2   *The vehicle*

Vehicle design affects both the probability and consequences of accidents. A distinction is made between the active safety features of the vehicle and its crashworthiness. The *active safety features* aim at making it possible for the driver to control the vehicle in a safe way and to avoid transportation conflicts (Englund et al. 2006). They include:

- Viewing conditions from the driver's position. These are determined by seating position and design of windows, windscreen wipers, rear mirrors, headlights, and so on.
- Design of controls, primarily steering wheel and brake pedal, clutch, and accelerator pedals. These must be designed so that the driver has rapid and error-free control of the speed and direction of the vehicle.
- Vehicle stability against skidding and overturning.
- Vehicle controllability (i.e. the way the vehicle follows the driver's intended course, acceleration, or retardation).
- Design of displays to keep the driver informed about important driving parameters such as speed and possible technical failure.
- Ergonomic conditions (i.e. seating position and conditions, climate control, noise, vibrations, and other factors that, if adequately designed, make the driver's conditions comfortable and reduce fatigue).

Increasingly, cars are equipped with in-vehicle driver support systems to reduce the mental load of the driving task. These include, for example, automatic speed control, anti-collision radar, anti-lock braking systems (ABSs), electronic stability programs (ESPs), and automatic driving.

The different active safety measures listed here typically will reduce the complexity of the driving task. As discussed earlier, changes in driver behaviour may offset the positive safety effects of these measures. The driver may, for example, utilise such features as a comfortable seating position, good acceleration, and low noise and vibration levels to prolong the driving time and to increase speed. Reduced feedback on driving speed due to low noise and vibration levels and good acceleration characteristics are of special concern.

The introduction of the concept of *crashworthiness* was determined by the needs to minimise injury to the driver and passengers in case of a collision. This is accomplished by allowing the colliding vehicles to absorb as much of the kinetic energy as possible and to reduce the energy transfer to the driver and passengers to a minimum. To accomplish this, vehicles are equipped with crash deformation zones in the front and rear and a stiff cabin, offering protection to the belted driver and passengers. During the first phase of a collision, the deformation zone is compressed by about half a metre. This phase lasts for about 100 ms for vehicles with a speed of approximately 50 km/hour. During the second phase, the body comes into contact with the protection system (seat belts, air bags) or with the interior of the cabin. Here, it is important to minimise injury by moderating the rate and concentration of the energy absorption.

Vehicle maintenance also is a contributing factor to the risk of accidents. Although research is not fully conclusive, some studies suggest that mechanical defects, and especially brake failures, are relatively common in transportation accidents and contribute to up to a third of all truck accidents in the United States (Loeb et al. 1994). Other 'technical' factors such as load characteristics and centre of gravity of the loaded truck also affect the risk of accidents.

27.2.2.3   The transportation environment

The transportation environment affects the exposure to transportation accident risks as well as the probability and consequences of transportation accidents. An important aspect of the transportation environment is technical design. This includes road standards (width, curvature, surface), design of crossings, signal regulation, speed limits, and so on (Englund et al. 2006).

The road network is divided into *road standard* categories to meet the needs of different users in relation to travelling distance and speed. Thus, it is common to distinguish among national, regional, and local roads. The following design recommendations apply, for example, to national roads in Sweden to support safe and efficient travel over long distances:

- Full or partial separation of pedestrians and cyclists from motor vehicle transportation
- Speed limit of 90 or 110 km/h on highways
- Specified minimum viewing distance and road curvature, requirements for a clear view for overtaking
- A limited number of intersections per kilometre
- Crossings in one plane or with overpass, roundabout, and signal regulation not allowed

The *speed limit* is one of the most basic measures for improving road safety. Although the significance of the vehicle speed as a determinant of transportation accident risk is indisputable, a trade-off has to be made between transportation safety and efficiency. Speed directly affects the probability of an accident through its influence on the stopping distance and other factors affecting the driver's ability to avoid a crash. In the case of a crash, the energy involved – and thus the extent of injury and damage – is to a large extent determined by the speed. Evidence from police-investigated fatal accidents in the United States suggests that 'driving too fast or in excess of posted speed limits' was a contributing factor in as much as 22% of the accidents (Loeb et al. 1994). Statistics show that a reduction in the average speed by 10 km/h on public highways reduces the fatal accident rate by almost 40% (Englund et al. 2006). Not only high speed in an individual vehicle but also lack of coordination of speed among different vehicles (speed variance) is a determinant of accident risk (Loeb et al. 1994). The reason is that vehicles moving at similar speeds are less likely to collide.

The aim of speed limits is to achieve an adequate speed adaptation on the part of the drivers. Research shows that, for speed limits to be effective, they have to be regarded as reasonable by the drivers, and there must be a fair chance of getting caught in case of speed violations (Englund et al. 2006). This issue points up the general concern about how transportation technical safety measures affect behaviour. Recent research aims at identifying some general principles for the design of technical safety measures that promote safe driving behaviour. Besides good speed adaptation, they include such issues as:

• Obvious rules for yielding in intersections
• Adequate viewing distances
• Simple and clear rules for interactions among road users
• Integration of unprotected road users where complete separation is not possible

Also *snow, rainfall, and lighting conditions* affect the risk of accidents. Compared to driving during ideal conditions (dry road surface, daylight), the accident frequency (per vehicle-kilometre) increases by a factor of two during night driving and by a factor of two to five during rain or snowfall (Englund et al. 2006). The combination of night driving and slippery roads results in an increase in the injury frequency rate (per vehicle-kilometre) by up to 10 times.

Emergency medical services after a crash aim at lowering accident fatalities and consequences to the health of the accident victims. The response time to delivery of the initial emergency medical services and the transportation time to hospital are considered critical for the probability of surviving a major trauma (Evanco 1999). This also applies to the quality of the emergency medical services, initially and at the hospital. Poor mobile phone coverage and inability of road users arriving at the accident site to take the correct action will delay response time. WHO estimates that, of the 1.2 million people worldwide who are killed each year in road transportation crashes, about half a million lives could have been saved by improvements in emergency medical services in low- and middle-income countries to the level of high-performing countries (WHO 2015). A recent study concludes that the potential effects of crash prevention and minimising injury severity as an immediate result of the crash have greater potential for mortality reduction than reductions in response time (Clark et al. 2013).

27.2.3   Road transportation safety management

Statistics presented in Section 27.1 showed that the risk of fatal road transportation crashes varies considerably among countries. This largely was explained by differences among the countries in national policies on road safety management. Figure 27.2 illustrates this relationship through a feedback system for the legislators' and the authorities' interventions to control road transportation safety.

Road safety legislation and enforcement varies significantly among countries (WHO 2015). Increasingly, new countries introduce road safety legislation based on experiences from the best-performing countries, especially within the areas of seat belt and motorcycle helmet use, control of speed, and reduction of drunk- and drugged-driving. There also are large differences among countries regarding requirements for the safety standards of vehicles and roads. Poor roads and vehicle safety standards are still the norm in many low- and middle-income countries (WHO 2015). Economic growth in these countries results in increased road fatalities

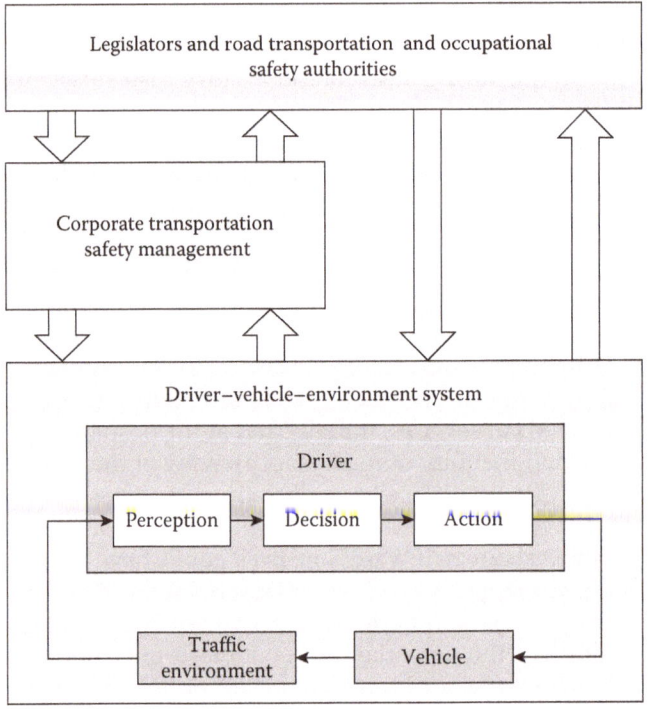

Figure 27.2 Simplified model of the safety management systems for work-related road transportation.

due to an increasing mismatch between the quality and capacity of the road infrastructure and the traffic volume (IRAP undated).

The corporate safety management systems for road transportation (Figure 27.2) are an intermediate factor with the potential to reduce the risk of work-related road transportation risks for company personnel, which is especially important in countries with weak national policies on road safety management. The span of control (see Ashby's Law of Requisite Variety, Chapter 7) and thus the potentials for effective accident countermeasures will vary depending on national legislation, whether transportation takes place on public or site roads, whether a company's own or contractor employees are involved, and so on.

The corporate road transportation safety management system does not differ in principal from a company's general health, safety, and environment (HSE) management system, but it is a complement focusing on an area with high potential for severe accidents. This system applies the general principles and elements of safety management as described in international standards and guidelines. An important example is the international standard on road traffic safety management systems, ISO 39001, which is rooted in the ISO 9001 and ISO 14001 quality and environmental standards (ISO 2012, 2015a, 2015b). It specifies requirements for a road transportation safety policy, objectives, and action plans.

Other examples are the recommended land transportation safety practices of the International Association of Oil & Gas Producers (OGP) that specify requirements for corporate road safety management system (OGP 2014b). Detailed requirements are given in the areas of:

- Driver training and qualifications
- Driver fitness and alertness, fatigue management
- Use of seat belts
- Driver use of mobile phone or other distracting activities
- Vehicle specification and outfitting, in-vehicle monitoring system
- Journey management

The research literature on the effects of corporate intervention strategies to manage safety in work-related transportation is sparse, and there is limited scientific evidence on the effects of such initiatives (Newman and Watson 2011; Mitchelle et al. 2012). Based on the general research literature on road transportation safety, we argue that a company has considerable scope for reducing the fatal crash risk involving its own and contractor road transportation through systematic management. This especially applies to transportation in countries with a generally high fatal risk level for road transportation. This conclusion is supported by evidence from an intervention program by a British firm, based on the Haddon matrix, which resulted in substantial reductions in road traffic crashes and associated costs (Murrey et al. 2014).

27.3   Case: Road safety management practices in a power company

This case is built on experiences by an international hydropower company (IHC) involved in developing and operating hydropower facilities in emerging markets. The company primarily invested in smaller, run-of-river plants. This involved construction and operation of power plants in mountainous areas, where the drop of the rivers is used to generate power. Transportation is challenging in these areas due to the topography, climatic conditions, and an often underdeveloped road system.

One of the company's projects, a 200 MW Greenfield run-of-river project in the Indian state of Himachal Pradesh, experienced seven fatalities in five separate accidents during a period of 15 months at an early stage in the project. Four of the accidents were caused by departures from site roads. This row of accidents made road transportation safety a focus of the corporate management agenda. A series of short and intermediate measures were introduced, primarily based on the principles of the Haddon matrix (Table 27.3).

The roadside barriers turned out to be the most visible life-saving measures by preventing road departures; see Figure 27.3. The totality of the different measures turned out to be effective. There were zero fatalities in the project during the subsequent 33 months until completion.

Table 27.3 Transportation safety measures applied in a project organised according to the Haddon matrix

Phase	Human (driver) factors	Vehicle/equipment factors	Road/environment factors
Pre-crash	Driver safety training Checking of driver qualification when entering the construction site	Regular preventive maintenance Safety inspection of vehicles, break and tier standard Checking of documentation of roadworthiness when entering the construction site	Control of traffic on site roads, including speed Stoppage of all traffic on site roads during high precipitation (rain, snow) or slippery road conditions
Crash	Not applicable	Not applicable	Roadside barriers
Post-crash	First-aid training	First aid, communications equipment	Increased availability of ambulances

Figure 27.3 Illustration of the use of roadside barriers in the project.

27.3.1   *Road transportation safety management*

Based on this experience, the IHC implemented systems to manage safety more systematically both for its own and for contractor road transportation.

A special procedure for road transportation during business travel by company employees was introduced. Because the road safety standard generally was beyond company control, the procedure focused on driver and vehicle standards and on journey management. The corporate procedure defined a company's minimum standard. It was the duty of the company's country-based managers to organise risk assessments of typical transportation routes used by company employees when in those countries with the aim of ensuring that all employees experience a tolerable risk level. The results were used to establish rules to compensate for the prevailing traffic safety standard in each country. The compensating measures focused on journey management of transportation in rural areas and were based on international best practice – for example, as illustrated by OGP's land transportation safety guidelines (OGP 2014b). Following are some examples:

- Documentation of the selected transportation route by a risk assessment, similar to a job safety analysis (JSA, see Chapter 22), covering the different segments of the road. This was used to determine the feasibility of the journey and alternative transportation means (e.g. by air).
- Assignment of a journey manager to assess and approve the journey.
- Documentation of the journey in a management plan, defining route selection, participants, safety precautions, schedule, and times for reporting back.
- Fatigue management, driver rest periods.
- Security escort, use of convoy (based on a security risk assessment).
- Restrictions in travel at night and during severe weather conditions.
- Pre-departure checks based on a checklist.

Contractor road transportation safety was managed in a different way. The main requirements were documented in a general specification for contractors' HSE management in construction, which specified requirements for a road transportation risk assessment, a road transportation management plan, and specific requirements regarding drivers' qualifications, vehicle safety standards – including preventive maintenance routines, and the safety standard for site roads. The principles for implementation of the requirements in the contracting process are illustrated by Figure 27.4.

Road transportation safety is addressed in the feasibility study phase of a project during site investigations. The project carries out specific road surveys to map the standard for the public access roads to the site. One purpose is to assess the feasibility of heavy transportation on public roads and need for upgrading. If upgrading is necessary, cooperation with the local road authorities will be required. The feasibility of site road construction due to topography and slope stability is also determined through site investigations undertaken in this phase. Road safety is included in the scope of the site investigations and road surveys, and the data are used in road transportation risk assessments.

The pre-qualification and tendering work follow the general practices for contracting, where transportation safety is integrated in HSE management of contractors. In this case, the company's project personnel verified the main contractor's transportation risk assessments and

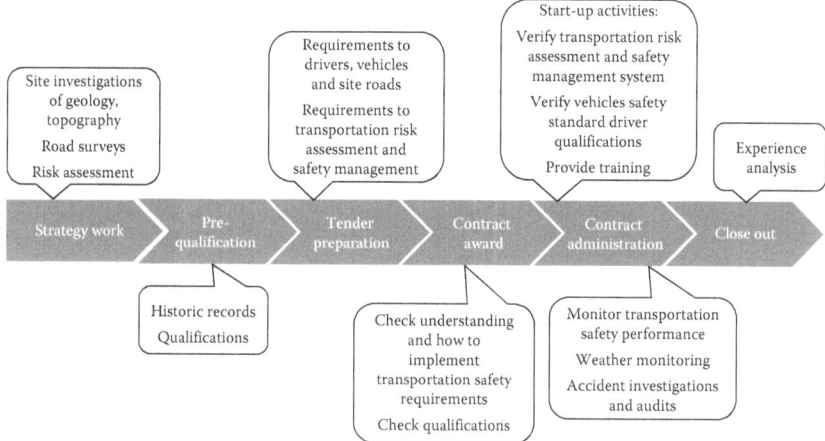

Figure 27.4 Management of contractor road transportation safety through the contracting process.

transportation safety management plan before site establishment. At start-up, contractors' routines for entry checks of vehicles and drivers were verified. Site driver licences were issued by the main contractor to those drivers that passed the tests. Site roads were subject to inspections and approval by the project team before they could be brought into use.

Monitoring of contractors' transportation safety performance involved regular inspections and checking of driver qualifications as well as vehicle and safety standards and adherence to site traffic rules. Special checklists were developed for use in accident investigations and transportation safety audits; see Table 27.4 for an example.

27.3.2 Road transportation risk assessment

According to Figure 27.4, risk assessments of road transportation are conducted by the project management and the contractors. Contractors' risk assessments normally are a variant of JSA, where the activities consist of driving different segments of the site roads with light vehicles and trucks. Here, we will focus on risk assessments carried out by one of the IHC's projects during the feasibility study phase. They covered transportation activities in two different phases in the project.

The first risk assessment aimed to assess the fatal road crash risk during transportation to and from the site and within the site during field investigation; this is carried out primarily during the feasibility study phase. Field investigation involves the collection of site- and area-specific data on, for example, hydrology, meteorology, typography, geology,

Table 27.4 Example of checklist for use in road transportation accident investigations and safety audits

Driver	*Knowledge, skills*
	Formal training, demonstrate skills, including hilly area experience
	Medical fitness (eyesight, hearing, health)
	Off-road/winter driving experience
	Heavy load, securing of load
	Behaviour
	Defensive driving
	Securing of passengers/load
	Use of alcohol/drugs
	Vigilance/fatigue
	Shift/work schedule
Vehicle	Maintenance standard, breaks, tires
	Structural integrity, stability
	Suitability for transportation of passengers/goods
	Tire standard
	Winterisation
	Safety standard
	Crashworthiness
	Securing of load, separate compartment from passenger cabin
	Securing of passengers, safety belts
	Rollover protection and so on
Roads	Road gradient
	Width, one- or two-way traffic
	Road stability
	Surface quality
	Viewing conditions
	Roadside conditions
	Road maintenance
	Roadside barriers
	Signage, marking of road edge
	Speed control road bumps
Environment	Geophysical hazards, falling rocks, landslide, fall out, flooding
	Reduced visibility due to rainfall (wind shield)
	Snow, ice, avalanche
Road users	Traffic density
	Variety of types of vehicles, speed (general public)
	Pedestrians

(Continued)

Table 27.4 (Continued) Example of checklist for use in road transportation accident investigations and safety audits

Traffic management	Driving safety rules (speed, daylight, no mobile phone)
	Journey management and monitoring system (including logistics officer)
	Restrictions in driving when high-intensity rain forecast
Emergency management	Turn-around time (including communication)
	Emergency hospital available
Security	Security escort

biology, and existing infrastructure including road surveys. In the feasibility study phase, the decision to realise the project is still pending, and minimum resources are spent on road improvements. The road standard can be very poor and, even if the amount of transportation is relatively limited during this phase, the risk – especially for individual drivers – may be very high.

The aim of the second risk assessment was to determine the transportation accident risks during construction. This is part of the constructability assessment of the project. The results are used in decisions on whether to proceed with construction or not, and if yes, on design, cost, and schedule.

Decisions regarding road transportation crash risk are based on two different types of acceptance criteria. The first focuses on workers at highest risk, where the individual frequency of a fatal accident shall not exceed one in 1000 years. It is assumed that full-time drivers on high-mountain roads define this category of workers, and their individual risk should meet this criterion. The second criterion relates to 'zero vision' with regard to fatal crashes in road transportation. This is interpreted as the likelihood of a fatality being well below one for a typical project year of field investigation or construction activity.

The method used in both risk assessments was a variation of the comparison risk assessment discussed in Chapter 24 (see Figure 27.5). Road transportation belongs to the domain characterised by high-frequency, 'small-scale' accidents, where an empirical risk management strategy based on statistical analysis is suitable (Rasmussen 1997).

The example shown in Figure 27.5 is based on a risk assessment for a field investigation project in the Indian state of Himachal Pradesh. In this example, the risk of fatality for a typical light-vehicle driver is assessed. In Step 1, the regional average for fatal crash risk in Himachal Pradesh was calculated based on statistics from the Indian Ministry of Road Transportation. The results are shown in Table 27.5.

In Step 2, the relative fatal crash risk for transportation from a hub in Himachal Pradesh to be used to go to a project site located in the same

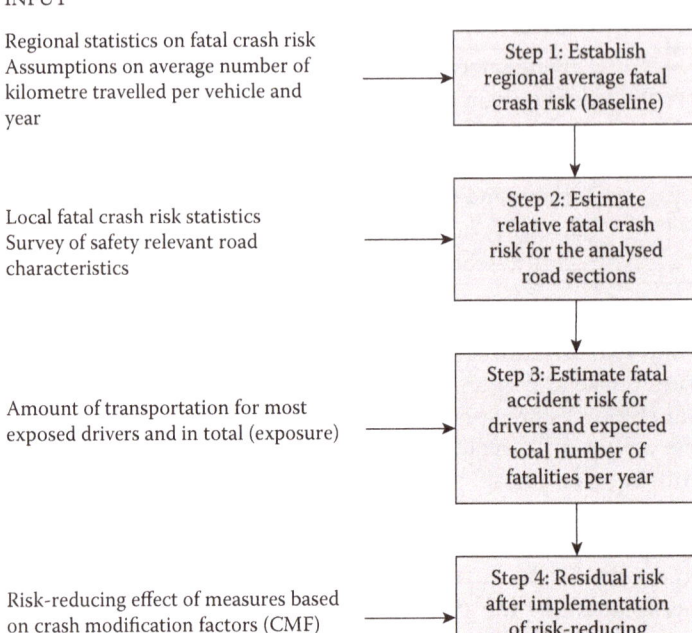

INPUT

Regional statistics on fatal crash risk
Assumptions on average number of
kilometre travelled per vehicle and
year

Step 1: Establish
regional average fatal
crash risk (baseline)

Local fatal crash risk statistics
Survey of safety relevant road
characteristics

Step 2: Estimate
relative fatal crash
risk for the analysed
road sections

Amount of transportation for most
exposed drivers and in total (exposure)

Step 3: Estimate fatal
accident risk for
drivers and expected
total number of
fatalities per year

Risk-reducing effect of measures based
on crash modification factors (CMF)

Step 4: Residual risk
after implementation
of risk-reducing
measures

Figure 27.5 The steps in the analysis of road transportation fatal crash risk.

Table 27.5 Estimated risk of fatality in road traffic in
Himachal Pradesh

Type of vehicle	Per million vehicle-kilometres	
	No. of fatalities	No. of fatal crashes
Car/light vehicle	0.19	0.14
Truck	0.11	0.08

state was estimated. The 315-km-long route was divided into five segments, and the fatal crash risk for each segment and for the route as a whole were estimated using two different methods, as shown in Table 27.6.

The coarse assessment was based on expert judgement by road engineers involved in a survey of the route. They identified road sections representing average regional risk, medium risk (two times the average), high risk (five times the average), and very high risk (ten times the average). The latter typically was represented by semi-tunnels carved out of a vertical wall, as shown in Figure 27.6. To verify these assumptions, statistics had been collected from the local police for road segments representing very

Table 27.6 Two different assessments of the number of fatal crashes per million vehicle-kilometres for the different route segments

Route segment (hub to project site)	Distance	Assessment based on detailed road survey and crash modification factor (CMF)			Coarse assessment based on expert judgement		
		Relative risk	Risk cars/light vehicles	Risk trucks	Relative risk	Risk cars/light vehicles	Risk trucks
Segment 1	42	1.0	0.14	0.08	1.0	0.14	0.08
Segment 2	47	2.4	0.33	0.18	2.9	0.40	0.22
Segment 3	57	2.3	0.32	0.18	2.9	0.40	0.22
Segment 4	47	2.5	0.34	0.19	3.2	0.45	0.26
Segment 5	122	3.9	0.55	0.30	4.2	0.57	0.32
Total route	315	2.8	0.39	0.22	3.2	0.44	0.25

Figure 27.6 Semi-tunnel road with vertical drop at the other side of the road.

high risk, and the results indicated a fatal crash risk per vehicle-kilometre of about 10 times the regional average.

The detailed assessment was based on a model of the mountainous parts of the route with respect to road safety characteristics such as transverse slopes (vertical: horizontal), hairpin bends, substandard bridges, road width, hard shoulder, and side guard rails. Each road characteristic was assigned a crash modification factor based on empirical data (Høye et al. 2012). Two examples illustrate the use of the crash modification factors (CMFs). Steep slopes beside the road had a considerable influence on the risk of road departure, and recovery is almost impossible for slopes steeper than 1:3. Transfer slopes 1:1 or steeper were given a CMF of 3.0 (i.e. an increase by the factor three in relative risk compared to the regional average). Side guard rails, on the other hand, decreased the risk of fatal road departures, and these roads were given a CMF of 0.5.

As we see from the table, the coarse assessment results in an approximately 40% higher assessment of the fatal crash risk but was considerably less time-consuming.

Table 27.7 shows the results of the risk assessment for drivers of light vehicles, which were assumed to be at highest risk due to their driving pattern (Step 3). A typical light-vehicle driver made daily trips from a village adjacent to the project office to the site, around the site, and then back to the project office. The driver also made weekly supply trips to the nearest town and monthly trips to the hub.

Table 27.7 Summary of risk estimates for drivers of light vehicles based on the detailed assessment of road crash risk compared to driving the same distance on roads representing the regional average safety standard

Type of trip	Distance	No. of tours per year	Per year (km)	Fatal crash frequency per driver and year		Remarks
				Detailed assessment	Regional average	
Daily, on site roads	35 km	240	8,400	0.0046	0.0012	Same relative risk assumed as for Segment 5
Weekly Segment 5	2 × 122 km	40	9,760	0.0054	0.0014	Weekly trips for supplies
Monthly complete route	2 × 315 km	10	6,300	0.0025	0.0009	Monthly trip to the hub with passengers at shift change
Sum			24,460	0.013	0.0035	

The results estimate that the light-vehicle drivers experience one fatal accident about every 80 years (i.e. 13 times the acceptance criterion of less than one fatality per 1000 years of exposure). Even the average regional fatal crash risk is about three times the acceptance criterion. Therefore, it is assumed here that the average driving safety standard is the same as for regular road users.

CMFs have also been established for other factors related to drivers and vehicles (Høye et al. 2012). The factors used in Step 4 of this study included:

- *Drivers:* Measures against drunk driving, increased seat belt use, measures against fatigue. The combined effect of these measures was expected to reduce the fatal crash frequency by 38%.
- *Vehicle:* Improved crashworthiness, improved tire conditions. The combined effect was assessed to be a reduction of the fatal crash frequency by 10%.

The combined effects of all assessed measures were estimated to reduce the fatal crash frequency by two-thirds under optimal conditions. This would still give a fatal crash frequency for light-vehicle drivers of about four times the acceptance criterion. Regulations for company transportation activities were also evaluated, but the effects were never assessed due to lack of data. These included restrictions on driving during winter and severe weather conditions, banning night-time driving, regulations about meeting traffic on single-lane roads, and improved emergency response.

The field investigation project was executed while introducing a combination of the measures described here, and this risk assessment was essential in planning risk management activities. The project did not experience any fatal road crashes involving project or contractor personnel in the course of three years of field investigation activities.

chapter twenty-eight

Epilogue

The objective of this book has been to present principles and methods for the prevention of accidents through experience feedback. These feedback systems are an important part of corporate management of safety. Statistics presented in this book show that significant reductions in accident risk may be achieved through the systematic application of such methods. Most of the examples presented are based on the experiences of large, resourceful companies, especially those in highly hazardous industries. This positive development in accident statistics has not been paralleled by small- and medium-sized companies. Norwegian statistics support this conclusion – that country's offshore oil and gas industry has a relatively low fatal accident frequency rate compared to other Norwegian industry sectors such as construction, transportation, and primary production. These sectors are dominated by small- and medium-sized companies and even solitary work as well as by a dynamic working environment.

The application of feedback control in safety management is not isolated from other corporate activities. Efficient management of safety can only be realised when it is an integrated part of the way a company does business. Health, safety, and environment (HSE) staff perform a support function that helps in developing and implementing a company's experience feedback systems. They are also responsible for managing those experience feedback activities that require expert competence. But the main responsibility for promoting and operating feedback systems lies with the line organisation. It must also be realised that the prevention of accidents is accomplished through many different means, of which safety management is one important part. We have used the word 'control climate' throughout this book to illustrate the positive relationship between a sound and controlled way of producing goods and services and the control of accident risk. Studies have shown that good safety performance is correlated with well-functioning general management, resulting in top performance within a range of fields including not only safety but productivity, progress, and quality as well. This is further illustrated through the close relationship between the principles and methods of quality management and those of HSE management. Total quality management (TQM) emphasises those efforts within organisations that integrate quality management into management systems and the entire company.

Since its humble beginnings in the 1920s, the area of systematic safety (and later HSE management) has developed through various phases – from a focus on people at the sharp end that err to a socio-technical view of safety as resulting from complex interactions among human, technical, and organisational factors. Is there an emerging new phase in the development of HSE management? Some claim that adaptive safety management approaches such as resilience engineering and theories on high reliability organisations represent a new phase for dealing with complexity in modern organisations and technology. Adaptive safety management can be a promising area for the prevention of incidents. To date, this area seems too conceptual for most safety practitioners. Another challenge is that these adaptive safety management approaches mainly focus on the prevention of major accidents having a very complex range of causal factors, which is a focus far beyond that of simple occupational accidents.

There are trends in society that influence HSE management. One of these trends is the globalisation of work. In high-income countries, foreign workers from middle- and low-income countries represent a safety challenge due to language and cultural issues. Western companies operating in developing countries also experience safety management challenges related to differences in culture, governance issues, and poor physical infrastructures. In addition, value chains in many industries have become more complex and multinational. This creates greater interaction among different companies and across international borders. Generally speaking, many systems have become more complex and dynamic, which in particular creates problems in predicting and managing future hazards and unwanted occurrences.

Another trend, especially after the events of September 11, 2001, is an increased attention to security (i.e. protection against acts intended to harm). This book has not dealt with security specifically, although many of the principles and methods for experience feedback presented here are applicable to this field as well. This is the case not only for the principles and methods of experience feedback but also for barrier principles and those dealing with emergency response. One difference is that security risk assessments involve a higher degree of uncertainty due to lack of knowledge about intentions to create harm and possible energy sources to be used, especially compared to random events that occur in connection with accidents.

In the first edition of this book, we emphasised the influence of information and communication technology (ICT) on the developments within accident prevention. Clearly, ICT has made a significant difference in the collection and capturing of experience data, in the retrieval and analysis of data, and in the distribution of experience reports. This development will, of course, continue. Cloud computing has already made collection and sharing of accident data even more efficient and also promotes multi-organisational sharing of data within industrial sectors. ICT also represents an opportunity to improve online monitoring and application of safety performance indicators. Software agents, or smart agents, have the potential to guide decision-makers more efficiently. These agents are similar to those applied in online stores that display 'items that you may also be interested in'. Techniques for analysis of big data represent another possibility but obviously require large data sets. More than technology, limits in human attention and information processing capacity will define the future as to what is possible.

Appendix A: Definitions

Acceptable risk Risk of accidents that is considered tolerable and further risk reducing measures are desirable but not mandatory (cf. HSE 2001b).

Acceptance criterion for the risk of accidents Criterion used to express a level of acceptable risk in an activity.

Accident A sequence of logically and chronologically related deviating events involving an incident that results in injury to personnel or damage to the environment or material assets.

Accident risks A collective term for any event or condition that is associated with an increased risk of accidents such as hazard, incident, deviation, contributing factor, or root cause.

Active barrier Barrier that is dependent on actions by an operator or on a technical control system to function as intended.

ALARP As low as reasonably practicable.

Analysis object The part of the workplace which defines the scope of the risk analysis (e.g. an area in coarse job safety analysis and a specific job in detailed job safety analysis).

Barrier A set of system elements (human, technical, organisational) that as a whole provide a barrier function with the ability to intervene into the energy flow to change the intensity or direction of it.

Barrier element A human, technological, or organisational element that contributes to the realisation of barrier function

Barrier function The ability of the barrier to intervene into the accident sequence to eliminate or reduce loss.

Barrier system The set of interacting human, technical, and organisational elements that make up the barrier function as an integrated whole.

Behaviour-based safety A process designed to reduce the frequency of work-related accidents by first reducing the frequency of negative or inappropriate employee behaviours (Janicak 2003).

Cause of accident Contributing factor or root cause.

Contributing factor More lasting risk-increasing condition at the workplace related to human, technical, and organisational sub-systems.

Danger zone Any zone within and/or around an energy source in which a person (or vulnerable object) is subject to a risk to his/her health or safety (or loss) by getting in contact with the energy flow from the source (European Council 2006).

Deviation Event or condition that departs from the norm for the faultless or planned-systems processes.

Diagnosis A decision cycle consisting of identification of symptoms, determination of causes, and prescription of remedy.

Experience carrier A document, database, or piece of hardware that represents the company's collective experiences and governs its activities and decisions.

Experience feedback The process by which information on the results of an activity is fed back to decision makers as new input to modify and improve subsequent activities (Melnick and Everitt 2008).

Experience, qualifying Experience that makes an individual or organisation able to understand real-world phenomena (know-why) and able to act efficiently (know-how).

Feedback control (negative) A regulating mechanism that produces corrective action.

Hazard A potential source of injury or damage to health of people or damage to the environment or material assets (cf. European Council 2006). In most cases, it is an energy source with the potential of creating injury to personnel or damage to the environment or material assets.

Hazardous event See Incident.

High-potential incident (HIPO) An incident, where the most serious probable outcome is a fatality or serious injury, resulting in permanent disability.

HSE Health, safety, and environment.

HSE audit A systematic and independent examination of a company's HSE management system (cf. ISO 2015). The aim is to determine that the elements within the system have been established and is effective and suitable for achieving stated HSE requirements and goals.

HSE management system The part of the management system of an organisation used to establish HSE policies, objectives, and processes to achieve those objectives (cf. ISO 2015).

Human error A subset of human actions that transgress a norm or limit of what is planned/intended, normal, or acceptable.

Incident Loss of control of energies in the system or body movements, resulting in a potential for exposure of personnel (or the environment/material assets) to the energy flow.

Job A series of inter-related manual activities that are performed to achieve organisational objectives.

Job safety analysis (JSA) A series of logical steps to enable systematic examination of the hazards associated with jobs.

Lagging safety performance indicator An indicator that changes after the safety performance in the activity has changed.

Leading safety performance indicator An indicator that predicts future safety performance, i.e. that changes before the safety performance changes (Janicak 2003; Kjellén 2009).

LTI rate Lost-time injury frequency rate, i.e. the number of lost-time injuries at work per one million hours of work.

Management of safety The totality of an organisation's efforts to control hazards to avoid accidental loss.

MTO Man–technology–organisation

Near accident A sequence of logically and chronologically related deviating events involving an incident that under slightly different circumstances could have resulted in injury to personnel or damage to the environment or material assets.

Nonconformity Non-fulfilment of specified requirement.

Occupational disease A disease or disorder that is caused by the work or working conditions.

OHS Occupational health and safety

Passive barrier Barrier that is not dependent on actions by an operator or on a technical control system to realise its barrier function.

Precision, degree of The proportion of the retrieved reports from a database that correspond with what was wanted.

Recordable injury Fatality, lost-time injury, medical treatment injury or injury resulting in loss of consciousness, transfer to another job, or restricted work.

Reliability in the reporting of incidents The probability that a reportable incident is reported. It is expressed as the number of reported incidents divided by the 'true' number of incidents, as defined by the reporting criterion.

Retrieval, degree of The share of the total number of wanted reports from a database that have been retrieved.

Risk analysis Establishing a risk picture by determining frequencies and consequences of unwanted incidents

Risk assessment The overall process of hazard identification, risk analysis, and risk evaluation (ISO 2009)

Risk evaluation Process of comparing the results of the risk analysis with acceptance criteria for the risk of accidents.

Risk of accidents A combined measure of the probability or frequency of accidents involving losses per unit of exposure to a specified activity, and the extent of the losses (consequences).

Root cause of accidents Most basic cause of an accident/incident, i.e. a lack of adequate management control, resulting in deviations and contributing factors.

RUO Report on unwanted occurrences.

Safety audit See HSE audit.

Safety culture The product of individual and group values, attitudes, perceptions, competencies, and patterns of behaviour that determine the commitment to, and the style and proficiency of, an organisation's health and safety management (ACSNI 1993).

Safety information system A system that provides the information needed for decisions and signalling relating to the prevention of accidents.

Safety performance measure/indicator Metric used to measure the organisation's safety performance in terms of its effectiveness in controlling the risk of accidents in its activities.

STEP analysis Sequential time event plot analysis. A tool for graphical presentation of the incident sequence.

Symptom A deviation of the system's behaviour from what is considered to be 'normal'.

Taxonomy A classification system made up of a complete set of mutually exclusive classes.

Total recordable injury (TRI) Work-related injury resulting from an accident which involves one or more of the following: fatality, lost workday(s), loss of consciousness, restriction of work or motion, transfer to another job, and medical treatment other than first aid.

TRI rate Total recordable injury rate, i.e. the number of total recordable injuries at work per one million hours of work.

Unsafe act Subset of human errors that increases the risk of accidents, e.g. by degrading a barrier.

Unsafe condition Deviating condition that increases the risk of accidents, e.g. by degrading a barrier.

Unwanted occurrence Near accident, unsafe act, or unsafe condition

Validity as indicator of the risk of accidents A measure of the degree to which the indicator represents the risk of accidents.

Variety The total number of states of a system.

Appendix B: SMORT checklists and questionnaire

B.1 Checklists

Tier 1: Sequence of events/risk situation

☐ 1.1 Work situation

☐   1.1.1 Human error, e.g. wrong action, wrong sequence, omission

☐   1.1.2 Technical failure, e.g. substandard equipment, break down, missing equipment or tools

☐   1.1.3 Disturbance in material flow, e.g. bad quality, delays

☐   1.1.4 Personnel deviations, e.g. absence, not qualified, indisposed

☐   1.1.5 Inadequate information, e.g. job procedure, permit to work, risk assessment, supervision

☐   1.1.6 Progress delay

☐ 1.2 The environment

☐   1.2.1 Intersecting or parallel activities, e.g. lack of co-ordination of work

☐   1.2.2 Poor housekeeping

☐   1.2.3 Poor physical environment, e.g. excessive noise, high temperature

☐   1.2.4 Substandard building and infrastructure, e.g. roads

☐ 1.3 Incident

☐   1.3.1 Loss of control of energy or person relative to energy flow

☐   1.3.2 Failure in active safety barriers

☐   1.3.3 Failure in fixed barriers

☐   1.3.4 Failure in personal protective equipment or cloths

☐   1.3.5 Persons in danger zone

☐ 1.4 Development of injury/damage

☐   1.4.1 Failure in alarm and mobilisation of emergency response team

☐   1.4.2 Failure in limiting injury/damage, e.g. medical treatment and evacuation

☐   1.4.3 Failure in management of information to internal and external stakeholders

Tier 2: Work system and department

☐ 2.1 HSE culture and values
☐   2.1.1 Line management commitment, risk perception
☐   2.1.2 Compliance culture, adherence to safety rules
☐   2.1.3 Communication on HSE
☐   2.1.4 Employee involvement, reporting culture
☐ 2.2 HSE management
☐   2.2.1 HSE organisation
☐   2.2.2 HSE goals, program, and action plans
☐   2.2.3 Safety routines and procedures
☐   2.2.4 Co-ordination of safety work at site
☐   2.2.5 Emergency preparedness
☐   2.2.6 Documentation
☐ 2.3 Management of operation
☐   2.3.1 Assignment of responsibilities
☐   2.3.2 Resources
☐   2.3.3 Production and activity plans, progress planning, co-ordination
☐   2.3.4 Planning of job execution, use of calculations, and risk assessment
☐   2.3.5 Work and safety instructions
☐   2.3.6 Supervision, instructions, monitoring of performance, and correction
☐   2.3.7 Co-ordination between activities, hand-over between shifts
☐   2.3.8 Planning for the management of disturbances
☐   2.3.9 Procurement
☐ 2.4 Management of contractors
☐   2.4.1 Pre-qualification
☐   2.4.2 Tendering and contract award
☐   2.4.3 Start-up
☐   2.4.4 Monitoring during execution
☐   2.4.5 Close-out and experience transfer
☐ 2.5 Human resources
☐   2.5.1 Resources, manning, qualifications
☐   2.5.2 Education, training
☐   2.5.3 Remuneration
☐ 2.6 Fixed assets, site, infrastructure
☐   2.6.1 Design of buildings and infrastructure, perimeter guarding
☐   2.6.2 Plant layout, access
☐   2.6.3 Physical barriers, safety systems
☐   2.6.4 Physical working environment
☐   2.6.5 Design of machinery/equipment
☐   2.6.6 Man–machine interface

Continued

Tier 2: Work system and department

☐   2.6.7 Availability of machinery and equipment
☐   2.6.8 Documentation
☐ 2.7 Maintenance of fixed assets
☐   2.7.1 Maintenance planning, budgeting, preventive maintenance vs. repair
☐   2.7.2 Resources
☐   2.7.3 Methods, procedures, use of risk assessment
☐   2.7.4 Co-ordination with operations
☐   2.7.5 Permit-to-work system
☐   2.7.6 Modifications, management of change, quality control
☐   2.7.7 Documentation
☐ 2.8 Material supply
☐   2.8.1 Composition
☐   2.8.2 Material transportation, storage
☐   2.8.3 Waste handling
☐   2.8.4 Documentation
☐   2.8.5 Quality control

Tier 3: Project management

☐ 3.1 HSE culture and values
☐   3.1.1 Project management commitment
☐   3.1.2 Communication on HSE
☐ 3.2 HSE management
☐   3.2.1 Organisation and responsibilities
☐   3.2.2 Project's HSE program
☐   3.2.3 Suppliers' HSE programs
☐   3.2.4 Experience transfer
☐   3.2.5 Regulations, standards, specifications
☐   3.2.6 Analyses and verifications
☐   3.2.7 Audits and management reviews
☐   3.2.8 Nonconformity handling
☐ 3.3 Project management
☐   3.3.1 Project scope, responsibilities, procedures, design basis
☐   3.3.2 Human resources, qualifications
☐   3.3.3 Budget
☐   3.3.4 Schedule
☐ 3.4 Relation to stakeholders
☐   3.4.1 Co-ordination of stakeholders
☐   3.4.2 Liaison with operations
☐   3.4.3 Contact with authorities, licenses, and permitting

Continued

Tier 3: Project management
☐ 3.5 Study phase
☐   3.5.1 Concept exploration and definition
☐   3.5.2 Demonstration and validation
☐   3.5.3 Definition of scope, change control
☐ 3.6 Project execution
☐   3.6.1 Engineering
☐   3.6.2 Procurement
☐   3.6.3 Follow-up of HSE in fabrication and construction
☐   3.6.4 Mechanical completion and commissioning
☐ 3.7 Take-over by operations
☐   3.7.1 Recruitment and training of operations personnel
☐   3.7.2 Development of procedures and work instructions
☐   3.7.3 Start-up

Tier 4: Higher management
☐ 4.1 HSE culture and values
☐   4.1.1 Leadership and commitment
☐   4.1.2 Compliance culture
☐   4.1.3 Co-ordination and communication
☐   4.1.4 Employee involvement
☐   4.1.5 Stakeholder management
☐ 4.2 HSE policy, goals, and action plans
☐   4.2.1 Policy
☐   4.2.2 Goals and acceptance criteria
☐   4.2.3 Action plans
☐ 4.3 Resource management
☐   4.3.1 Responsibilities
☐   4.3.2 Human resources, qualification
☐   4.3.3 Business planning and budgeting, risk management
☐   4.3.4 Fixed assets and materials
☐   4.3.5 HSE organisation
☐ 4.4 Identification and evaluation of risks
☐   4.4.1 Routines
☐   4.4.2 Data collection
☐   4.4.3 Storage, analysis, and distribution of information
☐ 4.5 Handling of governing HSE documents
☐   4.5.1 Routines
☐   4.5.2 Regulations, codes, standards

Continued

Tier 4: Higher management
☐   4.5.3   HSE management program
☐   4.5.4   Procedures
☐   4.6   Performance monitoring and auditing
☐   4.6.1   Performance indicators
☐   4.6.2   Follow-up of results
☐   4.6.3   HSE auditing and management reviews

B.2   Questionnaire

Tier 1: Sequence of events/risk situation

1.1   Work situation

     1.1.1   Human error

         1. Wrong work method or equipment used?

         2. Actions omitted, delayed, or out of sequence?

         3. Misunderstanding within work crew?

         4. Improvisations, horseplay?

         5. Inadequate error recovery?

     1.1.2   Technical failure

         1. Faulty machinery, equipment, or tools?

         2. Errors in technical control system?

         3. Faulty plant containment of hazardous substances?

         4. Machinery, equipment, or tools missing?

         5. Wrong location of machinery?

     1.1.3   Disturbance in material flow

         1. Deficiencies in the quality of the work material?

         2. Wrong dimensions or insufficient quantity of delivered materials?

         3. Delays in material delivery?

         4. Inadequate packaging or storage of materials?

     1.1.4   Personnel deviations

         1. Did persons other than the usual ones carry out the work?

         2. Temporary reductions in the work crew?

         3. Personnel in an unsatisfactory condition (due to inadequate sleep, drug use, etc.)?

     1.1.5   Inadequate information

         1. Insufficient or incomplete work instructions (written or oral)?

         2. Technical documentation (drawings, user instructions, certificates, etc.) incomplete or lacking?

         3. Did the personnel lack the required driver's license, certificate, etc.?

1.1.6 Progress delay
   1. Not getting started in time
   2. Slow progress

1.2 The environment
   1.2.1 Intersecting/parallel activities
      1. Disturbances from adjacent work crew?
      2. Delays affecting downstream activities?

   1.2.2 Poor housekeeping
      1. Waste from work materials or packaging materials present?
      2. Spillage from production process present?
      3. Unattended tools and equipment present?

   1.2.3 Poor physical environment
      1. Too high noise level?
      2. Too high vibration level?
      3. Inadequate lighting, glare?
      4. Too high wind speed?
      5. Too high/low temperature?
      6. Unusually heavy rainfall, snow?
      7. Contamination of the working atmosphere?
      8. Uneven or slippery floor?
      9. Unstable slope, unsecured rock, other unfavourable geological conditions?

   1.2.4 Substandard buildings and infrastructure
      1. Poor structural integrity
      2. Violation of building code on safety/fire protection
      3. Substandard road quality
      4. Inadequate roadside protection
      5. Inadequate road signage

1.3 Incident
   1.3.1 Loss of control
      1. Loss of control of own body movements?
      2. Loss of control of energies in the system?
         a. Operator loses control of manually operated tools/machinery equipment?
         b. Loss of control of machinery movements due to technical or control system failure?
         c. Accidental release of hazardous substances?
         d. Other?

   1.3.2 Failure in active safety barriers
      1. Energy release not automatically detected?
      2. Active safety barrier function not activated or not functioning?

1.3.3 Failure in fixed barriers
    1. Fixed barriers not present?
    2. Fixed barriers penetrated by energy flow?

1.3.4 Failure in personal protection, cloths
    1. Personal protection not used?
    2. Wrong personal protection in use?
    3. Personal protection did not give adequate protection?

1.3.5 Person(s) in danger zone
    1. Unauthorised person(s) present in danger zone?
    2. Too many people present in danger zone?

1.4 Development of injury/damage
  1.4.1 Failure in alarm and mobilisation of emergency response team
    1. Alarm not received or delayed?
    2. Delayed or incomplete mobilisation of emergency team?

  1.4.2 Failure in limiting injury/damage
    1. Inadequate evacuation or rescue of victims?
    2. Inadequate or inefficient combat of energy flow?
    3. Inadequate protection of emergency team?

  1.4.3 Failure in management of information
    1. Delayed or inadequate information to the authorities?
    2. Delayed or inadequate information to the next-of-kin?
    3. Delayed or inadequate information to higher management and/or the employees?
    4. Delayed or inadequate information to the media and public?

Tier 2: Work system and department
2.1 HSE culture and values
  2.1.1 Line management commitment, risk perception
    1. Do the supervisors promote HSE by their own vigour and example?
    2. Do the supervisors take active part in safety activities (accident and near-accident investigations, workplace inspections, job safety analyses)?

  2.1.2 Compliance culture, adherence to safety rules
    1. Do the supervisors monitor the safety behaviour of the workers and take action in case of non-compliance?
    2. Are the reactions on rule violations defined and made known and are they experienced as fair?
    3. Do the supervisors themselves act as role models by showing strict compliance with safety rules and procedures?

2.1.3 Communication on HSE
  1. Is communication between supervisors and higher manage-
     ment and workers on HSE adequate?
  2. Are there regular meetings between supervisors and workers
     where HSE issues are highlighted?

2.1.4 Employee involvement, reporting culture
  1. Is it ensured that the employees are involved in decisions on
     HSE, where they possess relevant knowledge and experience?
  2. Is ownership of HSE measures ensured through employee
     involvement?
  3. Do employees participate in accident and near-accident inves-
     tigations, workplace inspections, job safety analyses, toolbox
     meetings?
  4. Are employees sharing their experiences through an adequate
     reporting of unwanted occurrences?

2.2 HSE management
  2.2.1 HSE organisation
    1. Has an adequate HSE organisation been established
       with safety representatives and a working environment
       committee?
    2. Does this include contractors and hired labour?
    3. Has the division of responsibility between line management
       and the HSE organisation been adequately defined?

  2.2.2 HSE goals, program, and action plans
    1. Have adequate HSE goals and program been established?
    2. Is it ensured that they are implemented in action plans with
       clear responsibilities and deadlines?
    3. Are the plans adequately followed up?

  2.2.3 Safety routines and procedures
    1. Are there adequate routines for accident and near-accident
       reporting and investigations, workplace inspections, job-
       safety analyses?
    2. Do the routines for workplace inspections ensure that they are
       carried out with an acceptable frequency and scope?
    3. Do these routines function adequately in practice?
    4. Are the actions from these different activities adequately doc-
       umented and followed up?
    5. Are there adequate routines for the identification, analysis,
       and documentation of new jobs with respect to safety?
    6. Is the physical working environment monitored at regular
       intervals?

7. Are key characteristics of the operation and activities that can have a significant impact on the environment monitored and measured regularly?
8. Is compliance with relevant environmental legislation and regulations periodically evaluated?
9. Are there regular health checks of the personnel?

2.2.4 Co-ordination of safety work at the site
1. Have the responsibilities for co-ordination of safety work been adequately defined?
2. Are the different safety activities adequately co-ordinated with contractors on the premises?

2.2.5 Emergency preparedness
1. Does an adequate emergency preparedness plan exist that describes the responsibilities and routines?
2. Has the plan been established based on an analysis of the different possible accident scenarios with consequences to personnel, the environment, and material assets?
3. Are there adequate organisational and technical resources for emergency operations in relation to the potential accident scenarios? Are they available at all hours when they are needed?
4. Is a single line of command ensured in emergency situations?
5. Are there adequate evacuation and counting routines?
6. Is the plan for emergency training and drills adequate?
   a. Of the emergency organisation?
   b. Of all concerned personnel?
   c. Does it ensure that the emergency crew is familiar with the facilities?
   d. That the actions of internal and external resources are co-ordinated?
7. Does the plan ensure an adequate notification of accidents, mobilisation of emergency organisation, rescue operations, combat of fires, etc.?

2.2.6 Documentation
1. Has the HSE management system been adequately documented?
2. Does the documentation include:
   a. Policy and goals?
   b. Organisation and responsibilities?
   c. Identification of applicable procedures and instructions?
   d. Action plans?
3. Is the HSE management system available within the organisation? Has it been ensured that it is known?
4. Is the access to statutory regulations adequate?

2.3 Management of operation
  2.3.1 Assignment of responsibilities
    1. Have the safety responsibilities been adequately defined:
       a. Between first-line supervisors and middle management?
       b. Between line management and safety officers?
       c. Between operation and maintenance?
       d. In hand-over between shifts?
       e. Within plant areas that are used by more than one department?
       f. In work outside normal working hours?
       g. For housekeeping?
       h. For hired personnel?
    2. Has authority to make decisions been delegated to the lowest possible level?
    3. Have the operators been provided with adequate autonomy and well-defined operational goals and decision criteria to be able to handle critical disturbances in a satisfactory way?
    4. Do the authority and accountability match the responsibility?

  2.3.2 Resources
    1. Do the supervisors receive adequate support from the line organisation and staff officers?
    2. Do the supervisors receive the necessary information?
    3. Is selection and training of supervisors adequate?
    4. Have the supervisors received the required HSE education and training?
    5. Do the supervisors have adequate time at their disposal to follow-up on safety issues?
    6. Do the supervisors have an adequate budget for safety measures?

  2.3.3 Production and activity plans, progress planning, co-ordination
    1. Are there adequate production plans?
    2. Are new jobs adequately planned?
    3. Is it possible for the workers to control their own work pace?
    4. Have working hours, shift schedule, and breaks been adequately planned?
    5. Are the activity plans and schedule acceptable considering the age and skills of the workers and their needs of restitution?
    6. Do the activity plans allow for adequate variations in job execution and work pace?
    7. Is the co-ordination of simultaneous activities adequate to avoid hazardous conflicts?

2.3.4 Planning of job execution, use of calculations, and risk assessment
   1. Has the job been adequately planned before start-up through calculations and risk assessment, identification of safety requirements?
   2. Have identified measures been implemented through elimination of hazards, substitution, or barriers?
   3. Has the work crew been given necessary coaching and training, e.g. in toolbox meeting before start-up?
   4. Has the job been adequately documented in work and safety instructions?

2.3.5 Work and safety instructions
   1. Are there adequate work and safety instructions for the different jobs?
   2. Have persons with direct experience of the jobs participated in the development of the instructions?
   3. Are the instructions based on a systematic evaluation of the hazards?
   4. Are hazards adequately identified and necessary control actions described?
   5. Do the instructions coincide with the way work is actually carried out?
   6. Are the instructions checked and updated at regular intervals and also when the jobs are changed?
   7. Do the instructions cover start-up, shut-down, normal operation, handling of disturbances, cleaning, and repair?
   8. Have potential conflicts between safe operation and requirements for productivity been adequately evaluated?
   9. Are the responsibilities and routines for housekeeping adequately defined and documented?
  10. Do they ensure regular housekeeping to avoid accumulation of waste and litter at the workplaces?

2.3.6 Supervision, instructions, monitoring of performance
   1. Do the supervisors have required qualifications to make the necessary decisions?
   2. Do the supervisors act as role models on safety and communicate and coach to ensure safe work performance?
   3. Do the supervisors monitor hazardous conditions at the workplaces and take the necessary decisions on remedial actions?
   4. Do the supervisors balance the priorities for safety, quality, and progress?
   5. Do the supervisors adequately address HSE issues in management meetings?
   6. Is the co-ordination between different supervisors in different departments and shifts adequate?

2.3.7 Co-ordination between activities, hand-over between shifts
1. Is there adequate co-ordination of different activities to avoid interference with potential unwanted HSE effects?

2.3.8 Planning for the management of disturbances
1. Are there adequate provisions for the handling of deviations from activity plans due to disturbances?

2.3.9 Procurement
1. Have the relevant HSE requirements been implemented in the routines for procurement to ensure that equipment, materials, and services conform to the requirements?
2. Is it checked at delivery to site that the supply is in compliance?

2.4 Management of contractors
2.4.1 Pre-qualification
1. Have adequate criteria been defined for HSE resources, qualifications, and historical performance?
2. Are potential bidders evaluated in relationship to the criteria? Does this include auditing of the existing site?
3. Are potential bidders that do not meet the HSE criteria excluded from the bidder's list?

2.4.2 Tendering and contract award
1. Are company HSE requirements adequately specified in the tender (and contract)?
2. Are company HSE expert(s) participating in bid evaluation?
3. Are bidders' HSE qualifications and performance given adequate weight in bid evaluation?
4. Are bid qualification meetings used to align expectations on HSE performance?

2.4.3 Start-up
1. Has a person in company's organisation been assigned to follow-up on the contractor's work at site?
2. Are qualifications of contractor's personnel and standard of equipment brought to site checked before start of work?
3. Are all contractor personnel subject to HSE induction before start of work? Does this include an adequate review of hazards in the work and remedial actions?
4. Are contractor's HSE management program and safety systems (as specified in Contract) checked before start of work?

2.4.4 Monitoring during execution
1. Is safety performance a fixed point on the agenda in regular meetings with contractor?

2. Is contractor participating in regular safety inspections? Are contractor personnel reporting experienced accidents and unwanted occurrences?
3. Is it checked that contractor executes JSA as a part in planning of work?

2.4.5 Close-out, experience transfer
1. Is it checked that the workplace is reinstituted after completion of contract work?
2. Have the experiences with contractor on HSE been adequately documented and implemented in steering documents and experience databases?

2.5 Human resources
2.5.1 Resources, manning, qualifications
1. Have the requirements as to manning levels been adequately evaluated and defined?
2. Have the requirements as to competence and skills in the different positions in the organisation been adequately evaluated and defined?
3. Have the requirements as to manning and competence taken the needs of back-up from colleagues during disturbance-handling and error recovery adequately into account?
4. Is there adequate slack in the organisation to allow for consultation in the handling of new situations and to facilitate learning from experience?
5. Has it been possible to meet the requirements as to manning and competence?
6. Are there adequate extra resources (manning, competence) to fill positions in case of absenteeism?
7. Have the desires of the individual employees been adequately considered in the allocation of personnel?

2.5.2 Education, training
1. Is there a satisfactory education and training plan for the department?
2. Does the plan include an adequate general safety and emergency-preparedness training?
3. Does the plan ensure that the requirements as to skill and competence in the individual positions are met?
4. Have personnel that are required to perform critical recovery and emergency response actions been adequately trained until correct behaviour is automatic?
5. Does the plan ensure an adequate knowledge of the hazards at the workplace?

6. Have all personnel received adequate training according to the plan? Has it been documented?

7. Is it controlled and verified that the training is efficient?

8 Is there adequate practice and new training when required by changes in job content?

9. Does the education and training ensure correct attitudes towards safe work practices?

10. Does the training plan include service personnel, contractors, and temporary hired personnel?

2.5.3 Remuneration

1. Are there adequate remuneration systems that promote safe behaviour?

2. Are there adequate routines to punish violations of safety rules? Are these experienced as fair?

2.6 Fixed assets, site, infrastructure

2.6.1 Design of buildings and infrastructure, perimeter guarding

1. Are buildings adequately located, designed, and built considering their purpose and environmental conditions (climate, geological conditions, etc.)

2. Are roads adequately designed considering collision, road departure, and slope instability hazards?

3. Is infrastructure adequately protected against third-party accident?

4. Is the perimeter adequately fenced in and guarded?

5. Has the security risk level been adequately defined and implemented?

2.6.2 Plant layout, access

1. Have the size and structural integrity of the building been adequately designed in consideration of safety?

2. Are the different areas of the plan (production, maintenance and storage areas, transportation routes, rooms for personnel and administration, etc.) of an adequate size, and are they adequately located and segregated in relation to each other, considering fire hazards, transportation, communication, etc.?

3. Have the transportation routes been adequately designed and marked, considering traffic intensity, the separation of different types of traffic, sight lines, etc.?

4. Have access and escape ways been adequately designed and marked?

5. Is machinery and equipment adequately located? Is the access for operation, inspection, and maintenance adequate?

6. Are there adequate provisions for housekeeping and waste handling?
7. Are there adequate facilities for the storage of combustible and toxic materials?
8. Are there adequate work platforms and guard-rails?

2.6.3 Physical barriers, safety systems
1. Have the different plant areas been adequately classified in relation to the risk of fires and explosions?
2. Have ignition sources been adequately isolated?
3. Are buildings adequately ventilated to avoid build-up of explosive gases?
4. Have fire and explosion partitions been adequately established and equipment adequately insulated against fire?
5. Are buildings adequately designed to withstand the most severe fires and explosions during a specified time period in order to meet the defined risk-acceptance criteria?
6. Are there adequate detection systems for releases of gas, fires, and explosions?
7. Are there adequate systems for emergency shut-down?
8. Are there adequate provisions for firefighting and first aid?
9. Are there adequate communication systems in case of an emergency?

2.6.4 Physical working environment
1. Is the ventilation and temperature regulation adequate?
2. Are there adequate provisions to avoid contamination of the working atmosphere?
3. Are the noise and vibration levels acceptably low?
4. Is lighting adequate?
5. Are alarm and warning signals audible?

2.6.5 Design of machinery/equipment
1. Have the hazards been adequately eliminated in design?
2. Are there adequate barriers (guarding, etc.) to protect against hazards?
3. Are areas for operation, inspection, and maintenance adequately located outside the danger zone?
4. Is there adequate access for operation, inspection, and maintenance?
5. Is there adequate access to production and emergency-stop functions, and for isolation of the machinery from the energy sources?
6. Are work postures and methods of work adequate to avoid the risk of strain injuries?
7. Is the process enclosure of adequate quality to ensure containment of hazardous substances within the process?

2.6.6 Man–machine interfaces
1. Have controls and displays been designed in accordance with recognised ergonomic design standards?
2. Have controls and displays been located for adequate view and access in a normal working posture?
3. Does the operator have access to the necessary process parameters to be able to watch process safety performance adequately?
4. Are there adequate provisions to avoid information overload such as alarm filtration?
5. Have displays and controls been arranged in a way that is consistent with both normal and emergency response tasks?
6. Are there adequate provisions for error recovery? Are the consequences of actions transparent, are they traceable, and are errors reversible?
7. Are there adequate provisions for communication with other operators and supervisors?

2.6.7 Availability of machinery and equipment
1. Is the reliability of the machinery adequate?
2. Are machinery, equipment, and lifting aids available in a sufficient number?
3. Is the access to containers for waste disposal adequate?

2.6.8 Documentation
1. Are the routines for storage, retrieval, and updating of documentation on machinery and equipment (drawings, instructions, certificates, inspection protocols, etc.) adequate?
2. Are the hazards of the machinery during operation and maintenance adequately documented in manuals and work instructions?
3. Is the technical documentation adequate and kept up to date?
4. Is the documentation available to those who need it in the language that they understand?

2.7 Maintenance of fixed assets
2.7.1 Maintenance planning, budgeting, preventive maintenance versus repair
1. Has an adequate plan been established and implemented to ensure safe and reliable operation of processes and machinery? Is it based on risk assessment?
2. Is the budget adequate?

2.7.2 Resources
1. Are there adequate human resources (manning, qualifications) for inspection, maintenance, and scaffolding (see item 2.5)?

2. Is the access to tools and equipment for inspection, maintenance, material handling (lifting), and scaffolding adequate?
3. Are spare parts adequately available?

2.7.3 Methods, procedures, use of risk assessment
1. Has a maintenance philosophy been established that defines the balance between inspection, preventive maintenance, and repair?
2. Is it based on experiences with each type of equipment? On the criticality of the equipment from a safety point of view?
3. Are there adequate instructions and checklists for inspection and maintenance?
4. Is the inspection and testing frequency of safety barriers adequate?
5. Are the routines for inspection and testing of barriers adequate? Do they ensure that the total barrier efficiency is tested?
6. Are there adequate routines for evaluation of maintenance work through use of risk assessment before start of work?
7. Are there adequate routines to prioritise between different maintenance tasks?
8. Is the quality of the maintenance work ensured?

2.7.4 Co-ordination with operations
1. Are there adequate routines for the reporting of maintenance and repair needs?
2. Is the division of responsibility between operation and maintenance clear?
3. Are there adequate routines to co-ordinate operation and maintenance tasks to avoid conflicts? Are there adequate communication links between operation and maintenance personnel?
4. Is it ensured that the necessary safety precautions are taken before start of maintenance? Is it ensured that the equipment is secured and free from hazards? That the area is roped off?

2.7.5 Permit-to-work system
1. Have the needs of a formal work permit system been adequately evaluated, considering the hazards in the process?
2. Do the routines for work permits ensure:
   a. Adequate definition of the types of work for which a permit is required?
   b. Responsibilities for application and approval?
   c. When and where the work permit is valid?
   d. Identification and implementation of the necessary safety barriers before the equipment is released for maintenance?

3. That the work is performed at the right place?

4. That the equipment is brought back to an adequate operational status before start-up?

5. Overview of active work permits by the person responsible for operations?

2.7.6 Modifications, management of change

1. Have routines for all phases of the modification work been established?

2. Have responsibilities for HSE been adequately defined?

3. Have the HSE requirements as to design been adequately defined?

4. Do the routines ensure an adequate and timely definition of the safety scope of the modification work? Does it ensure that the area affected by the modification is brought up to today's HSE standard?

5. Have plans for risk analyses and HSE design reviews been adequately established?

   a. Do they cover all relevant hazards?

   b. Are the activities adequately scheduled in relation to the progress of design work?

   c. Is access to competent personnel including operations personnel ensured?

   d. Are the results adequately documented and followed up?

6. Have routines been established to verify that the modifications meet all relevant requirements?

2.7.7 Documentation

1. Are inspection, repair, and maintenance activities adequately documented? Does an overview of the status of the equipment exist that identifies deviations from an acceptable technical standard?

2. Are they adequately used as a basis in determining equipment status and in the planning of maintenance activities?

3. Are modifications of machinery and equipment adequate?

2.8 Material supply

2.8.1 Composition

1. Have the chemical hazards in connection with the handling of materials (raw materials, materials in the process, and end product) been adequately evaluated and controlled?

2. Is the weight and shape of the materials adequate for safe handling?

3. Is the material adequately wrapped up?

2.8.2 Material transportation, storage
   1. Are the provisions for material handling and storage (buildings, equipment, routines) adequate? Is manual handling avoided when this represents a hazard?
   2. Is the access to materials adequate to avoid production disturbances?

2.8.3 Waste handling
   1. Are the routines and facilities for waste removal and storage adequate?
   2. Is waste disposed of in an adequate way? By certified firm (hazardous materials)?

2.8.4 Documentation
   1. Are all products containing hazardous chemicals documented in approved safety data sheets? Are these available to the workers?
   2. Are all materials and products adequately marked?

2.8.5 Quality control
   1. Is the quality of the materials adequate to avoid production disturbances?
   2. Are there adequate routines to control and document the quality of the materials in relation to the specifications?

Tier 3: Project management
3.1 HSE culture and values
   3.1.1 Project management commitment
      1. Do the project managers demonstrate an adequate concern regarding HSE issues?
      2. Does project management take an active part in relevant HSE activities?
      3. Does project management demonstrate a proactive attitude by addressing potential HSE problem areas and making the necessary decisions in due time?
      4. Does project management take the necessary actions when substandard HSE conditions are brought to their attention?
      5. Is adequate priority given to HSE in relation to other project goals (budget, schedule, etc.)? Also during cost-cutting exercises?

   3.1.2 Communication on HSE
      1. Have adequate routines been established to ensure communication within the project on HSE issues?
      2. Do the routines ensure adequate communication to all project members of the project's HSE policy, goals, and requirements, and the client's expectations?

3.2 HSE management

    3.2.1 Organisation and responsibilities

        1. Have the responsibilities for the management, execution, and verification of work affecting HSE been adequately defined?

        2. Does it include responsibilities for the identification, implementation, and verification of requirements to HSE?

        3. Has the distribution of responsibilities between the project and the client (operations) been adequately defined?

        4. Has the distribution of responsibilities between the project and suppliers (contractors and vendors) been adequately defined?

        5. Have the responsibilities for review and acceptance of changes and nonconformities been adequately defined?

        6. Have the responsibilities been adequately documented in organisational charts and job descriptions?

    3.2.2 Project's HSE program

        1. Is it ensured that the project's HSE program is established at an early phase?

        2. Does the program adequately define:

           a. The project's HSE objectives, goals, and acceptance criteria?

           b. Governing documents relating to HSE?

           c. Responsibilities for the implementation and follow-up of HSE?

           d. Plans for control and verification activities?

        3. Does the program adequately cover:

           a. Management of major accident risks, working environment, and environmental care in design?

           b. Management of HSE risks in fabrication and construction?

        4. Do the routines ensure a timely update of the HSE program in new project phases and when the conditions change?

        5. Do the routines ensure an adequate verification of the HSE program by qualified personnel, project management, and the client?

    3.2.3 Suppliers' HSE programs

        1. Are requirements as to a documented HSE management system adequately defined in contracts and purchase orders?

        2. Do the requirements adequately cover aspects relating to HSE in design and fabrication/construction?

        3. Do the requirements also apply to contractors' subcontractors?

        4. Are the routines adequate to ensure a proper evaluation of the program before start of design and fabrication/construction activities?

3.2.4 Experience transfer
1. Are the routines to ensure experience transfer relating to HSE from earlier projects, operations, and other companies to the project adequate?
2. Does the experience transfer to the project include adequate information about accidents, incidents, and disturbances in similar types of production?
3. Does the experience transfer include information about previous HSE analyses and evaluations of similar plant concepts?
4. Does the experience transfer include adequate information about occupational health problems in similar types of production?
5. Does the experience transfer include adequate information about environmental pollution in similar types of production?
6. Is experience transfer adequately ensured through participation of personnel with operations experience in design work?
7. Does the experience transfer from the project to operations ensure adequate information on requirements to operational instructions and procedures to avoid hazards?
8. Does the experience transfer adequately cover accident experience in earlier projects of a similar type?

3.2.5 Regulations, standards, and specifications
1. Have applicable regulatory requirements been identified, evaluated, and implemented in the project's governing documents?
2. Have adequate and verifiable goals concerning HSE in design been established?
3. Have adequate and verifiable acceptance criteria for the risk of accidents with effects on personnel and environment been established?
4. Have adequate HSE design specifications been established?
5. Do the specifications cover all relevant HSE aspects concerning:
   a. The prevention of major accidents? Are they based on adequate intrinsic safety and defence-in-depth philosophies?
   b. The working environment, including requirements as to the prevention of occupational accidents and strain injuries, industrial hygiene, man–machine interfaces, and psychosocial working environment?
   c. Environmental care including prevention of emissions to the air, discharges to the sea, contamination of land, waste handling, energy conservation, and recycling?
6. Do the specifications adequately cover operation, maintenance, and transportation?

7. Have the client's (operation's) requirements and expectations in relation to HSE been adequately implemented in the specifications?
8. Is it ensured that the HSE requirements are unambiguous? That the order of priority between HSE requirements has been adequately defined?
9. Are all relevant requirements available and made known to the personnel with responsibility for implementation and verification?

3.2.6 Analyses and verifications
1. Have the requirements and responsibilities as to analyses and verifications of HSE in design and construction been adequately specified in a HSE management program and activity plan?
2. Do the activity plans ensure adequate analyses and verifications during concept selection and definition, engineering, procurement, construction, and commissioning in relation to all relevant HSE requirements?
3. Have the work processes for analysis and verification been adequately documented?
4. Do the plans ensure that the analyses and verifications are performed in due time, with the necessary design documentation available, by competent personnel, and with an acceptable quality?
5. Is it ensured that the analyses and verifications adequately cover:
   a. Risk analyses of accident scenarios in the plant in operation with the potential of major losses for comparison with the project's acceptance criteria?
   b. Working environment aspects in the operating plant including analyses of the risk of occupational accidents and strain injuries, man–machine interface evaluations, predictions of air quality, noise, vibration and illumination levels, and evaluation of psychosocial conditions?
   c. Environmental impact assessment including analysis of the risk of accidental pollution and predictions of emissions to the air, discharges to the sea and land contamination, energy analysis, waste-handling evaluations, and product-life-cycle analyses?
   d. Emergency-preparedness analysis?
   e. Analysis of design with respect to risk of accidents during construction (constructability analysis)?
6. Is it ensured that the assumptions and findings from the analyses and verification activities are adequately documented and followed up?

3.2.7 Audits and management reviews

1. Are the plans for audits and management reviews adequate to ensure that the project's and suppliers' HSE management systems satisfy the client and project requirements and expectations and that they are adequately implemented?
2. Are audits and management reviews performed at adequately regular intervals?
3. Are the audits and management reviews adequately comprehensive to cover critical aspects of the HSE management systems?
4. Are the audits and management reviews conducted by personnel with adequate competence?

3.2.8 Nonconformity handling

1. Has a procedure for the handling of nonconformities been established? Does it ensure that nonconformities relating to HSE are adequately identified, recorded, and reviewed?
2. Is it ensured that nonconformities relating to HSE are approved at the right level of the organisation and that adequate compensatory measures are taken?
3. Is it ensured that design not conforming to regulatory requirements is adequately evaluated and approved?

3.3 Project management

3.3.1 Project scope, responsibilities, design basis, procedures

1. Has the project's scope and interfaces been adequately defined?
2. Are the job descriptions for key personnel in the project influencing HSE adequately defined? Is the line responsibility for HSE adequately defined?
3. Has a HSE management position in the project been defined? Is it adequately co-ordinated with QA/QC to ensure proper management and verification of compliance?
4. Has a project design basis been defined that lists the applicable design criteria including regulations, standards, guidelines?
5. Have the procedures governing the execution of the project been adequately defined and documented?

3.3.2 Human resources, qualifications

1. Is it ensured that positions of significance to HSE are identified and staffed with personnel with adequate qualifications?
2. Is it ensured that the project members have adequate knowledge of their own responsibilities for the implementation and follow-up of HSE?

3. Is it ensured that the project members have adequate knowledge of the HSE requirements within their area of responsibility?
4. Is the project organisation adequately manned with personnel with expert knowledge of HSE?

3.3.3 Budget

1. Is it ensured that the control estimate is accurate enough to avoid cost overrun and subsequent cost cuts? Has the HSE scope been adequately implemented in the estimate?
2. Is the budget adequate to ensure the procurement of buildings and machinery of an adequate HSE standard?
3. Are adequate resources allocated to the amelioration of HSE problems identified during project work?
4. Does the man-hour budget ensure adequate access to the necessary HSE expertise?
5. Have adequate working hours been allocated to HSE-related activities?

3.3.4 Schedule

1. Has the schedule been subdivided into phases and milestones with adequate goals concerning HSE activities and documentation?
2. Does the schedule enable the project to generate sufficient information about design and to make the right decisions related to HSE as early as feasible with minimal expenditure of time and money?
3. Is it ensured that critical decisions related to HSE are not postponed until later phases when a proper solution will have major cost or schedule impact?
4. Is the schedule realistic? Does it allow for adequate evaluations of HSE and implementation of results?
5. Is the schedule adequately flexible to absorb possible delays without undue HSE effects?

3.4 Relation to stakeholders

3.4.1 Co-ordination of stakeholders

1. Is it ensured that the project's stakeholders have been adequately identified, including:
   a. The client?
   b. Business parties?
   c. The authorities?
   d. Employees and their organisations?
   e. Contractors and suppliers?

      f. Affected local population?

      g. External interest groups?

  2. Is it ensured that the stakeholders' needs and expectations relating to HSE are adequately identified and evaluated? That potential conflicts are adequately addressed?

3.4.2 Liaison with operations

  1. Is it ensured that the co-ordination between the project and operations is adequate?

  2. Does this co-ordination ensure that the project has adequate access to personnel with adequate operations and maintenance experience of significance to HSE?

  3. Does this co-ordination ensure that operational experience related to HSE is adequately considered:

      a. In the definition of scope?

      b. In choosing technical solutions?

      c. In the handling of changes and nonconformities?

      d. In developing documentation for operation and maintenance?

  4. Is it checked that the permit-to-work system is adequate for the needs during the tie-in of new systems to the existing plant and during commissioning and start-up of new systems?

  5. Is it ensured that the take-over of the plant by operations functions smoothly and with an acceptable level of safety?

3.4.3 Contact with the authorities

  1. Have the responsibilities for contacts with the authorities been adequately defined? Is the responsibility-split between the client and the project clear? Between project and contractors?

  2. Is it ensured that the required document deliveries to the authorities and the authority permits are identified and adequately scheduled and followed up?

  3. Are there adequate routines to ensure timely delivery to the authorities of the required documents? Delivery of documentation of acceptable quality?

  4. Is it ensured that the authorities are kept adequately updated on the progress of the project through status meetings, inspections, document reviews, etc.?

3.5 Study phases

3.5.1 Concept exploration and definition

  1. Is it ensured that the main HSE issues relating to the concept for the plant are adequately identified and addressed?

2. Is it ensured that HSE aspects relating to plant location are adequately evaluated such as:
   a. Security zones?
   b. Distance to residential areas, other plants, and transportation routes?
   c. Transportation of goods?
   d. Access to air and water?
   e. Cultural heritage?
   f. Environmental baseline data and proximity to sensitive environmental resources?
3. Is it ensured that the advantages and disadvantages of the different concept development alternatives are adequately evaluated based on HSE criteria?
4. That HSE characteristics disqualifying any of the alternatives are adequately identified and evaluated?
5. Is it ensured that technological uncertainties relating to HSE are adequately identified and evaluated? That back-up solutions are provided if the new technologies are found unfeasible?

3.5.2 Demonstration and validation
   1. Is it ensured that the evaluation of the economic and technical feasibility of the selected plant concept takes HSE acceptance criteria and requirements adequately into account?
   2. Is it ensured that basic decisions regarding safety systems are addressed, such as:
      a. Segregation into fire cells?
      b. Plant shut-down philosophy?
      c. Facilities for emergency response such as firefighting?
      d. Escape ways and evacuation?
   3. Is it ensured that basic decisions regarding the working environment are addressed, such as:
      a. Layout promoting safe and efficient operation, maintenance, and transportation? Separation of hazardous areas from non-hazardous areas?
      b. An acceptable working atmosphere (climate, pollution)?
      c. Storage and handling of hazardous substances?
      d. Location of major noise and vibration sources?
      e. Solitary work?
   4. Is it ensured that basic decisions regarding environmental care are addressed, such as:
      a. Prevention of accidental pollution of the environment?
      b. Control of emissions to the air, discharges to water recipients, and contamination of land?

       c. Waste disposal?

       d. Sampling and monitoring of effluent streams and exhaust gases?

    5. Is it ensured that the plant design has been defined and documented in adequate detail concerning HSE issues before making decisions to go ahead with realisation?

3.5.3 Definition of scope, change control

    1. Is it ensured that the scope of the project work is adequately defined and documented before deciding on realisation and entering into agreements with external parties (contractors, vendors)? Has the design and planning documentation been subject to a maturity evaluation?

    2. Is it ensured that the HSE-related parts of the scope are adequately defined? Does the scope include an upgrading of existing facilities to today's HSE standard?

    3. Has a procedure been established that defines the routines and responsibilities for the handling of changes in scope?

    4. Does this ensure an identification and evaluation of changes that are critical with respect to HSE? Is the authority to approve changes with safety impact placed at the right level of the organisation (project/client)?

3.6 Project execution

  3.6.1 Engineering

    1. Are the plans for identification, documentation, and review of design input (statutory requirements, client requirements, etc.) adequate?

    2. Is it ensured that the design concept has been adequately defined and qualified from a HSE point of view before placement of major contracts and purchase orders?

    3. Are the routines for implementation of HSE requirements in contracts and purchase orders adequate?

    4. Are the plans for documentation of design output for verification and validation adequate?

    5. Are the plans for design reviews adequate? Do they ensure reviews at regular intervals and with the participation of competent personnel?

    6. Are the routines for control of design changes adequate to ensure proper identification, review, and approval by authorised personnel?

    7. Are the routines for documentation and follow-up of actions adequate?

8. Are the routines for document control adequate? Do they ensure that the documents are traceable and that they are complete, updated, and accessible?
9. Are there adequate routines for documentation of design changes, nonconformities, and ameliorative actions?

3.6.2 Procurement

1. Is it ensured that the suppliers (contractors and vendors) are evaluated and selected in pre-qualifications and bid evaluations on the basis of their ability to meet the project's HSE goals, requirements, and acceptance criteria related to:
   a. HSE in design?
   b. HSE in fabrication and construction?
2. Is it ensured that the evaluation includes the tenders':
   a. Organisational resources and competence?
   b. Technical resources?
   c. Planned activities to meet the requirements?
   d. Documented procedures for control and verification?
   e. Experience data on the tenders' previous performance?
3. Is it ensured that the technical evaluation of the proposed design includes a qualification with respect to the project's HSE goals, acceptance criteria, and requirements?
4. Is it ensured that the tenders provide the necessary documentation (risk analyses, verifications) of the feasibility of the proposed design?
5. Is it ensured that the contract adequately defines:
   a. The HSE goals, acceptance criteria, and requirements?
   b. The scope of work related to HSE including activities to analyse, verify, and document compliance with the contractual requirements to HSE?
   c. The required delivery of documents including:
      i. Suppliers' and sub-suppliers' HSE programs for the project?
      ii. Reports on analyses and verifications?
      iii. Reports on HSE activities and results in fabrication and construction, including accident reports?
   d. The responsibilities for HSE including the split of responsibility between subcontractors and vendors?
6. Has the extent of control exercised over the suppliers been determined on the basis of the results of pre-qualifications and bid evaluations?
7. Is it ensured that the suppliers adequately understand the scope of work and HSE requirements and work according to these?

8. Is it ensured that the product's compliance with the contractual HSE requirements is adequately verified before delivery? Is the verification documented?

3.6.3 Follow-up of HSE in fabrication and construction
1. Is it ensured that the suppliers have established adequate HSE programs for fabrication and construction before start-up of work?
2. Are suppliers adequately followed up through monitoring of HSE performance?
3. Are the routines for reviews and audits of suppliers' HSE activities adequate?
4. Are HSE issues in fabrication and construction adequately brought up in management meetings with suppliers?
5. Are the routines adequate to implement corrective measures when suppliers' performance is substandard or deteriorates?

3.6.4 Mechanical completion and commissioning
1. Is it ensured that equipment and plant areas are adequately inspected to verify compliance with the contractual requirements to HSE, design specifications, drawings, and other engineering documents?
2. Is it ensured that the dynamic testing adequately verifies compliance with the contractual requirements as to HSE, design specifications, drawings, and other engineering documents?
3. Is it ensured that the findings from inspections and testing are adequately evaluated and followed up?
4. Is it ensured that the inspection and testing results during mechanical completion and commissioning and follow-up of nonconformities and deficiencies are adequately documented?

3.7 Take-over by operations
3.7.1 Recruitment and training of operations personnel
1. Have the requirements as to knowledge and skills been adequately defined for the different positions in the operations organisation?
2. Is the recruitment of personnel adequate, and does it take place in due time to allow for satisfactory education and training?
3. Does the recruitment include experienced personnel?
4. Are personnel recruited in due time to participate in commissioning and start-up?
5. Has the training program adequate contents and schedule to ensure that the personnel meet the requirements as to knowledge and skills?

3.7.2 Development of procedures and work instructions
    1. Are the routines to ensure timely development of work and safety instruction of acceptable quality adequate?
    2. Do the instructions adequately implement suppliers' instructions, experience from evaluations in design and commissioning, and experience from the operation of similar facilities?
    3. Are the routines to ensure evaluations of the instructions from a HSE and functionality point of view adequate?

3.7.3 Start-up
    1. Are the HSE criteria for start-up adequate? Do they adequately cover:
        a. Requirements as to the completeness of technical systems?
        b. The availability of safety systems?
        c. Requirements as to the availability and qualifications of personnel?
        d. Requirements as to the availability and quality of procedures?
    2. Is it ensured that the status of the plant is adequately evaluated and documented before the introduction of hazardous substances into the process?
    3. Are any nonconformities and changes adequately evaluated and accepted by persons with satisfactory competence and authority before start-up?
    4. Is it adequately ensured that the drawings and other design documentation with as-built status are adequately updated, verified, and approved before hand-over to operations?

Tier 4: Higher management and management of HSE
4.1 HSE culture and values
    4.1.1 Leadership and commitment
        1. Does company management promote HSE by demonstrating adequate leadership and commitment?
        2. Does management foster a culture of continuous improvements in the area of HSE?
        3. Is adequate priority given to HSE in relation to other company goals?
        4. Does management show adequate acceptance of its own responsibility and accountability for HSE activities, performance, and results?
        5. Does management take the necessary actions when substandard conditions and work practices are brought to its attention?
        6. Does management demonstrate a concern by actively querying possible new hazards?

4.1.2 Compliance culture
1. Has adequate company policy been established to ensure compliance at all levels with the HSE policy, procedures, and rules?
2. Has the policy been adequately implemented and followed up?
3. Does management adequately promote safe behaviour through its own vigour and example?

4.1.3 Co-ordination and communication
1. Have adequate routines been established to ensure communication within the company on HSE issues?
2. Do the routines ensure an adequate communication of company HSE policy, goals, and rules, as well as hazards to the employees?
3. Do the routines ensure adequate communication on the employees' HSE experiences and concerns?

4.1.4 Employee involvement
1. Have adequate routines been established to promote employee involvement in HSE activities at all levels of the company and to address employee expectations?
2. Does management promote employee involvement at all levels in HSE-related activities?
3. Do the routines ensure adequate use of employees' experience and their ownership of the results of the activities?

4.1.5 Stakeholder management
1. Have interested parties in addition to the employees (e.g. third party within the influence area of the company activities) been identified and their needs and expectation concerning HSE been identified as relevant?
2. Has a program been established to manage those needs and expectations?

4.2 HSE policy, goals, and action plans
4.2.1 Policy
1. Does a documented company HSE policy exist? Is it consistent with the company's other policies?
2. Does it include a management commitment to meet all relevant statutory requirements?
3. Does it commit management to continuous efforts to improve HSE performance and to reduce risks as low as reasonably practicable?

4.2.2 Goals and acceptance criteria
1. Has the company's HSE policy been translated into verifiable HSE goals and acceptance criteria for the risk of losses due to accidents?

2. Do the goals and acceptance criteria cover all relevant aspects of HSE?
3. Is it ensured that goals are established for all departments of the company?
4. Is it ensured that the line management and employees are involved in the development of goals and acceptance criteria?
5. Do the goals ensure continuous improvements in HSE performance?
6. Are the goals regularly reviewed and updated in line with new legislation and other changes?

4.2.3 Action plans
1. Are there adequate routines to ensure the development of action plans at all levels of the company to implement the HSE policy and goals?
2. Are adequate resources allocated to the execution of the plans?
3. Is progress adequately monitored and followed up?
4. Has coordination and allocation of responsibilities for HSE activities involving several parts of the organisation been adequately defined?

4.3 Resource management
4.3.1 Responsibilities
1. Have the responsibilities for HSE been clearly defined at all levels of the company?
2. Does adequate authority and accountability accompany these responsibilities?
3. Is the division of responsibilities between line management, HSE staff officers, and safety representatives of the workers on HSE made clear? Is it clearly stated that HSE is a line management responsibility?
4. Is it ensured that the line management knows about and accepts its responsibilities?

4.3.2 Human resources, qualifications
1. Have adequate resources been allocated to human resource management including the education and training of personnel?
2. Have personnel with adequate HSE competence and management skills been allocated to the HSE management and specialist positions?
3. Does management have access to adequate HSE expertise for advice and support in the handling of hazards connected with production?

4. Have sufficient person-hours been allocated to HSE activities?
5. Have sufficient resources been allocated to the follow-up and implementation of results of HSE activities?
6. Do the resources ensure adequate expert support in the identification and evaluation of accident risks, HSE education and training, advice on HSE regulations and on solutions to HSE problems, contacts with the authorities, etc.?
7. Have adequate resources been allocated to support the line management?
8. Have adequate resources been established for the development and maintenance of procedures and work instructions?
9. Do they ensure that procedures and instructions are complete and kept updated?
10. Are measures to avoid accidents adequately identified and defined?
11. Do the instructions represent best practice?
12. Do the employees know about, accept, and comply with the instructions?

4.3.3 Business planning and budgeting, risk management
1. Is risk assessment and management part of the regular (yearly) business planning and budgeting?
2. Is HSE integrated in the business planning and budgeting process? Is it ensured that the critical HSE risks are visualised and addressed?

4.3.4 Fixed assets and materials
1. Have adequate routines been established in the area of fixed asset management? Do they ensure that building, machinery, and equipment have an adequate HSE standard?
2. Are the routines adequate for the procurement of materials, equipment, and contract work?
   a. Do the routines ensure that suppliers with substandard HSE performance are avoided?
   b. Do the routines ensure that HSE requirements to the supply are defined in the contract and that the supply meets the requirements?

4.3.5 HSE organisation
1. Have safety representatives of the workers been elected in all departments?
2. Have the safety representatives been given adequate duties and status?
3. Have working environment committees been established?

4. Do the working environment committees function adequately in accordance with the intentions and requirements? Does management take an active part in the work?

5. Does the HSE organisation ensure adequate employee involvement and influence on decisions affecting HSE?

4.4 Identification and evaluation of risks

4.4.1 Routines

1. Are the routines for systematic identification and evaluation of risks to personnel, environment, and assets adequate?

2. Do they cover all phases of the life cycle of industrial systems?

3. Do the routines ensure an adequate distribution, follow-up, and implementation of results?

4. Do the routines ensure the implementation of corrective actions to prevent recurrence?

5. Do the routines ensure adequate experience transfer between departments and to projects for new plants and modifications?

6. Do the routines ensure involvement by the personnel concerned?

7. Are the routines adequately adapted to the different needs within the company?

8. Are the routines kept updated in line with changes in production, organisation, legislation, etc.?

9. Are the routines adequately documented?

4.4.2 Data collection

1. Do the routines ensure a systematic identification of all different types of hazard?

2. Do the routines ensure prompt detection of new hazards?

3. Do the routines ensure adequate reporting and investigation into accidents and near-accidents?

4. Are incidents relating to risks of personal injury, fires and explosions, and environmental releases reported?

5. Are the investigation resources adequately prioritised in relation to the degree of severity (actual/potential)? Have routines been established for investigations at different levels (e.g. supervisor's first investigation, problem-solving groups, and independent investigation commission)?

6. Do the routines ensure regular workplace inspections of a satisfactory quality?

7. Do the routines ensure an adequate identification of new hazards by means of risk analysis?

8. Are the routines for monitoring of the physical working environment adequate?

    9. Are the routines for health checks of personnel adequate?

   10. Is the quality of the data adequately checked?

4.4.3 Storage, analysis, distribution, and use of information

1. Are the results analysed and summarised in an adequate way for decision-making?

2. Are the incidents evaluated concerning potential for severe loss as a basis for prioritising?

3. Are the results compared with established goals and acceptance criteria and are gaps used adequately as a basis for decisions?

4. Is the information stored in an adequate way for information retrieval and experience transfer?

5. Is it possible for decision-makers to access the information when it is needed?

6. Are the routines for a periodic summary and presentation of the information adequate?

7. Is relevant information adequately distributed to the line management, HSE specialists, working environment committees, safety representatives?

8. Is it ensured that management and the HSE organisation gets the necessary information to evaluate the hazards at the workplaces and to take necessary preventive measures?

9. Is it possible to get an overview of all identified hazards and nonconformities that have not been resolved and the status of actions?

4.5 Handling of governing HSE documents

4.5.1 Routines

1. Are the routines for the handling of governing documents in the area of HSE adequate?

2. Do the routines ensure that the documents are complete and available to the personnel needing them?

3. Is it ensured that the documentation is kept updated in line with changes in legislation, organisation, or production?

4.5.2 Regulations, codes, and standards

1. Are all relevant regulatory requirements, codes, and standards available at the company?

2. Are changes adequately identified and implemented?

4.5.3 HSE management program

1. Have the different elements of the HSE management program been adequately documented?

2. Does the documentation cover established policy, goals and acceptance criteria, responsibilities, and activities?

3. Does the documentation cover operation and project work?
4. Is the program regularly reviewed and updated?

4.5.4 Procedures
1. Has the need for the documentation of HSE activities in procedures been adequately defined?
2. Do the procedures adequately cover daily operation, modifications, and building of new plants?
3. Do the procedures adequately cover the procurement of new materials and equipment and contract work?
4. Is it ensured that the procedures reflect best practice?
5. Is it ensured that the procedures are stated simply, unambiguously, and understandably?

4.6 Performance monitoring and auditing
4.6.1 Performance indicators
1. Have adequate performance indicators been established?
2. Do the indicators adequately cover the different areas of HSE?
3. Do the indicators adequately cover operation and project work?
4. Are the indicators adequately congruent with the HSE policy and goals?
5. Are the indicators adequately accepted within the organisation as fair measures of performance?

4.6.2 Follow-up of results
1. Are the performance indicators used adequately in performance monitoring and follow-up of HSE results?
2. Are adequate actions taken when performance indicators show substandard development?

4.6.3 HSE auditing and management reviews
1. Has an adequate plan been established for HSE audits and management reviews?
2. Has the plan been adequately implemented?
3. Does the plan ensure a systematic and independent examination of the HSE management system and activities?

Bibliography

AAAM. 1985. *Abbreviated injury scale.* American Association for Automotive Medicine, Committee on Injury Scale, Arlington Heights, IL.

Aasjord, H.L., Holmen, I.M. and Thorvaldsen, T. 2013. *Analyse av årsaksforhold ved dødsulykker og alvorlige personskader i norsk fiskeri* [Analysis of causes of fatal accident and severe occupational accidents in Norwegian fishery]. SINTEF report A23369, SINTEF, Trondheim. In Norwegian.

ACSNI. 1993. *Organising for safety.* Advisory Committee on the Safety of Nuclear Installations, HSE Books, Suffolk, UK.

Adams, N., Barlow, A. and Hiddlestone, J. 1981. Obtaining ergonomics information about industrial injuries – A five-year analysis. *Applied Ergonomics* 12:71–81.

Alteren, B. 1999. Implementation and evaluation of the safety element method at four mining sites. *Safety Science* 31:231–264.

ANSI. 2012. *American National Standard for Occupational Health and Safety Management Systems.* Standard No. ANSI Z10-2012, American National Standard Institute, Washington, DC.

Antonsen, S. 2009. Safety culture and the issue of power. *Safety Science* 47:183–191.

Antonsen, S., Almklov, P. and Fenstad, J. 2008. Reducing the gap between procedures and practice – Lessons from a successful safety intervention. *Safety Science Monitor* 12:1–16.

API. 2010. *Process safety performance indicators for refining and petrochemical industries.* ANSI/API Recommended Practice 754. API, Washington, DC.

Arbeidstilsynet. 1996. *Regulations relating to systematic health, environmental and safety activities in enterprises.* Regulations No. 1127. Oslo.

Arbeidstilsynet. 2015. *Arbeidsskadedødsfall i Norge – Utviklingstrekk 2009–2014 og analyse av årsakssammenhenger i fire næringer.* [Fatal work accidents in Norway 2009–2014, trends and causes in four sectors]. Report Kompass Tema nr. 3. Trondheim. In Norwegian.

Arbetarskyddsstyrelsen. 1996. *Internkontroll av arbetsmiljön. Arbetarskyddsstyrelsen författningssamling.* [Internal control of the working environment] Regulations No. AFS1996: 06. Stockholm. In Swedish.

Argyris, C. 1992. *On organizational learning.* Blackwell, Cambridge, MA.

Arocena, P. and Nunez, I. 2010. An empirical analysis of the effectiveness of occupational health and safety management systems in SMEs. *International Small Business Journal* 4:398–419.

ATSB. 2007. *Analysis, causality and proof in safety investigations*. Report AR-2007-053. Australian Transport Safety Bureau, Canberra.

Aven, T. and Renn. O. 2009. On risk defined as an event where the outcome is uncertain. *Journal of Risk Research* 12:1–11.

Aven, T. and Vinnem, J.E. 2007. *Risk management: With applications from the offshore petroleum industry*. Springer Science & Business Media, London.

Azaroff, L.S., Levenstein, C. and Wegman, D.H. 2002. Occupational injury and illness surveillance: Conceptual filters explaining underreporting. *American Journal of Public Health* 92:1421–1429.

Bahn, S. 2013. Workplace hazard identification and management: A case of an underground mining company. *Safety Science* 57:129–137.

Barbaranelli, C., Petitta, L. and Probst, T.M. 2015. Does safety climate predict safety performance in Italy and the USA? Cross-cultural validation of a theoretical model of safety climate. *Accident Analysis and Prevention* 77:35–44.

Benner, L. 1975. Accident investigations. Multilinear events sequencing methods. *Journal of Safety Research* 7:67–73.

Bento, J.P. 1999. *MTO-analys av händelsesrapporter. [MTO analysis of reports on incident occurrences]*. Report No. OD-00-2. Norwegian Petroleum Directorate, Stavanger. In Swedish.

Bird, F.E. and Germain, G.L. 1985. *Practical loss control leadership*. Institute Publishing, Division of International Loss Control Institute, Loganville, GA.

Blewett, V. and O'Keeffe, V. 2011. Weighting the pig never made it heavier: Auditing OHS, social auditing as verification of process in Australia. *Safety Science* 49:1014–1021.

Blindheim, G. and Lindtvedt, J. 1996. [Experience feedback in accident prevention by the incident database Synergi]. Project report, Norwegian University of Science and Technology, Trondheim. In Norwegian.

Boe, K. 1996. *Evaluering av HMS-styring i byggeprosjekter*. [Evaluation of HSE management in construction projects]. Master's thesis, Norwegian University of Science and Technology, Trondheim. In Norwegian

Bolman, L.G. and Deal, T.E. 1984. *Modern approaches to understanding and managing organizations*. Jossey-Bass, San Francisco, CA.

Booth, M.J. 1991. *The incident potential matrix*. Paper presented at the Society of Petroleum Engineers' First International Conference on Health, Safety and Environment, The Hague, 10–14th November.

Bråten, M., Ødegård, A.M. and Andersen, R.K. 2012. Samarbeid og HMS-utfordringer i bygg- og anleggsnæringen. [Collaboration and HSE-challenges in the construction industry]. FAFO-report 2012:52. Oslo. In Norwegian.

Briscoe, G.J. 1982. *Risk management guide*. Report No. SSDC-11. Systems Safety Development Center, EG&G Idaho, Idaho Falls, ID.

Briscoe, G.J. 1991. *MORT-based risk management*. Working paper No. 28. Systems Safety Development Center, EG&G Idaho, Idaho Falls, ID.

British Standards Institution. 2007. *Occupational health and safety management systems. Requirements*. BS OHSAS 18001:2007

Brown, R.L. and Holmes, H. 1986. The use of a factor-analysis procedure for assessing the validity of an employee safety climate model. *Accident Analysis and Prevention* 18:455–470.

Bureau of Labor Statistics. 2016. *Fatal occupational injuries, charts and table*. Washington, DC. http://www.bls.gov/iif/oshwc/cfoi/cfch0013.pdf. (Accessed 29 May, 2016).

Cambon, J. and Guarnieri, F. 2008. *Maitriser les defaillances des organisations en santeet securite au travail: La methode Tripod. Notes de synthese et de recherche.* Paris, France. In French.

Carlucci, D. 2010. Evaluating and selecting key performance indicators: An ANP-based model. *Measuring Business Excellence* 14:66–76.

Carter, N. and Menckel, E. 1985. Near-accident reporting: A review of Swedish research. *Journal of Occupational Accidents* 7:61–64.

Champoux, D. and Brun, J.-P. 2003. Occupational health and safety management in small size enterprises: And overview of the situation and avenues for intervention and research. *Safety Science* 41:301–318.

Chaplin, R. and Hale, A. 1998. An evaluation of the use of the International Safety Rating System (ISRS) as intervention to improve the organisation of safety. In *Safety management – The challenge of change*, eds. A.R. Hale and M. Baram, 165–185. Pergamon, London.

Christian, M.S., Bradley, J.C., Wallace, J.C. and Burke, M.J. 2009. Workplace safety: A meta-analysis of the roles of person and situational factors. *Journal of Applied Psychology* 94:1103–1127.

Clark, D.E., Winchell, R.J. and Betensky, R.A. 2013. Estimating the effect of emergency care on early survival after traffic crashes. *Accident Analysis and Prevention* 60:141–147.

Clarke, S. 2006. The relationship between safety climate and safety performance: A meta-analytic review. *Journal of Occupational Health Psychology* 11:315–327.

Cohen, A., Smith, M. and Cohen, H. 1975. *Safety program practices in high vs. low accident rate companies – An interim report.* National Institute for Occupational Safety and Health, Cincinnati, OH.

Connell, L.J. 2004. *Cross-industry applications of a confidential reporting model.* NASA ASRS, Publication 62, Moffett Field, CA.

Cooper, M.D. 2000. Towards a model of safety culture. *Safety Science* 36:111–136.

Cornelison, J.D. 1989. *MORT based root cause analysis.* Working paper No. 27, Systems Safety Development Center, EG&G Idaho, Idaho Falls, ID.

Cox, S. and Flin, R. 1998. Safety culture: Philosopher's stone or man of straw? *Work & Stress* 12:189–201.

Crutchfield, N. and Roughton, J. 2014. Developing the job hazard analysis. In *Safety culture: An innovative leadership approach*, eds. N. Crutchfield and J. Roughton, 235–248. Butterworth-Heinemann, Oxford.

Darby, P., Murray, W. and Raeside, R. 2009. Applying online fleet driver assessment to help identify, target and reduce occupational road safety risks. *Safety Science* 47:436–442.

Daugherty, J.E. 1999. *Industrial safety management.* Government Institutes, Rockville, MD.

Dedobbeleer, N. and Béland, F. 1991. A safety climate measure for construction sites. *Journal of Safety Research* 22:97–103.

Deepwater Horizon Study Group. 2011. *Final report of the investigation of the Macondo well blowout.* UC Berkeley, Center for Catastrophic Risk Management, Berkeley, CA.

DeJoy, D.M. 1994. Managing safety in the workplace: An attribution theory analysis and model. *Journal of Safety Research* 25:3–17.

Dekker, S. 2014. *The field guide to understanding 'Human Error'.* Ashgate, Farnham, UK.

Dekker, S. and Breakey, H. 2016. 'Just culture': Improving safety by achieving substantive, procedural and restorative justice. *Safety Science* 85:187–193.

Deming, W.E. 1993. *The new economics for industry, government and education.* MIT Press, Boston, MA.

Deming, W.E. 2000. *Out of the crisis.* MIT Press, Cambridge, MA.

Dien, Y., Dechy, N. and Guillaume, E. 2012. Accident investigation: From searching direct causes to finding in-depth causes – Problem of analysis and/or analyst? *Safety Science* 50:1398–1407.

Dien, Y., Llory, M. and Montmayeul, R. 2004. Organisational accidents investigation methodology and lessons learned. *Journal of Hazardous Materials* 111:147–153.

DNV GL. 2014. *ISRS For the health of your business.* DNV GL, Katy, TX.

DOE. 2000. *DOE workbook: Conducting accident investigations.* U.S. Department of Energy, Washington, DC.

Döös, M., Backström, T. and Sundström-Frisk, C. 2004. Human actions and errors in risk handling an empirically grounded discussion of cognitive action-regulation levels. *Safety Science* 42:185–204.

Doran, G., Miller, A. and Cunningham, J. 1981. There's a S.M.A.R.T. way to write management's goals and objectives. *Management Review* 70:35–36.

Drupsteen, L. and Guldenmund, F.W. 2014. What is learning? A review of the safety literature to define learning from incidents, accidents and disasters. *Journal of Contingencies and Crisis Management* 22:81–96.

Dzeng, R.J., Lin, C.T. and Fang, Y.C. 2016. Using eye-tracker to compare search patters between experienced and novice workers for site hazard identification. *Safety Science* 82:56–67.

Dyreborg, J., Lipscomb, H.J., Olsen, O., Törner, M., Nielsen, K., Lund, J., Kines, P., et al. 2015. *Safety interventions for the prevention of accidents at work.* Protocol ID no. SW2010-05. The Campbell Collaboration, Oslo.

Dyreborg, J., Nielsen, K., Kines, P., Dziekanska, A., Frydendall, K.B., Bengtsen, E. and Rasmussen, K. 2013. *Review af ulykkesforebyggelsen – review af den eksisterende videnskabelige litteratur om effekten af forskellige typer tiltag til forebyggelse af arbejdsulykker.* [Review of scientific literature on the effect of various safety interventions on the prevention of occupational accidents] Report, Det Nationale Forskningscenter for Arbejdsmiljø, Copenhagen. In Danish.

Eisner, H.S. and Leger, J.P. 1988. The international safety rating system in South African mining. *Journal of Occupational Accidents* 10:141–160.

Elmasri, R. and Navathe, S.B. 2006. *Fundamentals of database systems.* (5th ed.). Addison Wesley, Boston, MA.

Energy Institute. 2008a. *Guidance on investigating and analyzing human and organizational factors aspects of incidents and accidents.* Energy Institute, London.

Energy Institute. 2008b. *Managing rule-breaking.* Publication PO3141. Energy Institute, London.

Energy Institute. 2008c. *HSE – Understanding your culture.* Publication PO3069. Energy Institute, London.

Englund, A., Gregersen, N.P., Hydén, C., Lövsund, P. and Åberg, L. 2006. *Trafiksäkerhet – En kunskapsöversikt.* [Road transportation safety – An overview of current knowledge] Studentlitteratur, Lund. In Swedish.

EPA. 1990. *The Clean Air Act Amendments of 1990.* U.S. Environmental Protection Agency, Washington, DC.

European Council. 1989. *The introduction of measures to encourage improvements in the safety and health of workers at work.* Council Directive 89/391/EEC. European Council, Brussels.

European Council. 1998. *The protection of the health and safety of workers from the risks related to chemical agents at work.* Council Directive 98/24/EC. European Council, Brussels.

European Council. 2001. *Minimum safety and health requirements for the use of work equipment by workers at work.* Directive 2001/45/EC. European Council, Brussels.

European Council. 2006. *Machinery.* Council Directives 2006/42/EC. European Council, Brussels.

European Council. 2012. *Directive on the control of major-accident hazards involving dangerous substances.* Directive 2012/18/EU. European Council, Brussels.

Eurostat. 2013. *European statistics on accidents at work – Summary methodology.* European Union, Luxembourg.

Evanco, W.M. 1999. The potential impact of rural mayday systems on vehicular crash fatalities. *Accident Analysis and Prevention* 31:455–462.

Feng, Y., Zhang, S. and Wu, P. 2015. Factors influencing workplace accident costs of building projects. *Safety Science* 72:97–104.

Fernández-Muñiz, B., Montes-Peón, J.M. and Vázquez-Ordás, C.J. 2009. Relation between occupational safety management and firm performance. *Safety Science* 47:980–991.

Flin, R., Mearns, K., O'Connor, P. and Bryden, R. 2000. Measuring safety climate: Identifying the common features. *Safety Science* 34:177–192.

Fort, E., Pourcel, L., Davezies, P., Renaux, C., Chiron, M. and Charbotel, B. 2010. Road accidents, an occupational risk. *Safety Science* 48:1412–1420.

Gallagher, C., Underhill, E. and Rimmer, M. 2003. Occupational health and safety management systems in Australia: Barriers to success. *Policy and Practice in Health and Safety* 1:68–81.

Geurts, K. and Wets, G. 2003. *Black spot analysis methods: Literature review.* Steunpunt Verkeersveiligheid Report no. RA-2003-07. Steunpunt Verkeersveiligheid, Diepenbeek, Belgium

Ghahramani, A. and Khalkhali, H.R. 2015. Development and validation of a safety climate scale for manufacturing industry. *Safety and Health at Work* 6:97–103.

Gibson, J. 1961. The contribution of experimental psychology to the formulation of the problem of safety. In *Behavioral approaches to accident research*, ed. H.H. Jacobs, Association for the Aid of Crippled Children, New York.

Glenn, D.D. 2011. Job safety analysis: Its role today. *Professional Safety* 56:48–57.

Grimaldi, J.V. 1970. The measurement of safety engineering performance. *Journal of Safety Research* 2:137–159.

Grimaldi, J.V. and Simonds, R.H. 1975. *Safety management.* Richard D. Irwing, Homewood, IL.

Groeneweg, J. 1998. *Controlling the controllable. The management of safety.* (4th ed.). DSWO Press, Leiden University, Leiden.

Groth, K.M. and Mosleh, A. 2012. A data-informed PIF hierarchy for model-based human reliability analysis. *Reliability Engineering and Systems Safety* 108:154–174.

Guastello, S.J. 1993. Do we really know how well our occupational accident prevention programs work? *Safety Science* 16:445–464.

Haddon, W. 1968. The changing approach to epidemiology, prevention and amelioration of trauma. *American Journal of Public Health* 58:1431–1438.

Haddon, W. 1980. The basic strategies for reducing damage from hazards of all kinds. *Hazard Prevention* 16:8–12.

Hale, A.R. 2000. Editorial: Cultures confusions. *Safety Science* 34:1–14.

Hale, A.R. 2003. *Management of industrial safety.* Technical memo, TU Delft, the Netherlands.

Hale, A.R. and Borys, D. 2013. Working to rule, or working safely? Part 1: A state of the art review. *Safety Science* 55:207–221.

Hale, A.R. and Glendon, A.I. 1987. *Individual behaviour in the control of danger.* Elsevier, Amsterdam.

Hale, A.R. and Hovden, J. 1998. Management and culture: The third age of safety. A review of approaches to organizational aspects of safety, health and environment. In *Occupational injury: Risk, prevention and intervention,* eds. A.M. Feyer and A. Williamson. 129–165. CRC Press, London.

Hale, A.R., Wilpert, B. and Freitag, M. (eds.). 1997. *After the event – From accident to organisational learning.* Elsevier, Oxford.

Hansen, O.R., Wiik, A. and Wilkins, B. 2002. Suppression of secondary explosions in transformer rooms. *Journal de Physique IV France* 12:385–392.

Harms-Ringdahl, L. 2013. *Guide to safety analysis for accident prevention.* IRS Riskhandtering, Stockholm.

Harrison, F.E. 1999. *The managerial decision-making process.* (5th ed.). Houghton Mifflin, Boston, MA.

Haselton, M.G., Nettle, D. and Andrews, P.W. 2005. The evolution of cognitive bias. In *Handbook of evolutionary psychology,* eds. D.M. Buss. Wiley, Hoboken, NJ.

Haslam, R., Hide, S., Gibb, A., Gyi, D., Pavitt, T., Atkinson, S. and Duff, A. 2005. Contributing factors in construction accidents. *Applied Ergonomics* 36:401–451.

Haslam, R.A., Hide, S.A., Gibb, A.G.F., Gyi, D.E., Atkinson, S., Pavitt, T.C., Duff, R. and Suraji, A. 2003. *Causal factors in construction accidents,* HSE Report RR156, HMSO, Norwich.

Heinrich, H.W. 1959. *Industrial accident prevention – A scientific approach.* (4th ed.) McGraw-Hill, New York.

Hendrick, K. and Benner, L. 1987. *Investigating accidents with STEP.* Marcel Dekker, New York.

Hollnagel, E. 2004. *Barriers and accident prevention.* Ashgate, Aldershot, UK.

Hollnagel, E. 2011. Prologue: The scope of resilience engineering. In *Resilience engineering in practice. A guidebook,* eds. E. Hollnagel, J. Pariés, D.D. Woods, and J. Wreathall. 275–294. Ashgate, Aldershot, UK.

Hollnagel, E. 2014. *Safety-I and Safety-II: The past and future of safety management.* Ashgate, Aldershot, UK.

Hollnagel, E. and Speziali, J. 2008. *Study on developments in accident investigation methods: A survey of the "State-of-the-Art".* SKI Report 2008:50, Statens Kärnkraftinspektion, Stockholm.

Hollnagel, E., Woods, D.D. and Leveson, N.C. (eds.). 2007. *Resilience engineering: Concepts and precepts.* Ashgate, Aldershot, UK.

Hopkins, A. 2009. Thinking about process safety indicators. *Safety Science* 47:460–465.

Hovden, J. and Larsson, T.J. 1987. Risk: Culture and concepts. In: *Risk and decisions,* eds. W.T Singleton and J. Hovden. 47–66. Wiley, Aldershot, UK.

Høye, A. 2014. *Fatality risk in transportation for two alternative routes to Dugar HEP.* TØI Report 1375/2014. Institute of Transport Economics, Oslo.

Høye, A., Elvik, R., Sørensen, M. and Vaa, T. 2012. *Transportsikkerhetsboken.* [Transport safety handbook]. Institute of Transport Economics, Oslo. In Norwegian.

HSE. 2001a. *A guide to measure health and safety performance.* Health and Safety Executive, Sheffield, UK.

HSE. 2001b. *Reducing risks, protecting people. HSE's decision-making process.* Her Majesty's Stationery Office, Norwich, UK.

HSE. 2003. *Factoring the human into safety: Translating research into practice. The development and evaluation of a human factors accident and near miss reporting form for the offshore oil industry.* Health and Safety Executive, Research Report 60, Her Majesty's Stationery Office, Norwich, UK.

HSE. 2005. *Guidance on permit-to-work systems.* Health and Safety Executive, Sheffield.

HSE. 2006. *Developing process safety indicators.* Health and Safety Executive, Sheffield.

HSE. 2007. *An investigation of reporting of workplace accidents under RIDDOR using the Merseyside Accident Information Model.* Health and Safety Executive, Sheffield.

HSE. 2013. *Reporting accidents and incidents at work. A brief guide to the Reporting of Injuries, Diseases and Dangerous Occurrences Regulations 2013 (RIDDOR).* Health and Safety Executive, Sheffield.

HSE. 2015. *Statistics on fatal injuries in the workplace in Great Britain 2015.* Health and Safety Executive, Sheffield.

Hudson, P. 2007. Implementing a safety culture in a major multi-national. *Safety Science* 45:697–722.

IAEA. 1999. *Basic safety principles for nuclear power plants.* 75-INSAG-3, Rev. 1. International Atomic Energy Agency, Vienna.

IAEA. 2007. *Best practice in identifying, reporting and screening operating experience at nuclear power plants.* IAEA-TECDOC-1581. International Atomic Energy Agency, Vienna.

ICMM. 2012. *Overview of leading indicators for occupational health and safety in mining.* International Council of Mining and Metals, London.

ICOLD. 1995. *Dam failures statistical analysis.* Bulletin 99, International Commission on Large Dams, Paris.

ICOLD. 2005. *Risk assessment in dam safety management. A reconnaissance of benefits – Methods and current applications.* Bulletin 130, International Commission on Large Dams, Paris.

IEC. 2010. *Functional safety of electrical/electronic/programmable electronic safety-related systems.* Standard IEC 61508, Part 1 and 2, International Electrotechnical Commission, Geneva.

IFC. 2012. *Performance standards on environmental and social sustainability.* International Finance Corporation, Washington, DC.

ILO. 1996. *Recording and notification of occupational accidents and diseases.* International Labour Office, Geneva.

ILO. 1998. *Resolution concerning statistics of occupational injuries (resulting from occupational accidents), adopted by the Sixteenth International Conference of Labour Statisticians, 6–15 October, 1998.* The Sixteenth International Conference of Labour Statisticians, Geneva.

ILO. 2012. *Estimating the economic costs of occupational injuries and illnesses in developing countries: Essential information for decision-makers.* Working paper. International Labour Office, Geneva.

Ingason, H., Lönnermark, A., Frantzich, H. and Kumm, M. 2010. *Fire incidents during construction work of tunnels.* SP Report 2010:83. SP Technical Research Institute of Sweden, Borås, Sweden.

International Water Power & Dam Construction. 2010. *Sayano Shushenskaya accident – Presenting a possible direct cause*. http://www.waterpowermagazine. com/story.asp?storyCode=2058518 (Accessed 22 September, 2012).

IOGP. 2015. *Safety performance indicators – 2014 data*. International Association of Oil and Gas Producers, London.

IRAP. n.d. *Road deaths in developing countries: The challenge of dysfunctional roads*. International Road Assessment Program. http://www.irap.net/about-irap-3/research-and-technical-papers%3Fdownload%3D43:road-deaths-in-developing-countries-the-challenge-of-dysfunctional-roads (Accessed 16 January, 2016).

Ishikawa, K. 1976. *Guide to quality control*. Asian Productivity Organizations, Tokyo.

ISO. 2000. *Petroleum and natural gas industries – Offshore production installations – Guidelines on tools and techniques for hazard identification and risk assessment*. ISO 17776:2000. International Organization for Standardization, Geneva.

ISO. 2009. *Risk management. Principles and guidelines*. ISO 31000:2009. International Organization for Standardization, Geneva.

ISO. 2010. *Safety of machinery. General principles for design. Risk assessment and risk reduction*. International standard ISO 12100:2010. International Organization for Standardization, Geneva.

ISO. 2011. *Guidelines for auditing management systems*. ISO 19011:2011. International Organization for Standardization, Geneva.

ISO. 2012. *Road traffic safety (RTS) management systems – Requirements with guidance for use*. ISO 39001:2012. International Organization for Standardization, Geneva.

ISO. 2015a. *Quality management systems – Requirements*. ISO 9001:2015. International Organization for Standardization, Geneva.

ISO. 2015b. *Environmental management systems – Requirements with guidance for use*. ISO 14001: 2015. International Organization for Standardization, Geneva.

ISO. 2015c. *Quality management systems – Fundamentals and vocabulary. Standard*. ISO 9000:2015. International Organization for Standardization, Geneva.

ISO. 2016. *Occupational health and safety management system – Requirement with guidance for use*. ISO/DIS:45001, Draft international standard. Stage: 40.20 (2016-02-12). International Organization for Standardization, Geneva.

Jackson, P., Hilditch, C., Holmes, A., Reed, N., Merat, N. and Smith, L. 2011. *Fatigue and road safety: A critical analysis of recent evidence*. Road safety web publication no. 21, Department for Transport, London.

Jacobsson, A., Ek, Å. and Akselsson, R. 2012. Learning from incidents – A method for assessing the effectiveness of the learning cycle. *Journal of Loss Prevention in the Process Industry* 25:561–570.

Janicak, C.A. 2003. *Safety metrics. Tools and techniques for measuring safety performance*. Government Institutes, Lanham, MD.

Johnson, W.G. 1980. *MORT safety assurance system*. Marcel Dekker, New York.

Jubert, A. 1999. Developing an infrastructure for communities of practice. In *Proceedings of the 19th International Online Meeting*, ed. B. McKenna, 165–168. Learned Information, Hinksey Hill, UK.

Juran, J.M. 1989. *Juran on leadership for quality – An executive handbook*. The Free Press, New York.

Kaplan, S. and Garrick, J. 1981. On the quantitative definition of risk. *Risk Analysis* 1:11–27.

Khanzode, V.V., Maiti, J. and Ray, P.K. 2012. Occupational injury and accident research: A comprehensive review. *Safety Science* 50:1355–1367.

Kim, J.W. and Jung, W. 2003. A taxonomy of performance influencing factors for human reliability analysis of emergency tasks. *Journal of Loss Prevention in the Process Industries* 16:479–495.

Kjellén, U. 1982. An evaluation of safety information systems at six medium-sized and large firms. *Journal of Occupational Accidents* 3:273–288.

Kjellén, U. 1983. *Analysis and development of corporate practices for accident control.* Thesis, Report No. Trita AVE-0001. Royal Institute of Technology, Stockholm.

Kjellén, U. 1984. The role of deviations in accident causation and control. *Journal of Occupational Accidents* 6:117–126.

Kjellén, U. 1987a. A changing role of human actors in accident control. In Rasmussen, J., Duncan, K. and Leplat, J. (eds) 1987. New technology and human error. John Wiley and Sons, Chichester.

Kjellén, U. 1987b. Simulating the use of a computerized injury and near accident information system in decision making. *Journal of Occupational Accidents* 9:87–105.

Kjellén, U. 1990. Safety control in design – Experience from an offshore project. *Journal of Occupational Accidents* 12:49–61.

Kjellén, U. 1992. Arbeidsulykker. [Occupational accidents]. In: *Grunnbok i arbeidsmiljøopplæring.* Tiden Norsk Forlag, Oslo. In Norwegian.

Kjellén, U. 1993. *Skade- og hendelsesrapportering på Oseberg feltsenter og Oseberg C – Resultat av evaluering.* [Incident reporting at the Oseberg offshore installations]. Report No. NHT-F15-00026, Norsk Hydro, Oslo. In Norwegian.

Kjellén, U. 1995. Integrating analyses of the risk of accidents into the design process – Part II: Method for prediction of the LTI-rate. *Safety Science* 19:3–18.

Kjellén, U. 1997. Feedback control of accidents. In *The workplace*, eds. D. Brune, G. Gerhardsson, G.W. Crockford and D. D'Auria, Vol. 1, Fundamentals of Health, Safety and Welfare. Scandinavian Science Publisher, Oslo.

Kjellén, U. 1998. Adapting the application of risk analysis in offshore platform design to new framework conditions. *Reliability Engineering and System Safety* 60:143–151.

Kjellén, U. 2007. Safety in the design of offshore platforms: Integrated safety versus safety as an add-on characteristic. *Safety Science* 45:107–127.

Kjellén, U. 2008. Experience feedback. In: Melnick, E.L. and Everitt, B.S. (Eds.), *Encyclopedia of Quantitative Risk Analysis and Assessment*, Wiley, Hoboken, NJ

Kjellén, U. 2009. The safety measurement problem revisited. *Safety Science* 47:486–489.

Kjellén, U. 2012. Managing safety in hydropower projects in emerging markets – Experiences in developing from a reactive to a proactive approach. *Safety Science* 50:1941–1951.

Kjellén, U. 2016. Safety in hydropower development and operation. In *Alternative energy and shale gas encyclopedia*, eds. J.U.H Lehr and J. Keeley. 413–422. Wiley, Hoboken, NJ.

Kjellén, U., Boe, K. and Løge Hagen, H. 1997. Economic effects of implementing internal control of health, safety and environment: A retrospective case study of an aluminium plant. *Safety Science* 27:99–114.

Kjellén, U. and Hovden, J. 1993. Reducing risks by deviation control – A retrospection into a research strategy. *Safety Science* 16:417–438.

Kjellén, U. and Larsson, T.J. 1981. Investigating accidents and reducing risks – A dynamic approach. *Journal of Occupational Accidents* 3:129–140.

Kjellén, U., Menckel, E., Lauritzen, J. and Maijala, P. 1986. Utredning av arbetsolycksfall och tillbud som del i ett lokalt skyddsinformationssystem. [Study of occupational accidents]. Arbete och Hälsa, No. 3. Arbetarskyddsstyrelsen, Solna. In Swedish.

Kjellén, U., Tinmannsvik, R.K., Ulleberg, T., Olsen, P.E. and Saxvik, B. 1987. *SMORT – Sikkerhetsanalyse av industriell organisasjon, offshore versjon.* [SMORT – Safety analysis of industrial organisations]. Yrkeslitteratur, Oslo. In Norwegian.

Klein, G., Ross, K.G., Moon, B.M., Klein, D.E., Hoffman, R.R. and Hollnagel, E. 2003. Macrocognition. *IEEE Intelligent Systems* 18:81–85.

Kletz, T. 1994. *Learning from accidents.* Butterworth-Heinemann, Oxford.

Kletz, T. 2001. *An engineer's view of human error.* (3rd ed.). Institute of Chemical Engineers, Warwickshire, UK.

Knox, N.W. and Eicher, R.W. 1992. *MORT user's manual.* Report No. SSDC-4, Rev. 3. Systems Safety Development Centre, EG&G Idaho, Idaho Falls, ID.

Koeler, D.J. and Harvey, N. (eds.). 2004. *Blackwell handbook on judgement and decision making.* Blackwell, Victoria, Australia.

Komaki, J., Barwick, K.D. and Scott, L.R. 1978. A behavioral approach to occupational safety: Pinpointing and reinforcing safe performance in a food manufacturing plant. *Journal of Applied Psychology* 63:434–445.

Kongsvik, T., Almklov, P. and Fenstad, J. 2010. Organisational safety indicators: Some conceptual considerations and a supplementary qualitative approach. *Safety Science* 48:1402–1411.

Kongsvik, T., Fenstad, J. and Wendelborg, C. 2012. Between a rock and a hard place: Accident and near-miss reporting on offshore service vessels. *Safety Science* 50:1839–1846.

Koren, O.S. 2006. *Evaluation of the severity of findings in checks of the technical standard on offshore installations.* Master's thesis, NTNU, Trondheim.

Körvers, P.M.W. and Sonnemans, P.J.M. 2008. Accidents: A discrepancy between indicators and facts! *Safety Science* 46:1067–1077.

Kowalski-Trakofler, K.M. and Barrett, E.A. 2003. The concept of degraded images applied to hazard recognition training in mining for reduction of lost-time injuries. *Journal of Safety Research* 34:515–525.

Krause, T.R., Seymor, K.J. and Sloat, K.C.M. 1999. Long-term evaluation of a behavior-based method for improving safety performance: A meta-analysis of 73 interrupted time-series replications. *Safety Science* 32:1–18.

Lander, F. Nielsen, K.J. and Lauritsen, J. 2016. Work injury trends during the last three decades in the construction industry. *Safety Science* 85:60–66.

LaPorte, T.R. and Consolini, P.M. 1991. Working in practice but not in theory: Theoretical challenges of 'High-reliable organizations'. *Journal of Public Administration Research and Theory* 1:19–47.

Lebeau, M. and Duguay, P. 2013. *The costs of occupational injuries – A review of the literature.* Report R-787, IRSST, Montréal.

Leigh, J.P., Marcin, J.P. and Miller, T.R. 2004. An estimate of the U.S. Government's undercount of nonfatal occupational injuries. *Journal of Occupational and Environmental Medicine* 46:10–18.

Lekka, C. 2011. *Highly reliable organisations – A review of the literature.* Report No. RR899. Buxton, Health and Safety Executive, Derbyshire.

Leplat, J. 1978. Accident analyses and work analyses. *Journal of Occupational Accidents* 1:331–340.

Leveson, N. 2004. A new accident model for engineering safer systems. *Safety Science* 42: 237–270.

Leveson, N. 2011. *Engineering a safer world: Systems thinking applied to safety.* MIT Press, Cambridge, MA.

Levitt, R.E. and Samelson, N.M. 1993. *Construction safety management* (2nd ed.). Wiley, New York.

Lindberg, A.-K., Hansson, S.O. and Rollenhagen, C. 2010. Learning from accidents – What more do we need to know? *Safety Science* 48:714–721.

Lindblom, C.E. 1959. The science of "muddling through". *Public Administration Review* 19:79–88.

Loeb, P.D., Talley, W.K. and Zlatoper, T.J. 1994. *Causes and deterrents of transportation accidents – An analysis by mode.* Quorum Books, Westport, CT.

Lund, J. and Aarø, L.E. 2004. Accident prevention. Presentation of a model placing emphasis on human, structural and cultural factors. *Safety Science* 42:271–324.

Lundberg, J., Rollenhagen, C. and Hollnagel, E. 2009. What-You-Look-For-Is-What-You-Find–The consequences of underlying accident models in eight accident investigation manuals. *Safety Science* 47:1297–1311.

Lundberg, J., Rollenhagen, C. and Hollnagel, E. 2010. What you find is not always what you fix—How other aspects than causes of accidents decide recommendations for remedial actions. *Accident Analysis and Prevention* 42:2132–2139.

March, J.G. and Simon, H.A. 1958. *Organizations.* Wiley, New York.

Matson, E. 1988. *Beregningsprinsipper for kostnader ved yrkesulykker.* [Principles for the calculation of costs of occupational accidents]. SINTEF Report STF83 A88007, Trondheim. In Norwegian.

Maycock, G. 1997. Accident liability – The human perspective. In *Traffic and transport psychology*, eds. T. Rothengatter and E.C. Vaya. 65–76. Pergamon, Amsterdam.

McKennan, F. 1983. Accident proneness – A conceptual analysis. *Accident Analysis and Prevention* 15:65–71.

Melnick, E.L. and Everitt, B.S. (eds.). 2008. *Encyclopedia of quantitative risk analysis and assessment.* Wiley, New York.

Menckel, E. 1990. *Intervention and co-operation – Occupational health services and prevention of occupational injuries in Sweden.* Doctoral thesis, Arbete och hälsa.

Miller, D.P. and Swain, A.D. 1987. Human Error and Human Reliability. In Salvendy, G. (ed.) *Handbook of Human Factors.* Wiley, New York.

Mitchell, R., Friswell, R. and Mooren, L. 2012. Initial development of a practical safety audit tool to assess fleet safety management practices. *Accident Analysis and Prevention* 47:102–118.

Mintzberg, H., Raisinghani, D. and Thèorèt, A. 1976. The structure of 'unstructured' decision processes. *Administrative Science Quarterly* 21:246–275.

Murie, F. 2007. Building safety – An international perspective. *International Journal of Occupational Environment Health* 13:5–11.

Murray, W. 2007. *Worldwide occupational safety review project.* NIOSH, Morgantown, WV.

Murray, W., Watson, B., King, M., Pratt, S. and Darby, P. 2014. Applying the Haddon matrix in the context of work-related road safety. *Proceedings of the Occupational Safety in Transportation Conference*, 20–21 September, 2012, Gold Coast, Queensland University of Technology, Queensland.

Nenonen, N., Saarela, K.L., Takala, J., Kheng, L.G., Yong, E., Ling, L.S., Manickam, K. and Hämäläinen, K. 2014. *Global estimates of occupational accidents and work-related Illnesses 2014*. Workplace Safety & Health Institute, Singapore.

Newman, S. and Watson, B. 2011. Work-related driving safety in light vehicle fleets: A review of past research and the development of an intervention framework. *Safety Science* 49:369–381.

NFPA. 2015. *Recommended practice for fire protection for electric generating plants and high voltage direct current converter stations*. NFPA Standard 850, National Fire Protection Association, Quincy, MA.

Niskanen, T. 1994. Safety climate in the road administration. *Safety Science* 17:237–255.

Nonaka, I. 1994. A dynamic theory of organizational knowledge creation. *Organization Science* 5:14–37.

Nonaka, I. and Takeuchi, H. 1995. *The knowledge creating company*. Oxford University Press, New York.

Nonaka, I., Toyama, R. and Konno, H. 2000. SECI, Ba and leadership: A unified model of dynamic knowledge creation. *Long Range Planning* 33:5–34.

Norman, D.A. 1981. Categorization of action slips. *Psychological Review* 88:1–15.

Norwegian Oil and Gas Association. 2004. *Application of IEC 61508 and IEC 61511 in the Norwegian petroleum industry*. Guidelines NOROG GL 070, Norwegian Oil and Gas, Stavanger.

Norwegian Oil and Gas Association. 2011. *Anbefalte retningslinjer for felles modell for sikker jobb analyse*. [Recommended guidelines for a common model for safe job analysis]. Report No. 090, Norwegian Oil and Gas Association, Stavanger. In Norwegian.

Norwegian Oil and Gas Association. 2015. *Recommended guidelines for common model for work permits*. Report No. 088, Norwegian Oil and Gas Association, Stavanger.

OECD. 2008. *Guidance on developing safety performance indicators related to chemical accident prevention, preparedness and response*. OECD Series on Chemical Accidents No. 18, OECD, Paris.

OGP. 2010. *HSE management – Guidelines for working together in a contract environment*. Report No. 423. International Association of Oil and Gas Producers, London.

OGP. 2011. *Process safety – Recommended practice and key performance indicators*. Report No. 456. International Association of Oil and Gas Producers, London.

OGP. 2014a. *Safety performance indicators – 2013 data*. International Association of Oil and Gas Producers, London.

OGP. 2014b. *Land transportation safety recommended practice*. OGP Report No. 365, International Association of Oil and Gas Producers, London.

OHSAS Project Group. 2007. *Occupational health and safety management systems – Requirements*. OHSAS 18001:2007, British Standards Institute, London.

Okstad, E., Jersin, E. and Tinmannsvik, R.K. 2012. Accident investigation in the Norwegian petroleum industry – Common features and future challenges. *Safety Science* 50:1408–1414.

OSHA. 1971/2015. *Recording and reporting occupational injuries and illness. Regulations*, 29 CFR, Part 1904, Occupational Safety & Health Administration, U.S. Department of Labor, Washington, DC.

OSHA. 1989. *Safety and health program management guidelines; Issuance of voluntary guidelines*. Federal Register No. 54: 3904–3916, Occupational Safety & Health Administration, U.S. Department of Labor, Washington, DC.

OSHA. 2000. *Process safety management*. Report OSHA-3132, Occupational Safety & Health Administration, U.S. Department of Labor, Washington, DC.

O'Sullivan, A. 2003. *Economics: Principles in action*. Pearson Prentice Hall, Needham, MA.

Oxford University Press. 2012. *Oxford English dictionary*. Oxford University Press, Oxford.

Paltrinieri, N., Dechy, N., Salzano, E., Wardman, M. and Cozzani V. 2012. Lessons learned from Toulouse and Buncefield disasters: From risk analysis failures to the identification of atypical scenarios through a better knowledge management. *Risk Analysis* 32:1404–1419.

Pearson, K. 2009. *The causes of incidence of occupational accidents and ill-health across the globe*. British Safety Council, London.

Perlman, A., Sacks, R. and Barak, R. 2014. Hazard recognition and risk perception in construction. *Safety Science* 64:22–31.

Perrow, C. 1984/1999. *Normal accidents: Living with high-risk technologies*. Basic Books, New York.

Phillips, R.O. and Frislid Meyer, S. 2012. *Kartlegging av arbeidsrelaterte trafikkulykker. Analyse av dødsulykker i Norge fra 2005 til 2010*. [Study of work-related fatal accident in road transportation 2005–2010]. TØI report 1188/2012, The Institute of Transport Economics, Oslo. In Norwegian.

Podgórski, D. 2015. Measuring operational performance of OSH management system – A demonstration of AHP-based selection of leading key performance indicators. *Safety Science* 73:146–166.

PSA. 2013. *Prinsipper for barrierestyring i petroleumsvirksomentet*. [Principles for barrier management in the petroleum industry]. Note from Petroleum Safety Authority Norway, Stavanger. In Norwegian.

PSA. 2014. *Trends in risk level in the petroleum activities*. Petroleum Safety Authority Norway, Stavanger.

PSA. 2015a. *Personal injuries on the Norwegian continental shelf 2004–2014*. Petroleum Safety Authority, Stavanger. http://www.psa.no/facts-and-statistics/personal-injuries-on-the-norwegian-continental-shelf-2004-2014-article11489-921.html (Accessed 6 February, 2016).

PSA. 2015b. *Regulations relating to management and the duty to provide information in the petroleum activities and at certain onshore facilities (the management regulations)*. Petroleum Safety Authorities Norway, Stavanger.

Rasmussen, J. 1982. Human errors. A taxonomy for describing human malfunction in industrial installations. *Journal of Occupational Accidents* 4:311–333.

Rasmussen, J. 1993. Learning from experience? How? Some research issues in industrial risk management. In *Reliability and safety in hazardous work systems*, eds. B. Wilpert and T. Qvale, 43–66. Lawrence Erlbaum Associates, Hove, UK.

Rasmussen, J. 1997. Risk management in a dynamic society: A modelling problem. *Safety Science* 27:183–213.

Rausand, M. 2011. *Risk assessment. Theory, methods and application*. Wiley, Hoboken, NJ.

Rausand, M. and Høyland, A. 2004. *System reliability theory: Models, statistical methods, and applications*. Wiley, Hoboken, NJ.

Reason, J. 1990. *Human error*. Cambridge University Press, New York.

Reason, J. 1991. Too little and too late: A commentary on accident and incident reporting systems. In *Near-miss reporting as a safety tool*, eds. T.W. van der Schaaf, D.A. Lucas and A.R. Hale, 109–120. Butterworth-Heinemann, Oxford.

Reason, J. 1997. *Managing the risks of organizational accidents*. Ashgate, Aldershot, UK.

Reason, J. 1998. Achieving a safe culture: Theory and practice. *Work & Stress* 12:293–306.

Reason, J., Hollnagel, E. and Pariés, J. 2006. *Revisiting the "Swiss Cheese" model of accidents*. Eurocontrol note, EEC Note no 13/06. Eurocontrol, Bruxelles.

Reynard, W.D. 1986. *The development of the NASA aviation safety reporting system*. Reference Publication No. 1114, NASA, Mountain View, CA.

Rikhardsson, P.M. and Impgaard, M. 2004. Corporate cost of occupational accidents: an activity-based analysis. *Accident Analysis and Prevention* 36:173–182.

Robson, L.S., Clarke, J.A., Cullen, K., Bielecky, A., Severin, C., Bigelow, P.L., Irvin, E., Culyer, A. and Mahood, Q. 2007. The effectiveness of occupational health and safety management system interventions: A systematic review. *Safety Science* 45:329–353.

Robson, L.S., Macdonald, S., Gray, G.C., Van Eerd, D.L. and Bigelow, P.L. 2012. A descriptive study of the OHS management auditing methods used by public sector organizations conducting audits of workplaces: Implications for audit reliability and validity. *Safety Science* 50:181–189.

Rockwell, T.H. 1959. Safety performance measurement. *Journal of Industrial Engineering* 10:12–16.

Rognstad, K. 1993. Costs of occupational accidents and diseases in Norway. *European Journal of Operational Research* 75:553–566.

Rosness, R. 1995. *Kostnadseffektiv prioritering av HMS-tiltak*. [Cost-efficient prioritization of HSE measures]. SINTEF Report STF75 A95031. SINTEF, Trondheim. In Norwegian.

Rosness, R., Blakstad, H.C., Forseth, U., Dahle, I.B. and Wiig, S. 2012. Environmental conditions for safety work–theoretical foundations. *Safety Science* 50:1967–1976.

Roy, M., Cadieux, J., Fortier, L. and Leclerc, L. 2008. *Validation d'un outil d'autodiagnostic et d'un modèle de progression de la mesure en santé et sécurité du travail*. Rapport R-584. IRST, Montreal, QC. In French.

Rozenfeld, O., Sacks, R., Rosenfeld, Y. and Baum, H. 2010. Construction job safety analysis. *Safety Science* 48:491–498.

Rundmo, T. 1990. *Atferdsvitenskaplig sikkerhetsforskning*. [Safety research on behaviour]. SINTEF report no STF75A9007. Trondheim. In Norwegian.

Runyan, C.W. 2015. Using the Haddon matrix: Introducing the third dimension. *Injury Prevention* 21:126–130.

Ruuhilehto, K. 1993. The management oversight and risk tree (MORT). In *Quality management of safety and risk analysis*, eds. J. Suokas and V. Rouhiainen, 125–146. Elsevier, Amsterdam.

Saari, J. 1998. *Participatory workplace improvement process*. ILO Encyclopaedia of Occupational Health and Safety, Geneva.

Saari, J. and Näsänen, M. 1989. The effect of positive feedback on industrial housekeeping and accidents; a long-term study at a shipyard. *International Journal of Industrial Ergonomics* 4:201–211.

Salguero-Caparros, F., Suarez-Cebador, M. and Rubio-Romero, J.C. 2015. Analysis of investigation reports on occupational accidents. *Safety Science* 72:329–336.

Salminen, S., Saari, J., Saarela, K.L. and Rásánen, T. 1992. Fatal and non-fatal occupational accidents: Identical versus differential causation. *Safety Science* 15:109–118.

Salmon, P.M., Cornelissen, M. and Trotter, M.J. 2012. System-based accident analysis methods: A comparison of Accimap, HFACS and STAMP. *Safety Science* 50:1158–1170.

Sanders, M.S. and McCormick, E.J. 1992. *Human factors in engineering and design.* McGraw-Hill, New York.

Schmidt, M. 2007. *Tolerable risk.* Chemical Engineering, September 2007. www.che. com (Accessed 3 November, 2012).

SCS (Step Change in Safety). 2003. *Leading performance indicators – Guidance for effective use.* Step Change in Safety, Aberdeen, UK.

Senneck, C.R. 1973. Over 3-day absence and safety. *Applied Ergonomics* 6:147–153.

Sepeda, A.L. 2006. Lessons learned from process incident databases and the process safety incident database (PSID) approach sponsored by the Center for Chemical Process Safety. *Journal of Hazardous Materials* 130:9–14.

Shannon, H.S. and Manning, O.P. 1980. Differences between lost-time and non-lost time industrial accidents. *Journal of Occupational Accidents* 2:265–272.

Shappell, S.A. and Wiegmann, D.A. 2000. *The human factors analysis and classification system (HFACS).* Report Number DOT/FAA/AM-00/7. Federal Aviation Administration, Washington, DC.

Sidenfuss, T. 2006. *Collapses in tunnelling.* Master thesis, Stuttgart University of Applied Science, Stuttgart.

Simonds, R. and Shafai-Sahrai, Y. 1977. Factors apparently affecting injury frequency in eleven matched pairs of companies. *Journal of Safety Research* 9:120–127.

Skaar, S. 1994. *Internkontroll – ørkenvandring eller veien til det forjettede land?* [Internal control – Wandering in the wilderness or on the road to the promised land?] SINTEF Report STF 82 A94002. SINTEF, Trondheim. In Norwegian.

Skinner, B.F. 1969. *Contingencies of reinforcement.* Appleton-Century-Crofts, New York.

Sklet, S. 2004. Comparison of some selected methods for accident investigation. *Journal of Hazardous Materials* 111:29–37.

Sklet, S. 2006. Safety barriers: Definition, classification, and performance. *Journal of Loss Prevention in the Process Industries* 19:494–506.

Sklet, S. and Mostue, B.A. 1993. *Kostnader ved arbeidsulykker i prosess- og verkstedsindustrien.* [Costs associated with occupational accidents in the process industry]. SINTEF Report STF75 A92032. SINTEF, Trondheim. In Norwegian.

Smith, M., Cohen, H., Cohen, A. and Cleveland, R. 1978. Characteristics of successful safety programs. *Journal of Safety Research* 10:5–15.

Solem, A. and Kongsvik, T. 2013. Facilitating for cultural change: Lessons learned from a 12-year safety improvement programme. *Safety Science Monitor* 17:Article 4.

Standards Norway. 1999. *Machinery – Working environment assessment and documentation.* Norsok Standard S-005. Standards Norway, Oslo.

Standards Norway. 2004. *Working environment.* Norsok Standard S-002. Standards Norway, Oslo.

Standards Norway. 2005. *Environmental care.* Norsok Standard S-003. Standards Norway, Oslo.

Standards Norway. 2008a. *Technical safety.* Norsok Standard S-001. Standards Norway, Oslo.

Standards Norway. 2008b. *Krav til risikovurderinger.* [Requirements for risk assessment] *NS5814:2008.* Standards Norway, Oslo. In Norwegian.

Standards Norway. 2010. *Risk and emergency preparedness assessment.* Norsok Standard Z-013. Standard Norway, Oslo.

Stanton, N.A. and Salmon, P.M. 2009. Human error taxonomies applied to driving: A generic driver error taxonomy and its implications for intelligent transport systems. *Safety Science* 47:227–237.

Stellman, J.M. (ed.). 1998. *Encyclopaedia of occupational health and safety.* (4th ed.). International Labour Office, Geneva.

Stephans, R.A. 2004. *Safety for the 21st century: The updated and revised edition of systems safety 2000.* Wiley, New York.

Sulzer-Azaroff, B., Haris, T.C. and McCann, K.B. 1994. Beyond training; Organizational performance management techniques. *Occupational Medicine: State Art Review* 9:321–339.

Surry, J., 1974. *Industrial accident research. A human engineering appraisal.* Labour Safety Council, Ontario Ministry of Labour, Toronto.

Svedung, I. and Rasmussen, J. 2002. Graphic representation of accident scenarios: Mapping system structure and the causation of accidents. *Safety Science* 40:397–417.

Swain, A.D. 1974. *The human element in systems safety.* Industrial and Commercial Techniques, Surrey.

Swedish Work Environment Authority. 2005. *ISA – The Swedish information system on occupational accidents and work-related diseases.* The Swedish Work Environment Authority, Solna.

Tague, N.R. 2005. *The quality toolbook.* ASQ Quality Press, Milwaukee, WI.

Tappura, S., Sievänen, M., Heikkilä, J., Jussila, A. and Nenonen, N. 2015. A management accounting perspective on safety. *Safety Science* 71:151–159.

Tarrants, W.E. 1980. *The measurement of safety performance.* Garland STPM Press, New York.

Taylor, F.W. 1911. *The principles of scientific management.* Harper & Brothers, New York.

Tinmannsvik, R.K. 1991. *Bruk av diagnoseverktøy i sikkerhetsstyring.* [Use of safety diagnosis tools in safety management]. Doctoral dissertation, Norges Tekniske Høgskole, Trondheim. In Norwegian.

Tinnmansvik, R.K. and Hovden, J. 2003. Safety diagnosis criteria – Development and testing. *Safety Science* 41:575–590.

Trist, E. 1981. *The evolution of socio-technical systems: A conceptual framework and an action research program.* Ontario Quality of Working Life Centre, Toronto, Canada.

Trost, W.A. and Nurtney, R.J. 1995. *Barrier analysis.* Report No. SCIE-DOE-01-TRAC-29-95. Technical Research and Analysis Center, Idaho Falls, ID.

Tuncel, S., Lotlikar, H., Salem, S. and Daraiseh, N. 2006. Effectiveness of behaviour based safety interventions to reduce accidents and injuries in workplaces: Critical appraisal and meta-analysis. *Theoretical Issues in Ergonomics Science* 7:191–209.

Tuominen, R. and Saari, J. 1982. A model for analysis of accidents and its applications. *Journal of Occupational Accidents* 4:263–273.

Tversky, A. and Kahneman, D. 1974. Judgment under uncertainty: Heuristics and biases. *Science* 185:1124–1131.

Underwood, P. and Waterson, P. 2014. Systems thinking, the Swiss Cheese Model and accident analysis: A comparative systemic analysis of the Grayrigg train derailment using the ATSB, AcciMap and STAMP models. *Accident Analysis & Prevention* 68:75–94.

Van Court Hare. 1967. *System analysis: A diagnostic approach.* Harcourt Brace & World, New York.

Van der Schaaf, T.W., Lucas, D.A. and Hale, A.R. (eds.). 1991. *Near-miss reporting as a safety tool.* Butterworth-Heinemann, Oxford.

Van der Want, P.D.G. 1997. Tripod incident analysis methodology. In *Safety performance measurement,* ed. J. van Steen. Institution of Chemical Engineers, Warwickshire, UK.

Vinnem, J.E. 2014. *Offshore risk assessment.* Springer-Verlag, London.

Visser, J.P. 1998. Developments in HSE management in oil and gas exploration and production. In *Safety management – The challenge of change,* eds. A.R. Hale and M.S. Baram, 43–66. Pergamon, Bingley, UK.

Weddle, M.G. 1996. Reporting occupational injuries: The first step. *Journal of Safety Research* 27:217–223.

Weick, K. and Sutcliffe, K. 2007. *Managing the unexpected: Resilient performance in an age of uncertainty.* Wiley, Hoboken, NJ.

Wendel, E. 1998. *SHE experience transfer from operations to projects.* Master's thesis, Norwegian University of Science and Technology, Trondheim.

Whaley, A.M, Xing, J., Boring, R.L., Hendrickson, S.M.L., Joe, J.C., Le Blanc, K. and Morrow, S.L. 2016. *Cognitive basis for human reliability analysis.* U.S. Nuclear Regulatory Commission Report, NUREG-2114.

WHO. 2015. *Global status report on road safety.* World Health Organization, Geneva.

Wiegmann, D.A. and Shappell, S.A. 2003. *A human error approach to aviation accident analysis. The human factors analysis and classification system.* Ashgate, Aldershot.

Wilde, G.J.S. 1982. The theory of risk homeostasis: Implications for safety and health. *Risk Analysis* 2:209–225.

Wilde, G.J.S. 1986. Beyond the concept of risk homeostasis: Suggestions for research and application towards the prevention of accidents and lifestyle-related diseases. *Accident Analysis & Prevention* 18:377–401.

Wilson, J.R. 2014. Fundamentals of systems ergonomics/human factors. *Applied Ergonomics* 45:5–13.

Wold, T. and Laumann, K. 2015. Safety management systems as communication in an oil and gas producing company. *Safety Science* 72:23–30.

Woods, D.D, Dekker, S., Cook, R. and Johannesen, L. 2010. *Behind human error.* Ashgate, Aldershot.

Zohar, D. 1980. Safety climate in industrial organizations: Theoretical and applied implications. *Journal of Applied Psychology* 65:96–102.

Index